SYNTHETIC FOSSIL FUEL TECHNOLOGY

Potential Health and Environmental Effects

Edited by

K. E. Cowser

C. R. Richmond

Coeditors:

J. M. Evans	R. B. Gammage
R. A. Lewis	M. R. Guerin
J. L. Epler	C. W. Gehrs
	J. R. Hightower

Proceedings of the
First Annual Oak Ridge National Laboratory
Life Sciences Symposium
September 25-28, 1978

Sponsored by the
Oak Ridge National Laboratory
Department of Energy

ANN ARBOR SCIENCE
PUBLISHERS INC / THE BUTTERWORTH GROUP

Contents

Kenneth E. Cowser is Program Manager, Life Sciences Synthetic Fuels Program, Oak Ridge National Laboratory. He is a graduate of the Oak Ridge School of Reactor Technology and holds an MS in Industrial Management from the University of Tennessee and a BS in Sanitary Engineering from the University of Illinois. He is a registered Professional Engineer in the state of Illinois.

Mr. Cowser's previous research areas include radioactive waste disposal, medical physics and internal dosimetry, and environmental health physics. A member of the United States' delegation to the International Atomic Energy Agency international symposia, he has also served on scientific committees of the National Committee on Radiation Protection and Measurement, and on sub-committees of the American Standards Association on radioactive waste disposal and management.

A former section editor for the Nuclear Safety Journal and assistant news editor for the Health Physics Journal, Mr. Cowser has authored or co-authored more than 45 articles in professional journals. He is a member of the Health Physics Society, the American Public Health Association, the American Society of Civil Engineers, Beta Gamma, and Sigma Xi.

Chester R. Richmond is Associate Laboratory Director for Biomedical and Environmental Sciences at Oak Ridge National Laboratory, Oak Ridge, Tennessee.

He was previously Alternate Health Division Leader, Los Alamos Scientific Laboratory, Los Alamos, New Mexico, and earlier was with the Medical Research Branch, Division of Biology and Medicine, U.S. Atomic Energy Commission, Washington, DC. He earned his PhD and MS degrees in Biology-Physiology from the University of New Mexico and his BA from New Jersey State College, Montclair.

Mr. Richmond is a Professor, the University of Tennessee—Oak Ridge Graduate School of Biomedical Sciences at Oak Ridge. He has published a number of papers, is a Fellow of the American Association for the Advancement of Science, and a member of several other professional societies reflecting his interests in experimental biology, public health, radiation protection and risk assessment. He was honored by the U.S. Atomic Energy Commission in 1974 with the E. O. Lawrence Award, and in 1976 with the G. Failla Lecture and Award by the Radiation Research Society.

FOREWORD

During the past three decades Oak Ridge National Laboratory has hosted numerous symposia on very specific topics in the Life Sciences Program. These activities often addressed scientific problems and research areas at the forefront of particular disciplines, always seeking out the fundamental truths by which various postulates and hypotheses may be accepted, revised, or rejected. Now with the national requirements of incorporating environmental- and health-protection goals in the implementation of energy programs, a new mechanism is necessary to ferret out the principles by which extremely complex and involved situations can be resolved.

We believe the proper setting is a multidisciplinary environment in which the life scientist and the technologist jointly can explore the questions of technical feasibility, economic viability, and environmental acceptability. This first ORNL Life Sciences Symposium brings together various disciplines and institutions concerned with the development of environmentally acceptable synthetic fossil fuel technologies. Emphasis is placed on an information exchange to illustrate the current state of knowledge concerning potential health and environmental effects of these technologies. Related symposia are planned for the future, and we expect this symposium to be an annual event.

Topics selected for inclusion in the 1978 symposium are those believed to be of special current interest and major importance in contribution of the life sciences to the future, successful implementation of various fossil energy processes. The authors of the invited papers present current work and recent experimental findings related to the existing understanding of potential impacts and their reduction. The scientist, the engineer, and the decision maker will find this new data and information of direct benefit as program implementation continues, and the summary and discussion included with each session will suggest those areas of continued uncertainty.

K. E. Cowser, Manager
Life Sciences Synthetic Fuels Program

C. R. Richmond, Associate Director
Biomedical and Environmental Sciences
Oak Ridge National Laboratory

ACKNOWLEDGMENTS

This is the first in a continuing series of Life Sciences Symposia of the Oak Ridge National Laboratory which consider broad topical areas of research and development of a multidisciplinary nature. Primary sponsorship is by the Office of the Assistant Secretary for Environment of the Department of Energy.

Plans for the symposium were developed by the Symposium Committee: K. E. Cowser and C. R. Richmond, General Chairmen; and J. L. Epler, R. B. Gammage, C. W. Gehrs, M. R. Guerin, and J. R. Hightower, committee members. Special acknowledgement is due the session chairmen and cochairmen who aptly led the discussion periods and summarized the salient points of the papers; these included J. M. Evans, Enviro Control, Inc.; R. A. Lewis, Department of Energy; and J. L. Epler, R. B. Gammage, M. R. Guerin, and J. R. Hightower, ORNL.

Requests for assistance with technical documents and related information during the symposium were provided through an information center staffed by L. W. Rickert, B. C. Talmi, and G. A. Dailey. The significant contribution of J. D. Mason and I. H. Brogden in providing editorial assistance to the authors and in coordinating the preparation of these proceedings is gratefully acknowledged.

Arrangements for the symposium were under the capable direction of Wendell Martin, and the many details of symposium mechanics were handled flawlessly by Sharon Clark, Lois McGinnis, Barbara Palmer, Brenda Roberts, Janice Shannon, and Pamela Valliant.

WELCOMING REMARKS TO THE SYMPOSIUM ON POTENTIAL HEALTH AND ENVIRONMENTAL EFFECTS OF SYNTHETIC FOSSIL FUEL TECHNOLOGIES

Chester R. Richmond*

Welcome to this symposium on "Potential Health and Environmental Effects of Synthetic Fossil Fuel Technologies." I am very honored to be able to welcome you to this first annual Oak Ridge National Laboratory Life Sciences Symposium. We hope that this will be the first of a continuing series. As some of you know, the Oak Ridge National Laboratory and its Biology Division have sponsored numerous symposia both here at Gatlinburg and at other locations for the last 31 years. These have been quite specific in content and scope and have most often addressed scientific problems and research areas that were at the forefront of the discipline. This spring, for example, I had the privilege of delivering the introductory remarks to the 31st Biology Division Symposium entitled "Genetic Mosaics and Chimeras in Mammals."

The previous symposium, which marked the 30th Biology Division Symposium, also held in Gatlinburg, was on the subject of "Mechanisms of Tumor Promotion and Cocarcinogenesis." At that time, on the occasion of 30 years of Biology Division symposia—a full generation—I pointed out the timeliness of many of the previous symposia. For example, in 1950 the topic was "Biochemistry of Nucleic Acids"; in 1955 the topic was related to the "Structure of Enzymes and Proteins"; "Mammalian Genetics and Reproduction" was the subject in 1960, followed by "Hormonal Control of Protein Biosynthesis" in 1965.

This particular symposium that we are convening this morning is an attempt to move away from the very specific research areas covered in the Biology Division symposia to those much broader areas of research and development which encompass multidisciplinary programmatic efforts that are also typical of much of what we do at the Oak Ridge National Laboratory in the Life Sciences Program.

I firmly believe we need both kinds of symposia. Times change, as do public and governmental attitudes toward research and development support and funding. No research and development institution is the same as it was ten or twenty years ago. Times and people change, as do the nature of technical and scientific problems and their priorities on the local, national, or global scales. Multidisciplinary research activities at large national multipurpose research and development institutions play a key role in the science and technology of today. We feel that we can organize our total resources to address large problems that cut across many disciplines and require the application of numerous disciplines varying from engineering to the biological and physical sciences and also including input from those grappling with the complexities of the socioeconomic arena. About two years ago, we formed at ORNL a coordinated program to ensure the development of life sciences support for developing coal conversion technologies. We have since expanded this task somewhat to include other materials such as shale and we now consider the program to be more representative of synthetic fossil fuels in the generic sense.

*Associate Director, Biomedical and Environmental Sciences, Oak Ridge National Laboratory, Oak Ridge, Tennessee 37830.

Many of us firmly believe that the development and demonstration of new energy technologies must proceed in concert with research supporting engineering and process design. Further, research designed to ensure protection of environment and human health should be initiated during the early stages of process conception and continued throughout the operation of the demonstration plants. A most serious concern is that in the haste of developing new demonstration units, technologists may not consider environmental issues to be of significance until the licensing procedure has to be initiated. When I refer to environmental issues, I include the working environment as well as the offsite environment. Very often environmental concerns are sometimes viewed as obstacles and roadblocks to be overcome rather than as a partner in the design and construction of new facilities. I believe that environmental, health, and safety research needs cannot be left to any one cadre of individuals. I believe we have learned this lesson (or should have) from one of our most recently developed new technologies, that is, electrical energy production from nuclear fission. Society now demands that those who develop the technologies work closely together as a team to help ensure that technology will not provide more stress than benefit to society in general. I firmly believe that the development of any new, large modern technology should include the following components:

- Determination of technical feasibility,
- Determination of economic viability, and
- Determination of environmental acceptability.

This troika approach means that all the players must pull together as a team to make the system work properly.

Research on environmental and health problems needs to be closely knitted to the developing technology. Changes in process or pollution control technology will modify anticipated pollutant release levels and may possibly shift environmental and health research priorities. These integrated research activities should incorporate both laboratory studies utilizing well-characterized materials and field studies and small-scale conversion facilities or similar industrial processes to ensure development of an environmentally—or perhaps I should say—socially acceptable synthetic fuel industry. It is most important that a holistic approach to solving this problem be adopted.

It is also quite clear that resources available to help solve these problems are by no means infinite. There are many chemical compounds that will require testing in ecological and biological systems. We must develop cost-effective procedures whereby we can test materials efficiently and economically by devising prescreening and screening tests and using the tier approach so that we can find those materials which are most likely to be potential problems. If we can identify those materials, we can concentrate on them rather than on materials that will be of much lesser importance or perhaps even benign to man and his environment. In this way we can avoid false negatives and design experiments which will necessitate the commitment of people, money, and large animal facilities for long-term toxicity testing. The name of the game is the conservation of resources. One of these, time, may over the long haul be the most important resource.

Another important concept in our multidisciplinary approach to working in support of developing synthetic fuel technologies is that of iteration. We develop systems whereby the analytical chemists interact iteratively with environmental and biological scientists since some classes of compounds, depending upon the biological results, will require repeated fractionation and recharacterization prior to additional biological testing. There are benefits because the process engineer is also one of the key members of this integrated team. Data obtained through the iteration process is made available so that he is aware of more than just the engineering and economic activities and problems throughout the developmental aspects of the evolving technology. Hopefully, this iterative process will increase the probability that we may be in a position to modify processes and systems and anticipate potential problems during, rather than after, the completion of a given facility or process.

I am pleased to see another important development that one might consider as a spin-off benefit arising from the integrated life sciences program in support of synthetic fossil fuels at ORNL. Basic researchers are working together with others who function under the "mission-oriented" or "applied" research banner. I believe more of our research personnel are convinced that first-class biological research can be practiced within the context of a real world problem. Again, the taxpayer benefits as well as the nation's storehouse of knowledge (hopefully, the latter will provide dividends in the future).

I look forward to an exciting several days here with you at Gatlinburg during the first annual life sciences symposium.

I also look forward to the coming years when some of us and others will return to discuss other important topics such as the global impact of carbon dioxide, bioengineering technologies or bioprocess developments or health and environmental assessment of energy technologies, all of which are examples of activities which involve multidisciplinary attacks on the respective problems.

While you are in the area, you may wish to see some of the dams and other TVA facilities. The Oak Ridge National Laboratory and TVA are now in the process of developing cooperative programs in several important areas of R&D related to energy production. We at ORNL look forward to this important coupling of our research expertise with the development and demonstration projects of TVA. Both the taxpayer and the nation in general should benefit from these interactions.

Again, on behalf of the Oak Ridge National Laboratory, I welcome you to this symposium, to Gatlinburg, and to the Great Smoky Mountains.

SESSION I: TECHNOLOGY AND CONTROLS

Chairman: J. R. Hightower, Jr.
Oak Ridge National Laboratory

SESSION I: TECHNOLOGY AND CONTROLS

SUMMARY

J. R. Hightower, Jr.*

The first session in this symposium deals with the technology for contending with harmful effluents primarily from coal conversion processes and has two objectives:

1. to serve as an introduction to the symposium by summarizing the effluent problems presented by coal conversion processes and

2. to provide an opportunity for new experimental results from researchers who are active in developing environmental control technology.

The first paper in the session presents an overview of the problems posed by liquid, gaseous, and solid effluents from coal gasification and liquefaction processes and thus meets the first objective of the session. Besides serving as the introduction to the symposium, this paper by L. E. Bostwick describes work performed by Pullman Kellogg under an EPA contract to define in as much detail as possible (1) the environmental control problems that must be solved so that coal conversion processes can operate in an environmentally acceptable manner and (2) the effectiveness and costs of controls that may be applied to process streams. Available information indicates that technology can be evolved to render each effluent stream acceptable for release, but this approach is likely to be costly and difficult to execute.

The remaining papers contain new experimental results from researchers who are currently active in developing environmental control technology. Development of this control technology for treating aqueous, gaseous, and solid wastes from coal conversion technologies is discussed separately. In addition, the final paper discusses the special problems arising when pollutant distribution or discharge information

obtained from small pilot plants must be applied to full-scale commercial plants for environmental characterization, design of effluent controls, or other reasons.

J. A. Klein of Oak Ridge National Laboratory (ORNL) categorizes major U.S. research programs aimed at developing technology for treating aqueous wastes and presents recent experimental results from ORNL's aqueous waste treatment research and development activities. Field studies to determine effectiveness of installed control technologies to meet conventional cleanup standards and tests of integrated process schemes using advanced technologies for meeting future standards, which are anticipated to be more stringent, are needs that are identified. Experimental data are presented on biological oxidation of phenolics and thiocyanate by use of a tapered fluid-bed bioreactor and removal of residual polyaromatic hydrocarbons (PAHs) by adsorption and ozonation.

W. J. Boegly, Jr., also of ORNL, describes new experiments to determine physical and chemical properties and leaching characteristics of solid wastes typical of those which will be generated by the coal conversion demonstration programs. Laboratory leaching studies and field-scale lysimeter studies will provide information with which to design solid waste landfills for the demonstration plants. Field validation of the acceptability of these designs at the demonstration plants will be conducted subsequently.

There appear to be only small efforts at developing processes that treat gaseous effluents from coal conversion for removal of nonsulfurous pollutants such

---

*Oak Ridge National Laboratory.

3

as carbon monoxide or unburned hydrocarbons. Research and development of environmental control technology for cleaning gaseous emissions from coal conversion processes is predominantly concerned with removal of sulfur as H_2S from high-temperature process gas streams, as much for economic practicality as for environmental reasons. J. T. Schrodt of the University of Kentucky describes the development of a process for removing H_2S from hot gas streams by gasifier ash sorbents. Capacities and rates of H_2S sorption onto four gasifier coal ashes were measured, and computer models were developed for the data used to design fixed-bed H_2S sorption equipment. This process is believed to offer a superior alternative to low-temperature desulfurization processes.

Many coal conversion pilot plants are being monitored and analyzed to determine effluent production characteristics of the various conversion processes. The information obtained from such studies will be used to design effluent controls, to assess the environmental acceptability of the processes, or for other purposes, but special problems arise when the information obtained in the pilot plants must be applied to large commercial plants. J. C. Craun and M. J. Massey of Environmental Research and Technology, Inc. discuss the methodologies and data needed to perform this type of scaleable environmental data analysis. Examples from the CO_2-Acceptor gasification process are used to demonstrate the needs.

POTENTIAL EMISSION, EFFLUENT, AND WASTE PROBLEMS IN COAL CONVERSION PROCESSES

Louis E. Bostwick*

ABSTRACT

Application of control technology to liquid effluents, gaseous emissions, and solid wastes from coal conversion processes is essential for protection of the environment. Government-funded studies are in progress to determine the components of the streams and their hazardous or toxic nature for establishing limits for each component at environmentally acceptable levels. Available information indicates that technology can be evolved for treating each stream to reduce its components to acceptable levels, but this approach may be costly and difficult to execute.

Pullman Kellogg, operating under an EPA contract, has studied the problems and, in view of available information, has concluded that liquid process effluents may be treated for recycling within the processes to avoid discharge to receiving waters, that gaseous and particulate emissions may be controlled to meet present environmental standards, and that disposal of solid wastes may be managed to avoid deleterious environmental effects. This engineering approach to solving the control problems considers the conversion process operator's desire for maximum effectiveness with minimum cost, but does not imply that efforts in assessment of possible effects on the environment of conversion process establishment and operation can be reduced or eliminated.

This paper presents an overview of the problems and descriptions of the proposed solutions that were developed in the EPA-funded study.

INTRODUCTION

Proposals for development of America's natural resources to help satisfy America's energy needs invariably give coal high-priority consideration. Unfortunately, coal is by no means a direct and satisfactory replacement for oil and natural gas, but conversion of coal into clean synthetic liquid or gaseous fuels promises to solve most of the problems of end use in industrial processes. This promise has spurred the development of numerous processes for production of synthetic fuels. A few of them have received commercial status, and the rest are in various stages of development in laboratories, pilot plants, and demonstration plants.

The primary advantage of synthetic fuels is the transfer of environmental problems associated with direct coal use from individual and often small end users to conversion processes. Further, control technology for conversion processes may differ considerably from control technology for conventional combustion.

The objective of synthetic fuels development is maintenance and improvement of the quality of life by supplying energy from our natural resources without unacceptable deterioration of the environment. The Environmental Protection Agency (EPA) is responsible for the assessment of environmental factors of energy technologies and for assistance in the development of controls to protect the environment. The EPA has adopted a rational approach in following the development of energy systems which begins with a low level of environmental concern during the bench-scale phase of the process investigation and continues with increasing awareness to realization of a comprehensive program during pilot plant and larger operations. Control technology development thus keeps pace with conversion process development.

*Pullman Kellogg, Division of Pullman Incorporated, Research and Development Center, 16200 Park Row, Industrial Park Ten, Houston, Texas 77084.

Pullman Kellogg is concluding 18 months of operation under an EPA contract. In the course of the study, data on the quantity and composition of the various emissions, effluents, and waste streams from coal conversion processes were gathered by both literature searches and contacts with conversion process operators. The study defines, in as much detail as possible, the problems that must be solved if the conversion processes are to operate successfully without unacceptable deterioration of the environment. The emission streams from coal gasification processes are shown in Fig. 1 and from coal liquefaction processes in Fig. 2.

To apply emissions control technology efficiently, goals must be set for the pollutant residuals. A major part of the total effort in the Pullman Kellogg program was the gathering and synopsizing of present and proposed environmental regulations and standards for federal, state, regional, and international jurisdictions. A summary of the most stringent of these regulations was developed for use as a standard for comparing the efficiencies

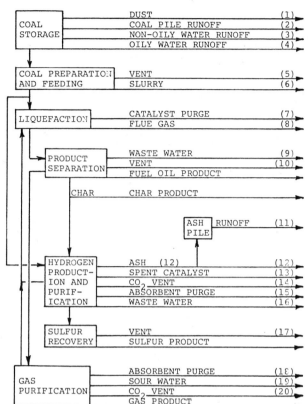

Fig. 2. Emission streams from coal liquefaction processes.

of emissions control processes, on the premise that a conversion plant with emissions equal to or lower than the most stringent standards could be built anywhere in the United States, Mexico, or Canada.

With problems scoped and objectives defined, data were gathered on available and developing control technologies. The goal was to define effectiveness and costs of controls for application to the conversion processes to enable the final streams leaving the process site to meet environmental standards.

GASEOUS EMISSIONS

Sulfur Compounds

About 84 to 97% of the sulfur species in the coal conversion process feed are reduced during reaction to hydrogen sulfide and organic sulfur compounds. Almost all of these are stripped from liquid streams in the conversion processes and are recovered as elemental sulfur, usually by the Claus or Stretford processes. Tail gases from these processes usually

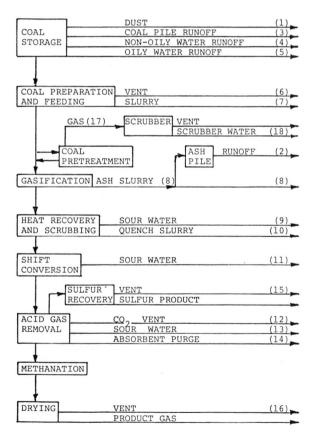

Fig. 1. Emission streams from coal gasification processes.

require further treatment to meet the most stringent environmental standards.* The commercial Beavon process is an example of tail gas treatment in which additional sulfur is recovered and an acceptable vent gas is released. Alternatively, the tail gas may be incinerated and scrubbed either to recover sulfur or to discard it as desulfurization sludge.

Process steam is raised in coal fired boilers or in incinerator/boilers in which part of the fuel is supplied by waste gas and liquid process streams. To reduce sulfur dioxide emissions from the boilers,† hydrodesulfurization of liquid waste streams may be possible, but investigation is required to establish operating conditions, sulfur removal efficiency, possible process problems and costs.

Fluidized bed combustion may be possible in an incinerator/boiler. Experimentation is needed to determine both the effect of adding liquid fuels and any problems in feeding waste gases.

Commercial nonregenerable flue gas desulfurization processes using lime or limestone slurry will scrub about 95 to 97% of the sulfur dioxide from flue gases. Capital and operating costs are low, but large quantities of raw materials and sludge must be handled.

Such processes as the citrate process recover elemental sulfur from flue gas by reaction of sulfur dioxide with hydrogen sulfide. Part of the feed stream to sulfur recovery can supply the hydrogen sulfide; the overall sulfur yield is increased, and the sulfur recovery unit size is reduced. Large-scale proof of the process is needed, even though the pilot plant has shown 95 to 98% removal efficiency.

Nitrogen Oxides

In the incinerator/boiler, combustion produces a mixture of nitrogen oxides consisting of about 95% nitric oxide and 5% nitrogen dioxide. Both are toxic and may react with hydrocarbons in the atmosphere to produce photochemical smog.

New Mexico, with 338 ppm maximum nitrogen oxides from coal fired boilers, has the most stringent current standards.‡ The EPA Combustion Research Branch has a Maximum Stationary Source Technology program for development of control methods, which the EPA believes can demonstrate combustion modifications in the field to reduce coal fired boiler emissions to 220 ppm nitrogen oxides as a 1980 goal. To attain the program's longer-term goal of 100 ppm in 1985, flue gas treatment will probably be required. Data from this research and development program are needed.

Hydrodenitrogenation of light liquid fuels, similar to hydrodesulfurization but at more severe conditions, remove about 80% of the fuel nitrogen, but proof of the technology is needed.

Removal of more than 90% of nitrogen oxides from flue gases by catalytic reaction with ammonia has been demonstrated commercially. At least one of the processes can remove sulfur dioxide simultaneously. Capital and operating costs are low, the processes are simple, and no waste streams are generated. A continuous ammonia supply is needed, however, and high operating temperature is critical.

Wet processes are available, many of them effective on both sulfur and nitrogen oxides. Nitrogen oxide removal efficiency is generally lower than for dry processes, but capital and operating costs are higher. Soluble nitrates are formed in the wastewater and must be removed prior to recycling or discharging the water. These findings, in a 1977 study for a client, led Pullman Kellogg to the conclusion that dry nitrogen oxides removal processes are superior to other processes.

Particulates

Current methods for the control of dust during storage, reclamation, and transport of coal include water sprays at car unloading points, reduction of stacker conveyor height to minimize the free fall onto the storage pile, and completely enclosing conveyors, crushers, and screens, with venting to baghouses. The actual dust losses from the unloading, storage, and reclamation operations have not been evaluated; control is by rule of thumb. It is presumed that only particles of less than 30-μm diameter will drift more than 1000 ft and that the percentage of the coal in storage in this size range will be small. Investigation is needed, however, on the actual amounts of dust to be expected and the drift potential because coal dust

---

*New Mexico, for gasification plants: 100 ppm by volume in effluent gas for any combination of H_2S, CS_2, and COS and 10 ppm by volume maximum of H_2S.

†Most stringent SO_2 standards for coal firing: 0.2 lb per million Btu (Wyoming); for residual oil firing: 0.34 lb per million Btu (New Mexico); for gas firing: 0.13 lb per million Btu (Montana).

‡Coal-fired boilers, 0.45 lb per million Btu, equivalent to 338 ppm (volume); oil-fired boilers, 0.30 lb per million Btu, equivalent to 225 ppm (volume); and gas-fired boilers, 0.20 lb per million Btu, equivalent to 150 ppm (volume).

accumulations can become a fire hazard, a physical hazard, and a toxic factor in the environment.

Commercially proved dust collection equipment includes cyclones, wet scrubbers, fabric filters, and electrostatic precipitators in order of increasing collection efficiency and cost. Combinations may be used for better economics.

Collector technology being developed for reducing energy consumption and increasing collection efficiency includes electrostatic charging of scrubbing water particles, controlled condensation of water vapor as an agglomerator, and many others. It is expected that these technologies will take their places with the well-proved control methods, or they may replace some of them as equipment is fitted to the job.

Trace Elements

Emissions from combustion of coal can contain a variety of toxic trace elements as vapors or particulates, which, in sufficient quantity, can cause adverse environmental and health effects. In a coal conversion plant, such emissions originate in coal fired boilers. Flue gas denitrification and desulfurization processes will collect both the vapors and the particulates with the exception of dry fly ash collection followed by dry oxides removal. This capability is an important consideration in selecting gaseous emission control systems for incinerator/boilers.

Trace elements will appear in coal gasification processes in oil, tar, and phenol condensates; in sour water; and eventually in sludges and blowdowns. In addition to the condensates, trace elements may appear in the fuel oil product from coal liquefaction, to be released when the fuel is burned or to be concentrated in residuals when the oil is refined. More investigation into the fate of these trace elements is needed.

Cooling Tower Drift

Compared with the conventional two-pass drift eliminators, new high-efficiency mist eliminators have been shown to reduce the drift from forced draft cooling towers by about 90%. Tests of these high-efficiency eliminators show that drift is largely confined to plant limits. Consideration must be given, however, to the composition of the cooling water as well as the effects the drift components may have on the conversion plant equipment and personnel. Economic studies are needed to contrast forced-draft and natural-draft (hyperbolic) cooling towers with respect to capital and operating costs of the two types and their drift potentials.

Lock Hopper Vent

Residual gas in coal lock hoppers usually contains hydrogen sulfide, organic sulfur compounds, and hydrocarbons and cannot be discharged to the atmosphere. Of the several schemes in operation or proposed for control of the vent stream, the best and the most practical from an environmental standpoint appears to be displacement of the residual gas with nitrogen from the oxygen plant or with carbon dioxide from the acid gas removal step and subsequent direction of the purge stream to the fuel system or to incineration.

Ash Quench Vent

The steam generated during quenching of hot gasifier ash may carry some particulates as well as noncondensable hydrocarbons formed from organic matter in the water and unreacted carbon in the ash. Direct-contact condensers have been effective in condensing the steam for disposal, and the uncondensed hydrocarbons may be sent to the incinerator. Molten slag from high-temperature gasifiers is usually quenched at system pressure by recirculating water; any noncondensables pass into the reactor. These methods appear to be effective.

LIQUID EFFLUENTS

In view of the stringency of environmental regulations concerning release of effluents to receiving bodies of water, consideration of the number of effluent streams from coal conversion processes, the flow quantities involved, and the levels of contaminants in these streams led to several conclusions:

• Proven technology appears to be available for treatment of conversion process effluent streams for reduction of contaminants to levels required by current environmental standards for releases to receiving waters.

• The application of treatment technology to individual conversion process effluent streams leads to maximum capital investment and operating costs for effluent treatment, to maximum raw water use, and to maximum problems in disposal of the residual sludges and inorganic salt concentrates from the treatment processes. All of these effects eventually increase the manufacturing costs of conversion process products.

- Present and future problems in meeting environmental regulations can be avoided if no effluents from the conversion processes are disposed of by either direct discharge or percolation into subsurface waters.

A simple twofold philosophy evolved from these conclusions: water treatment should be considered as the means of preparing individual conversion process effluent streams for recycle as process water within the conversion process battery limits; only the economically irreducible minimum of water should be released to receiving bodies of waters.

Following this philosophy has such advantages as:

- Severity of water treatment is minimized because treated water must now meet only the standards for process consumption rather than human (or fauna and flora) consumption.

- Water treatment costs are minimized.

- Raw water use is minimized.

- Problems in the disposal of water treatment process residuals are minimized.

- Problems in meeting possible future, more stringent, environmental standards are minimized.

No disadvantages are apparent in adoption of this philosophy. Accordingly, water treatment methods for application to conversion process effluents apparently need not be considered individually but rather as parts of a complete treatment scheme with the single objective of minimizing either environmental or economic problems.

Close examination of water treatment schemes as well as assessment of the relative merits of alternatives for parts of the schemes was difficult because of the lack of detailed information on quantities and compositions of the coal conversion process effluent streams. The scope of information is slowly increasing through the efforts of EPA and Department of Energy (DOE) contractors in specifying the methods of collecting and preserving samples, taking samples, analyzing them, and evaluating the methods of scaling up to full plant size the information gathered from pilot and demonstration plants. This developed information should be supplemented by continued efforts in experimentation on actual process streams with treatment methods that are now in commercial use and those that are developing.

Water Treatment for Lurgi Gasification

The Lurgi process was selected to represent low-temperature gasification processes because these processes produce phenols, oils, and tars and because all water issuing from the processes must be treated. Based on published information and private communications, analyses and quantities for the effluent streams were estimated.

The base-case study was derived from a conceptual design by the C. F. Braun Company. Process steps are oil separation, phenol extraction, stripping, ammonia recovery, dissolved air flotation, biological oxidation, sludge filtration, and evaporation. Oils and tars from separation and flotation are incinerated. Sludge from biological oxidation is also incinerated but may be sent to ash disposal. Clean water either becomes cooling tower makeup or substitutes for raw water. Boilers blow down to the cooling tower, and the cooling tower blowdown joins the inorganic sludge from raw water treatment as feed for the evaporator. Evaporator bottoms join cooled boiler ash, which joins cooled gasifier ash to be sent to disposal containing about 25% water. No liquid effluent streams leave the plant for entry into receiving bodies of water.

In an alternative study, inorganic sludges and blowdowns were considered for treatment by reverse osmosis followed by demineralization, with a small evaporator handling the reverse osmosis and demineralizer rejects. Contrasted with the base-case method of evaporating the complete stream, estimates showed little improvement in capital costs but revealed a significant steam reduction. Further investigation should be made into the operation of reverse osmosis on sludges to prove the practicality of the alternative.

Other studies indicated that chlorides, boron compounds, sulfates, silicates, and nitrates may be removed to an acceptable degree by reverse osmosis and demineralization, sending reject streams to evaporation. The same system might be used for side-stream treatment of cooling tower blowdown.

All of these various treatment methods appear to be effective in treatment of water for recycling, but this conclusion is based for the most part on estimated performance of treatment processes on streams of estimated composition. Performance testing of control process steps on actual effluents from coal conversion is required for assessment and evaluation of the treatment schemes and for optimization of treatment in view of coal characteristics, gasification process characteristics, and overall economic analysis.

It is noteworthy that in these schemes the inorganically contaminated waters are evaporated to recover clean water, and the concentrated bottoms are mixed with quenched ash and other solids. The damp mass is then transported to the disposal area. According to EPA, obtaining zero water discharge by this method has not been established as being the best option for effluent control; conceivably it may not be acceptable following further environmental consideration and analysis. Further study on this and other effluent control options is warranted.

Water Treatment for Bi-Gas Gasification

The Bi-Gas process from C. F. Braun's conceptual engineering design, supplemented by data from the Koppers-Totzek process, was selected to represent high-temperature gasification in which no phenols, oils, or tars are produced. The wastewater treatment scheme is considerably simplified and in the base-case scheme includes sour water stripping, ammonia recovery, and evaporation of inorganic sludge. No effluent stream is discharged to receiving waters.

Although the quantity of ammonia is much smaller than in the Lurgi case, stripping is required to avoid interference with sulfur recovery. Ammonia may be recovered for sale or incinerated as economics dictate.

Ammonia stripping is critical because there is no biological oxidation step to consume residuals, and ammonia fixation may occur. The Bethlehem Steel Company's proved procedure of liming before stripping avoids fixation, aids stripping, and removes particulates in the sludge. It is probable that virtually all trace metals as well as other inorganic compounds would also precipitate, but this point needs confirmation.

As pointed out in the discussion of Lurgi gasification, combinations of reverse osmosis, demineralization, and evaporation may be necessary to prepare the water for recycling.

Best estimates of effluent stream compositions and quantities and best estimates of control technology performance have led to the conclusions that process waters can be cleaned sufficiently for recycling and that the water in the damp slag discharge may be the only water leaving the process. Proof of these conclusions may be supplied by experimentation on

actual waters to establish operating methods, combinations, and economics. Such methods as ozonation, activated carbon adsorption, or alkaline chlorination may be needed to remove residual ammonia, sulfides, or cyanides.

Water Treatment for Liquefaction

The Solvent Refined Coal process was selected as representative of liquefaction processes. The water treatment scheme proposed in the R. M. Parsons Company conceptual engineering design for the process was chosen as a base case.

In the Parsons design, stripped sour water is returned to the steam generator at the Bi-Gas process gasifier for hydrogen production, and all organic matter is assumed to be vaporized and destroyed by combustion at the 3000°F gasifier temperature. Blowdown from the steam generator is combined with gas stream quench water and thus recycles to stripping. It appears that experimentation in a demonstration plant should be conducted to determine the fate of the inorganic compounds and whether or not their presence could lead to unacceptable scaling, clogging, or catalyst poisoning. The Parsons design also contemplates a considerable discharge of treated wastewater to receiving waters.

In an alternative case study, zero discharge appears practical when gasifier condensate is stripped with lime addition and the stripped water goes to cooling tower makeup. Liquefaction wastewater becomes cooling tower makeup after passing through oil separation, phenol extraction, ammonia stripping, flocculation, flotation, biological oxidation in two stages with activated carbon addition, and multimedia filtration. Biological oxidation sludges are concentrated and sent to the incinerator. Boilers blow down to the cooling tower, whose blowdown is treated by reverse osmosis of a side stream and evaporation of the reject together with the inorganic sludges from water treatment. Evaporator bottoms are combined with quenched slag and sent to disposal.

As in the other water treatment schemes, the bases for these treatment steps that are integrated into a complete scheme are estimations of effluent stream quantity and composition and estimations of treatment efficiency. Experimentation with actual effluent waters is needed for final process selection and specification and for optimization of costs. The cautions noted previously on attaining zero discharge by sending damp ash to disposal also apply here.

SOLID WASTES

The problem of fugitive dusts from coal handling, storage, and reclamation has been discussed earlier from the standpoint of dust control. If the dust were eliminated from the coal, however, there would be no fugitive dust problem anywhere from the mine to the conversion process pulverizer. While total elimination may not be possible, an effective means of reduction could be separation at the mine of particles of coal smaller than 100 U.S. mesh, agglomeration to 30 mesh or larger, and recombination with the main stream.

Treatment of about 0.1 to 0.5% of the coal might be involved. Agglomeration could be by pressure rolls, by briquetting, or by granulation with an additive. Of these methods, briquetting of fines to large sizes in preparation of Lurgi gasifier feed has been demonstrated and has been shown to be practical. Agglomeration of small sizes by compaction or granulation is practiced widely in other fields and might be economically applied to coal dust.

Solid wastes from coal conversion consist of ash, spent catalysts, and sludges from water treatment, including evaporator sludge and flue gas desulfurization sludge. As noted in the discussion of water treatment schemes, the total waste stream will probably be a damp mass containing about 25% water. The solids handling and disposal problem is large because, with a feed coal containing 10% ash, a gasification plant producing 250 billion Btu of synthetic natural gas per day will also produce about 16 to 20 million ft^3 of dry ash per year, to which must be added the other solids for overall disposal.

The most stringent environmental standards that concern the leaching of soluble salts from solid wastes state that the waste disposal area shall have a permeability no greater than 10^{-7} cm·s, or about 3 cm/year. This degree of impermeability can be attained by carefully choosing the disposal site, by incorporating clays or other such materials into the soil, by chemical stabilization of all or at least the bottom layer of the storage pile, by laying down a layer of impervious material such as asphalt aggregate, or by laying down a flexible film of impervious material. All of these methods are satisfactory for flat and level ground, and some of them may be adapted for pit disposal of the solids.

Drainage of the disposal area to a sump is needed so that runoff water can be returned to the conversion plant for treatment and reuse. A network of secondary drainage piping under the protective layer may be provided for monitoring purposes.

Lining the disposal area with an impervious layer, providing drainage collection, and returning the waters to the conversion plant are means of completely isolating the solids from the environment.

Disposal of the mass of waste material in open piles leads to problems of fugitive dust creation and leaching of soluble salts. Returning the solids to the mine is practical if the mine is adjacent to the conversion plant, is a strip mine, and no leaching problems are anticipated. Returning the solids to underground mines is considered to be costly and incompatible with mining techniques.

A method that has been in use in the power industry involves digging a series of trenches, filling them with ash, and covering them with the stockpiled earth. Such a method may be adaptable for conversion plant waste disposal where an area of about 1000 to 1100 ft square would be needed each year, which, when covered, would form a mound 25 to 30 ft high.

A second method to store one year's production of solids involves forming a pit about 800 to 900 ft square by moving earth from the center to form walls about 20 ft above grade and extending about 3 to 4 ft below grade, filling the pit with solids to a total depth of about 21 ft, and covering the filled pit with earth from the walls.

Strip mines may be usable disposal sites if their walls can be sealed against leaching, if retaining dams can be built to allow confinement of the solids, if drainage of the impounded material can be arranged, and if the drainage can be either neutralized or pumped back to the conversion plant. In addition, the economics of ash transport to the mine must be considered.

When the solids are covered, rainfall penetration, leaching, and the danger of contamination of groundwater are reduced. Revegetation for land reclamation is possible.

The choice of the most economical solutions to problems of both disposal and contamination of groundwater requires consideration of such factors as the geology and hydrology of the area, the terrain, the location, and regulatory requirements for monitoring and reclamation.

CONCLUSION

Available and developing coal conversion processes promise to satisfy at least part of our energy needs. Environmental standards and projected goals exist to aid in design and operation of these conversion processes. Control technology is available and developing to help meet the standards and goals. The challenge is to put these together in a manner that is practical, workable, and economically viable. To meet the challenge, it is essential that control technology be developed in an orderly manner and at the same pace as conversion process technology, so that the control technology to protect the environment is available when the conversion processes are ready for full-scale engineering and construction.

This project has been funded at least in part with federal funds from the U.S. Environmental Protection Agency under Contract Number 68-02-2198. The content of this publication does not necessarily reflect the views or policies of the U.S. Environmental Protection Agency, nor does mention of trade names, commercial products, or organizations imply endorsement by the U.S. government.

BIBLIOGRAPHY

For the Pullman Kellogg study, a technical reference file of 922 documents was assembled, which was supplemented by many other documentary sources, including magazines, brochures, and private communications. The environmental standards compilation used publications from federal, state, regional, Mexican, and Canadian jurisdictions. A list of representative references for the headings in this paper is given.

Sulfur Compounds

D. K. Beavon, "Four Years' Experience with the Beavon Sulfur Removal Process," paper presented at the 70th Annual Meeting of APCA, Toronto, June 1977.

S. Vasan, "Holmes-Stretford H_2S Removal Process Proved in Use," *Oil Gas J.* **72**(35), 56 (1974).

B. G. Goar, "Tighter Control of Claus Plants," *Oil Gas J.* **75**(34), 134 (1977).

T. Dowdy, *Summary Evaluation of Atmospheric Pressure Fluidized Bed Combustion Applied to Electric Utility Large Steam Generators*, EPRI FP-308, Electric Power Research Institute, Palo Alto, Calif. (October 1976).

P. S. Madenburg, "Citrate Process Demonstration Plant: A Progress Report," paper presented at *EPA Flue Gas Desulfurization Symposium, Miami, November 1977*. Available from N.T.I.S., U.S. Department of Commerce, Springfield, Va.

Nitrogen Oxides

H. L. Faucett, *Technical Assessment of NO_x Removal Process for Utility Applications*, EPA-600/7-77-127, Tennessee Valley Authority, November 1977. Available from N.T.I.S., U.S. Department of Commerce, Springfield, Va.

L. J. Ricci, "EPA Sets Its Sights on Nixing CPI's NO_x Emission," *Chem. Eng.* **84**(4), 33–36 (1977).

Particulates

H. F. Lund, *Industrial Pollution Control Handbook*, McGraw-Hill, New York (1971).

J. Sinor, *Evaluation of Background Data Relating to New Source Performance Standards for Lurgi Gasification*, EPA 600/7-77-057, U.S. Environmental Protection Agency (June 1977). Available from N.T.I.S., U.S. Department of Commerce, Springfield, Va.

Cooling Tower Drift

G. K. Wistrom, *Cooling Tower Drift, Its Measurement, Control and Environmental Effects,* paper presented at Cooling Tower Institute Annual Meeting, Houston, January 1973, Ecodyne Cooling Products Co., Santa Rosa, Calif.

Water Treatment

R. Detman, *Factored Estimates for Western Coal Commercial Concepts*, FE-2240-5 (October 1976). Available from N.T.I.S., U.S. Department of Commerce, Springfield, Va.

Woodall-Duckham Ltd., *Trials of American Coals in a Lurgi Gasifier, Westfield, Scotland*, FE-105 (1974). Available from N.T.I.S., U.S. Department of Commerce, Springfield, Va.

F. G. Glazer, *Emissions from Processes Producing Clean Fuel*, Booz, Allen and Hamilton Co., EPA 450/3-75-028 (1974). Available from N.T.I.S., U.S. Department of Commerce, Springfield, Va.

J. B. O'Hara, *Oil/Gas Complex Conceptual Design/ Economic Analysis: Oil and SNG Production*, FE-1775-8, (March 1977). Available from N.T.I.S., U.S. Department of Commerce, Springfield, Va.

R. Cooke, "Biological Purification of Effluent from a Lurgi Plant Gasifying Bituminous Coals," *Int. J. Air Water Pollut.* **9**, 97–112 (1965).

H. E. Knowlton, "Why Not Use A Rotating Disc?", *Hydrocarbon Process* **56**(9), 227 (1977).

Water Quality and Treatment, 3rd ed., American Water Works Association, Denver, Colo. (1971).

Water Purification Associates, *Water Conservation and Pollution Control in Coal Conversion Processes*, EPA 600/7-77-065, U.S. Environmental Protection Agency (June 1977). Available from N.T.I.S., U.S. Department of Commerce, Springfield, Va.

Water Purification Associates, *Innovative Technologies for Water Pollution Abatement*, NTIS 247390, U.S. Department of Commerce, Springfield, Va. (December 1975).

C. T. Lawson, "Limitations of Activated Carbon Adsorption for Upgrading Petrochemical Effluents," *AIChE Symp. Ser. Water–1973* **70**, 136 (1973).

Allied Chemical Co., Industrial Chemicals Division, *Alkaline Chlorination of Cyanide Waste Liquors*, Tech. Serv. Appl. Bull. 102, Morristown, N.J. (n.d.).

Solid Wastes

R. L. Sanks, *A Survey of Suitability of Clay Beds for Storage of Industrial Solid Wastes*, paper presented at the National Conference on Treatment and Disposal of Industrial Wastewaters and Residuals, Houston, Tex., April 1977. Available from Information Transfer, Inc., 6110 Executive Blvd., Rockville, Md.

J. M. Barrier, "Economics of FGD Disposal," *Proc. EPA Flue Gas Desulfurization Symposium, Miami, November 1977*. Available from N.T.I.S., U.S. Department of Commerce, Springfield, Va.

Institute of Gas Technology, *Preparation of a Coal Conversion Systems Technical Data Book*, FE-2286-4, (July 1976). Available from N.T.I.S., U.S. Department of Commerce, Springfield, Va.

J. J. Davis, *Coal Preparation Environmental Engineering Manual*, EPA 600/2-76-138, U.S. Environmental Protection Agency (May 1976). Available from N.T.I.S., U.S. Department of Commerce, Springfield, Va.

C. W. Matthews, "Chemical Binders: One Solution to Dust Suppression," *Rock Products* **69**(1), 82 (January 1966).

DISCUSSION

Bill Rhodes, Environmental Protection Agency: I guess more of a comment than a question. Usually people are reluctant to say something at the beginning of symposiums, so I'll try it. I would like to underline the caution that you exhibit in the latter half of your paper that I feel maybe was not there in the beginning, especially in terms of some of the liquid treatment. I think it is very difficult to indicate the adequacy of control technology, when, in fact, you indicated we don't know what's in the streams and the control technology hasn't been tried on these streams either. I think there are people in the audience who can go a long way towards solving some of the lack of data that we have on these systems, and I wonder if you might comment on some of the needs that you see in those areas.

Louis E. Bostwick: Nothing that I have said in this paper and the conclusions that were reached in this study should lead anyone to believe that there are firm, complete, and economically viable solutions to all of the problems connected with coal conversion. What we have done in our study is gather together the available information and assemble the best estimates of stream compositions and quantities. From this information and with the aid of consultations with process licensors and equipment vendors, we can determine whether or not there was even a possibility of treating these streams to meet present environmental standards. Now, throughout the whole study, we were continually reminded, and I do hope this has been made clear in this presentation, that there is not enough information available to make firm engineering decisions. However, this information is being developed and as it becomes available, it must be expressed in terms of chemical engineering processes that can be applied to the conversion processes themselves, always with the conversion plant operator in mind. After all, he is attempting to operate a conversion process at a profit, but at the same time he is recognizing that he must meet certain considerations with regard to the environment. He wants to do this as economically as possible. The objectives of process development, control technology development, and environmental standards are all part of the same problem, and they all must be considered together.

Jim Evans, Enviro Control: I was interested in the position that you put the cleanup from the utilities, the steam and power plants, that you mentioned this first, and that occupied a good third of your paper. It sounded like this may be one of the big problems.

Louis Bostwick: I think that we can say that it certainly is a problem and that we had not originally intended in our study to become deeply involved with combustion problems. Because, particularly in Lurgi gasification plants, much of the waste material and waste streams can be incinerated and, in many cases, around half the total Btu requirements for steam generation may come from waste streams, which contain all manner of contaminants, we did become involved in combustion processes. Therefore, we did begin to consider the sulfur compounds, SO_2 removal or conversion, and the nitrogen oxides problem. It can be considered a totally different study, of course, but the problem is that a steam generator in a coal conversion plant rarely is isolated completely from the coal conversion process itself. It is too convenient an incinerator for waste gas streams or waste liquid streams that have some heating value, and it should be considered with them so there must be involvement.

Ron Neufeld, University of Pittsburgh: In your opinion, are there sufficient environmental controls for the demonstration plants that are being considered today for the very near future?

Louis Bostwick: Yes, to the best of our knowledge, but I surely must hedge on that one. The point is that we cannot make firm statements on something like this without complete knowledge of the waste streams we are talking about, and we didn't have that in our study. So what it comes down to is the necessity for sampling and analysis and determination of the components of the various effluent emission and waste streams. Now, of course, an analysis like this could be carried on for years and years with determinations down to the last parts per billion of whatever is in a stream. If these streams, and I am speaking particularly of the liquid streams, are to be recycled within the process and are to be treated only sufficiently to allow recycling as process water, it is, of course, an entirely different problem from releasing them to the environment. Given the major components and their quantities, from the information that we have available at this time, we feel that it is possible to evolve combinations of treatment processes so that the water can be recycled within the process.

NATIONAL PROGRESS IN CONTROL OF WASTEWATER FROM COAL CONVERSION PROCESSES*

J. A. Klein[†]

ABSTRACT

Several ongoing research programs in the area of environmental control technology for coal conversion aqueous wastes have demonstrated that, although current technologies can meet today's effluent regulations, advanced technology will be needed for the future. The various national programs are described, including ORNL's program in wastewater control. A major gap in the testing of technology on actual waste streams is identified, and a program to perform field evaluation studies is proposed.

INTRODUCTION

In recent years, both the federal government and industry have begun the extensive effort needed to develop processes for converting this country's abundant coal resources into substitute gaseous and liquid fuels. Water is an essential raw material for these coal conversion processes. The most fundamental of the many reasons for this dependency is that water is a source of the hydrogen ($C + H_2O \rightarrow CO + H_2$; $CO + H_2O \rightarrow CO_2 + H_2$) needed to raise the atomic H:C ratio of coal from less than 1.0 to the approximate 1.5 and 3.5 values found in crude oil and natural gas respectively. Overall material balances for many coal conversion processes indicate that they are net users of water. Some gasifiers, especially those designed for the production of hydrogen, may in effect have no water discharge except for accidental releases. In practice, however, wastewater will normally be discharged in large amounts from coal conversion facilities. It has been estimated that by the year 2000 wastewater discharges from all types of coal conversion facilities will total nearly 700 million tons/year.[1]

Environmental and economic reasons will compel developers of coal conversion processes to maximize reuse of water and to reduce pollutant levels in aqueous effluents. Because of the economic incentives favoring location of a coal conversion facility close to the mine mouth, water resources available to such a plant will be either relatively scarce and highly competed for by nonindustrial users such as farmers, municipalities, and recreationists (as in Colorado, Wyoming, Montana, North Dakota, and New Mexico), or they will be relatively abundant but heavily used by other industrial facilities (as in Kentucky, West Virginia, and Illinois). Environmental considerations aside, recovery of dissolved "pollutants" as process by-products (e.g., phenol) may become economically desirable because of raw material scarcities. Thus, ensuring the availability of adequate technology for treating the wastewaters from coal conversion processes may be as necessary a part of fostering the commercialization of a synthetic fuels industry as is the development of the coal conversion technology itself.

Recent studies[2-5] of several coal gasification and liquefaction processes have identified the following operations as sources of wastewater: (1) separation of excess water from raw liquid products, (2) quenching and scrubbing of raw fuel gas, and (3) desulfurization and demetallization of raw liquid products. Other

*Research sponsored by the Division of Environmental Control Engineering, U.S. Department of Energy, under contract W-7405-eng-26 with the Union Carbide Corporation.

†Oak Ridge National Laboratory, Oak Ridge, Tennessee 37830.

parts of the overall processes will provide additional sources of wastewater: (1) coal storage and preparation (storm runoff from coal and tailing piles, wash water from coal grinding and screening); (2) waste treatment facilities (ash ponds, sludge from biological oxidation); and (3) utility and other auxiliary operations (cooling tower blowdown, raw and boiler feedwater treatment, sanitary sewage, plant storm sewers).

COMPOSITION OF COAL CONVERSION WASTEWATER

Although wide variations exist in the composition of coal conversion wastewaters, generalizations can be formulated, depending on the type of process, operating conditions, and nature of the coal used. The major constituents and concentrations of a representative wastewater are listed in Table 1.[2] The high phenol and ammonia concentrations are typical

Table 1. Composition of a typical coal conversion wastewater

Component	Concentration in wastewater ($\mu g/cm^3$)	Limits of anticipated regulations ($\mu g/cm^3$)
Phenol	6,000	0.03–0.3
NH₃	10,000	0.8–5.0
H₂S	1,000	0.02–0.2
CN⁻	100	0.02–0.1
SCN⁻	500	
PAH	10	
TOC	20,000	BOD 4–30
		COD 20–350

of coal conversion process wastewaters, whether they are condensation liquors or scrubber waters. Also listed are the concentration ranges expected in future federal regulations for a variety of components. Standards for related industries such as coking and petroleum refining were used to develop these values[2,6-7] because there are as yet no regulations for the coal conversion industry. As shown, the levels of all the components listed will far exceed the limits of the anticipated regulations.

Individual variations do exist in the component concentrations between various wastewaters and, although they are not shown, they may be quite important in the consideration of specific control technologies. As an obvious example, the presence of a significant quantity of a metallic compound could alter the effectiveness of a biological or catalytic oxidation treatment system.

DEVELOPMENT PROGRAM IN ENVIRONMENTAL CONTROL TECHNOLOGY

Development of the necessary control technology for the cleanup of coal conversion wastewaters was theoretically initiated with the first coal conversion processes at the turn of the century. Only recently, however, has this development gained the necessary emphasis to consider its integration into the coal conversion implementation program. Proposed and ongoing work in the area of control technology can be classified as to whether the proposed assessment is experimental or nonexperimental, whether the investigation is specific to one technology or one waste stream, or whether actual aqueous wastes are treated. A logical classification of the various assessment programs includes:

- Characterization of aqueous streams,
- Process stream modification studies,
- Assessment of control needs,
- Acquisition of field data,
- Laboratory assessment of available technology,
- Development of advanced technologies, and
- Field testing of integrated control technologies.

The chemical characterization of the various streams is of obvious importance and is needed before any of the other assessments can be implemented. Although new compounds are continually being identified in coal conversion wastewaters, recent characterization studies have given the developers of control technologies the major compounds of interest. Work is still needed on actual wastewater streams for processes that have not yet been characterized. Additional work will also be needed to develop analytical methods for implementing control technology.

Most control technology studies are considered to be add-ons for a particular coal conversion process. Unfortunately, this viewpoint is encouraged by the separation of those charged with the development of the technology (DOE's branch of Fossil Energy) and those that are to look at the technology available for the cleanup of aqueous streams (DOE's branch of Environmental Control Engineering and EPA). In many cases, judicious design of the various individual conversion process modules could allow minimal control modifications to clean up much of the wastewater problem.

Given that the discharge of a wastewater stream is unavoidable, the determination of the amount of cleanup needed is a function of both present and future effluent regulations and of the proposed use of the effluent water. The development of future regulations is presently causing a good deal of concern to those within EPA and to other investigators in the field.

The need for field studies to determine the effectiveness of installed control processes cannot be overemphasized. Current and future programs are being funded by EPA and DOE's Assistant Secretary for the Environment; Radian Corporation's low/medium-Btu gasification project, TRW's high-Btu project, Hitman's liquefaction project, and ORNL's Gasifiers in Industry Program are examples of such

work. Of obvious concern is the fact that the control technologies being assessed are those designed to meet only today's standards and not the more stringent regulations of the 1980s.

The last three areas (laboratory assessment, advanced technology development, and field testing) include the various experimental programs designed to test both available and advanced technologies on actual wastewaters or at least on synthetic waste streams that approximate wastewater. Although the number of investigators doing work in this area is constantly expanding, only 13 programs have been identified. These programs are listed in Table 2, along with the principal investigators and funding agencies.

In the area of laboratory assessment of available technology, some investigators are studying the amenability of various types of coal conversion wastewater to conventional treatment. Technologies being tested include solvent extraction, charcoal

Table 2. Experimental programs in environmental control technology

Organization	Principal investigator	Program	Funding agency[a]	Research type
University of North Carolina	P. C. Singer	Biodegradation Studies of Synthetic Wastewater	EPA	Laboratory assessment of available technology
Pacific Northwest Laboratory	G. W. Dawson	Biodegradation Studies of In Situ Oil Shale Wastewater	DOE ECE	Laboratory assessment of available technology
Water Purification Association	D. J. Goldstein	Biodegradation of Oil Shale Retort Water	DOE FE	Laboratory assessment of available technology
University of Pittsburgh	R. D. Neufeld	Treatability of Coal Gasification Wastewater	DOE FE	Laboratory assessment of available technology
University of Missouri-Columbia	R. H. Luecke	Solvent Extraction of Coal Gasification Wastewater	DOE ECE	Laboratory assessment of available technology; development of advanced technology
Pittsburgh Energy Technology Center	W. P. Haynes G. E. Johnson	Treatability of Synthane Wastewater	DOE FE ECE	Laboratory assessment of available technology; development of advanced technology
Boeing Corporation		Solvent Extraction of Coal Gasification Wastewater	In-house	Laboratory assessment of available technology; development of advanced technology
Laramie Energy Technology Center	R. E. Poulson	Treatability of Oil Shale Wastewater	DOE FE	Laboratory assessment of available technology; development of advanced technology
Oak Ridge National Laboratory	J. A. Klein D. D. Lee	Treatability of Coal Conversion Wastewater	DOE ECE PCSR	Laboratory assessment of available technology; development of advanced technology
University of Missouri-Columbia	S. E. Manahan	Use of Coal Humic Substances to Treat Wastewater	DOE FE and DOI OWRT	Development of advanced technology
Colorado State University	S. L. Klemetson	Ultrafiltration of Coal Gasification Wastewater	NSF	Development of advanced technology
Concentration Specialists, Inc.	W. Killilea	Product Recovery by Reverse Osmosis	DOE ECE	Development of advanced technology
Lawrence Berkeley Laboratory	P. Fox	Anaerobic Fermentation of Oil Shale Retort Water	DOE FE	Development of advanced technology

[a]ECE Division of Environmental Control Engineering; FE Fossil Energy; PCSR Division of Pollutant Characterization and Safety Research; OWRT Office of Water Research Technology.

adsorption, ozonation, and several biological degradation systems. All these programs are directed toward checking the tractability of a particular wastewater to a conventional treatment technique or toward evaluating the reaction rates and intermediate compounds for the various treatments. These studies have generally shown that biological treatment is probably sufficient to meet current effluent regulations, but that 10 to 15% of the organic material and most of the carcinogenic PAH are normally biologically refractive and will cause concerns with future standards.[8-10] Addition of an adsorption or ozonation step will probably remove a considerable amount of any residual compounds.[10,11] Major remaining questions concern the most efficient types of biological reactors, whether solvent extraction can be applied economically, and the nature of the intermediate compounds formed during ozonation.

Advanced technologies include new configurations and solvents for solvent extraction, new types of adsorption materials, and novel biological contactors. Other programs in this area include ongoing work in ultrafiltration, reverse osmosis freezing, and anaerobic fermentation. It is quite obvious that existing technologies can only be stretched a certain degree in the treatment of effluent streams and that

new advanced technologies will need to be developed.

Very little progress is being made in the area of field testing of integrated control schemes. The testing of oil and grease removal, biological oxidation, and adsorption on synthane condensate wastes at the Pittsburgh Energy Technology Center is as close as any project approaches an integrated control process.[10] With the number of demonstration plants being considered in the coal conversion field, it is reasonable to conclude that this area should receive more emphasis in the future.

ORNL'S WASTEWATER TREATMENT PROGRAM

A specific example of an experimental program is the ORNL involvement in an assessment and screening test evaluation of the amenability of coal hydrocarbonization wastewater and other coal conversion wastewaters to various treatment technologies. As part of this program, a flowsheet was proposed that takes advantage of several available unit operations, and would address the complete treatment of coal conversion wastewaters. This processing scheme is shown in Fig. 1 and includes physical, chemical, and

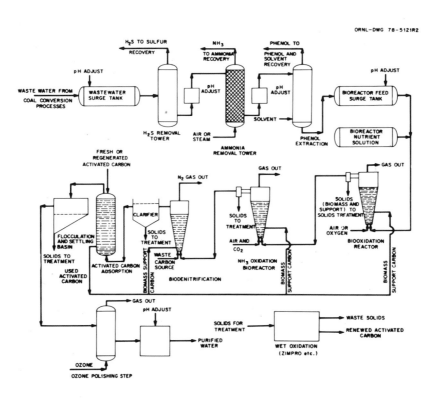

Fig. 1. Proposed flowsheet for treatment of aqueous coal conversion wastes.

biological treatment of the waste stream. The unit operations that are being investigated at ORNL include the bioreactor, the adsorption column, and the ozonator.

A tapered fluidized-bed bioreactor that is adapted to the treatment of coal conversion wastewater is used in the biological step. This tapered fluidized-bed bioreactor system, shown in Fig. 2, consists of a tapered (2.5 to 7.6 cm) section containing a solid support to which bacteria can adhere. The solid support, which is generally a substance such as coal or sand of about 30/60-mesh particle size, is fluidized by the flow of the waste stream to be treated. The tapered bed permits a wide range of fluidizing conditions and allows for expansion of the bed as biomass accumulates on the particles. Furthermore, the tapered section produces few large eddies and tends to minimize backmixing when used for two-phase (i.e., liquid-solid) systems.

The chief advantage of the fluidized-bed system is the high concentration of bacteria,[12] which is made possible by the large surface area available on the small particles. Also, particles with excess cells can be easily removed and fresh particles added to the reactor while operation continues. In cases where the reactor operates as a three-phase fluidized bed, the oxygen needed for metabolism is supplied by sparged oxygen.

The unit, which has been in operation for more than 1.5 years, has treated the aqueous waste produced by the ORNL Coal Hydrocarbonization Bench-Scale Facility. Some preliminary results indicate that the tapered fluidized-bed bioreactor in a single contacting stage can reduce the phenol and thiocyanate levels in the aqueous waste to $\leqslant 1 \ \mu g/cm^3$ and 1 to 5 $\mu g/cm^3$, respectively, at rates up to 5 to 10 kg of phenol per m^3 of reactor volume per day, and 0.5 to 1 kg of thiocyanate per m^3 of reactor volume per day, at a flow rate of 500 cm^3/min. In the fluidized-bed bioreactor for phenol degradation, the total soluble organic content is reduced an average of 95%, whereas phenol is reduced an average of >99.5%, using the hydrocarbonization scrubber water as a reactor feed. The concentrations of other organic components (PAHs, xylenols, etc.) are also decreased to some extent, depending on the chemical species involved and the residence time available in the reactor.[8]

As a final polishing step to remove any biologically refractive compounds, ORNL is assessing the feasibility of using adsorptive materials other than expensive activated charcoal. Current investigations have centered on several highly promising lignitic coals. These materials, although not having the very high adsorptive capacity of activated charcoal, do have relatively high surface areas of about 250 m^2/g.

Some column screening tests have recently been performed with two Texas and one North Dakota lignite coals on synthetic wastewater and actual wastewater from the ORNL Coal Hydrocarbonization Facility and a coal tar chemical plant. For the synthetic solutions, both naphthalene and phenanthrene were used as representative, but noncarcinogenic, PAH.

These tests were performed in a packed bed, upflow mode with approximately 33 g of coal loaded into a standard 50-ml burette. The results of one series of tests are shown in Table 3. In the synthetic feed runs, the phenol and PAH concentrations were reduced an average of 99 and 95% respectively.

As an alternate choice for a final cleanup step, ozonation is being investigated. In this case, ozonation, which is normally considered an expensive choice for effluent cleanup, would be used to destroy only those refractory materials that pass through biological treatment. For these screening tests, a

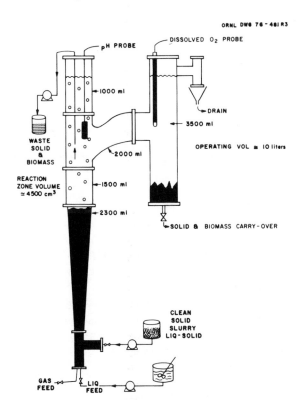

ORNL DWG 76-481 R3

Fig. 2. Schematic drawing of the tapered fluidized-bed bioreactor system.

Table 3. Adsorptive capacities of lignitic coals

Lignite coal origin	Solution tested	Concentration (ppm)	Flow rate (cc/min)	Coal loading (mg/g coal)
North Dakota	Naphthalene	3.0	6.0	1.6
Texas	Naphthalene	3.0	5.0	2.2
North Dakota	Phenanthrene	1.16	5.0	1.0
Texas	Phenanthrene	1.16	1.2	0.4
Texas	Phenol	200	6.0	2.1
Texas	Phenol	50	1.2	1.8
North Dakota	Coal tar Wastewater	55 Total carbon 1 PAH	1.2	0.6

small 1-liter batch ozone reactor was used to treat ORNL coal hydrocarbonization scrubber water.

Some preliminary results indicate that ozonation can reduce the phenol and PAH levels in the aqueous wastes to $\leqslant 0.1$ $\mu g/cm^3$ and $\leqslant 0.65$ $\mu g/cm^3$, respectively, with contact times of 15 min or less. Operating conditions were 21°C; pH, 11.2; gas flow, approximately 0.3 liter/min; and ozone concentration, 70 to 150 mg O_3 per liter of O_2. In the ozonation reactor, the phenol and PAH levels were reduced by 95 to 98%. Figures 3 and 4 show some sample results.

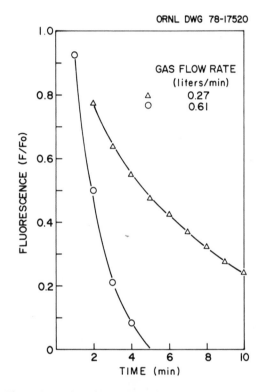

Fig. 4. **Ozonation of hydrocarbonization wastewater.** Effect of gas flow rate.

Even though high phenol and PAH removals are achieved, the soluble organic content remains unchanged. This indicates the formation of new compounds and incomplete oxidation to CO_2. Future work will focus on the identification and biodegradability of any intermediate compounds formed. Conversion of the system to a continuous mode of operation will also be performed.

CONCLUSIONS

Although several gasifiers now being developed are theoretically capable of being operated under conditions in which they should produce little, if any,

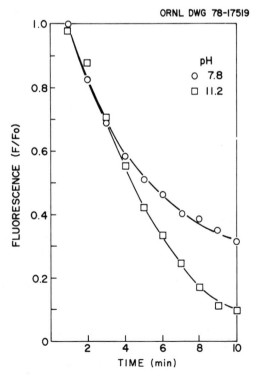

Fig. 3. **Ozonation of hydrocarbonization wastewater.** Effect of pH.

harmful pollutants, most of the developing coal conversion processes will, in all likelihood, produce large quantities of effluents. Both gaseous and aqueous effluents are usually present, but any sour-gas cleanup or wet scrubbing can transfer the pollutants to a water stream. Thus the aqueous effluent streams will be of great concern and will contain sizable amounts of known toxic and carcinogenic materials. Present standards do not call for specific levels of treatment; however, future standards will surely be more stringent and will mandate extensive control technology.

Several coal conversion processes are presently being considered for demonstration level projects. If, indeed, the purpose of these projects is to demonstrate the commercial feasibility of these processes, then it should also be desirable to demonstrate their environmental acceptability.

Wastewater processing trains incorporating some of both presently available and advanced techniques should be constructed and used to treat various coal conversion wastewaters. Although it is not practical to design these pilot treatment processes to treat all the wastewater from any of the various demonstration plants, they should provide data on the treatability of actual wastewater streams and allow evaluations of the various technologies.

REFERENCES

1. J. G. Cleland, "Summary of Multi-Media Standards," *Symposium Proceedings: Environmental Aspects of Fuel Conversion Technology, II*, EPA-600/2-76-149, U.S. Environmental Protection Agency, Washington, D.C. (July 1976).

2. J. A. Klein and R. E. Barker, *Assessment of Environmental Control Technology for Coal Conversion Aqueous Wastes*, ORNL/TM-6263, Oak Ridge National Laboratory, Oak Ridge, Tenn. (July 1978).

3. E. M. Magee, *Evaluation of Pollution Control in Fossil Fuel Conversion Processes: Final Report*, EPA-650/2-74-009n, U.S. Environmental Protection Agency, Washington, D.C. (January 1976).

4. F. Glaser, A. Hershaft, and R. Shaw, *Emissions from Processes Producing Clean Fuels*, Applied Research Report BA 9075-015, Booz-Allen, Bethesda, Md. (March 1974).

5. D. B. Emerson, "Liquefaction Environmental Assessment," U.S. Environmental Protection Agency, Washington, D.C., in *Symposium Proceedings: Environmental Aspects of Fuel Conversion Technology, III*, EPA-600/7-78-063 (April 1978).

6. U.S. Environmental Protection Agency, *Effluent Guidelines and Standards for Iron and Steel Manufacturing Point Source Category 39FR24114* (June 28, 1974).

7. U.S. Environmental Protection Agency, *Effluent Guidelines and Standards for Petroleum Refining Point Source Category 39FR16560* (May 9, 1974); *39FR32614* (Sept. 10, 1974); and *40FR21939* (May 20, 1975).

8. J. A. Klein and D. D. Lee, "Biological Treatment of Aqueous Wastes from Coal Conversion Processes," presented at Symposium on Biotechnology in Energy Production and Conservation, Gatlinburg, Tenn., May 10–12, 1978.

9. R. D. Neufeld, C. J. Drummond, and G. E. Johnson, "Biokinetics of Activated Sludge Treatment of Synthane Fluidized-Bed Gasification Wastewaters," presented at the 175th National Meeting of the American Chemical Society, Anaheim, Calif., Mar. 12–17, 1978.

10. G. E. Johnson, et al., *Treatability Studies of Condensate Water for Synthane Coal Gasification*, PERC/RI-77/13, Pittsburgh Energy Research Center (Nov. 1977).

11. R. D. Neufeld and A. A. Spinola, "Ozonation of Coal Gasification Plant Wastewater," *Environ. Sci. Technol.* 12(4), 470 (1978).

12. J. S. Jeris, et al., "Biological Fluidized-Bed Treatment for BOD and Nitrogen Removal," *J. Water Pollut. Control Fed.* 49, 816 (1977).

DISCUSSION

Phil Singer, University of North Carolina: With respect to the hydrocarbonization treatment, what kind of pretreating steps did you have prior to the biological, carbon, and ozone?

Jerry Klein: We had relatively few; we tried not to involve pretreatment very much. We had an adjustment to relatively low pH—2 or 2½—until we got a lot of settling; we had ammonia stripping and a filtration step in there, too, to get rid of the solids.

Phil Singer: You bring the pH back down for the removal of settleable material and then bring it back up?

Jerry Klein: Yes, you have to bring it up to a neutral condition for the bacteria.

Phil Singer: For ammonia stripping you are going much higher than neutral?

Jerry Klein: You can do it either way. To deal with ammonia, you have to strip it at a high pH, and to settle out the sludge, you go to low pH, but whatever order you use, you need to end up with a neutral condition before you go into the biological step.

Phil Singer: But you are having to acidify the solution and also alkalize the solution and then bring it back down to neutral?

Jerry Klein: It depends what the initial pH is. It may be better off doing it the other way first.

Phil Singer: Okay, but you are talking about major pH changes for a fairly well-buffered solution if I'm not mistaken.

Jerry Klein: Yes, there is a real problem in that you are obviously adding additional ions at the time when you change pH. How much of this you can add and yet maintain a system that the bacteria can survive in is a real concern.

R. H. Filby, Washington State University: The degradation products of your ozone treatment on PNAs—what are they? You have a reduction of fluorescence, which indicates a loss of aromaticity.

Jerry Klein: Yes, we looked at fluorescence, phenolic content, and total organic carbon. What we saw was a very high reduction in PNAs and in phenolic content but essentially no change in total organic carbon. So we are not going all the way to carbon dioxide, far from it. We are in the process of trying to identify, at least roughly, what we are forming as intermediates and whether these things are amenable to additional biological treatment. There are a lot of similarities between initial ozonation and initial biological treatment, and if we can get the first steps going by ozonation, additional biological treatment may do the job at a much cheaper cost than additional amounts of ozonation. We have flow schemes that show this material being recycled back into the biological reactor to let it have at least one pass through the biological reactor.

Audience: Are you trying to characterize these things?

Jerry Klein: Are we trying to characterize the ozonation products? We have not gone into that extensively yet. Initial tests were run this summer, and

starting after the first of next month, we will do the additional work on ozonation that we have not done yet on characterization.

Jim Evans, Enviro Control: Have you looked at the possibility of using fly ash as a filter medium?

Jerry Klein: Yes, we have. We have used fly ash from three or four of TVA's steam plants that are in the neighborhood of Oak Ridge. They all look excellent as far as the absorption capacity, mainly because they are very finely divided, with very high surface areas. They are very difficult to handle, and the fact is that we were not able to handle them in a very simple column mode; columns plugged very fast. The only way we were able to study them was in a batch mode. They did have very high capacity, so we have that information, but we were trying to figure how to handle these materials. But fly ash does work, yes.

Fred Witmer, Department of Energy: Jerry, you might want to say something about the relationship this program has to biological assessments; at least it is my understanding that you will be screening some of these waste materials.

Jerry Klein: Right, some of this will be brought up in additional sessions later on. Because of our diversity at ORNL, we have an ongoing program for the complete characterization of one of these steps, the biological reactor. We are taking the aqueous effluent from our hydrocarbonization pilot plant and doing a complete characterization on it. Oak Ridge National Laboratory's Analytical Chemistry Division is providing fractions of these materials to the Environmental Sciences Division and the Biology Division to do toxicity—both acute toxicity and carcinogenic-mutagenic—studies on this material before and after treatment, before any kind of pretreatment, after a minimal amount of pretreatment, prior to biological treatment, and after biological treatment. The initial part of that work, the before biological treatment portion, has been done. The after-biological-treatment portion has not yet been completed, but it will be available whenever it is finished.

SOLID WASTE MANAGEMENT*

William J. Boegly, Jr.†

ABSTRACT

Solid wastes produced in all stages of the coal fuel cycle (mining, processing, beneficiation, and utilization) may represent a major fraction of the total solid waste produced in the United States each year. Passage of the 1976 Resource Conservation and Recovery Act (RCRA) can produce significant impacts on generators of large volumes of solid waste. One of the key features of RCRA is the designation of wastes as being "hazardous" or "nonhazardous" depending on the results obtained by a standardized leaching test now under development. "Hazardous" solid wastes must be disposed of in landfills that ensure that no contamination results from leachates entering potential or existing water supplies. These landfills are likely to cost an order of magnitude more than "nonhazardous" landfills.

To date, only limited experimental work has been performed to determine the physical and chemical properties and the leaching characteristics of these residues. This paper reviews most of the experimental studies performed to date. It also describes the ORNL Stored Solid Study, which is directed to the disposal of solids resulting from coal gasification.

INTRODUCTION

Coal conversion, as well as coal combusion, produces solid wastes in the form of bottom ash, fly ash, chars, and sulfur residues. Solid wastes (sludges) are also generated during the treatment of aqueous wastes. Most of these wastes consist of inorganic compounds, which are of questionable value, and would ultimately be buried in landfills (probably located on some of the land surrounding the conversion or combustion plant site). Since most coals have an inorganic content (ash) of about 10% of their total weight, a plant processing 5000 tons of coal per day could be expected to dispose of 500 tons/day of residues. If the waste is landfilled at a density of 80 lb/ft³ (1300 kg/m³), about 12,000 ft³/day (340 m³/day) of storage space would be required [an area 30 ft by 30 ft (9.1 m by 9.1 m) filled to a depth of 14 ft (4.3 m)]. Assuming the disposal trench is 30 ft wide by 14 ft deep (9.1 m wide by 4.3 m deep), one year of continuous operation would require 10,950 linear feet (3340 m) of trench [~7 acres (~2.8 ha)]. Thus the final disposal of residue could result in a significant impact in terms of land requirements in addition to the potential environmental impact of leachate from the landfilled solids.

The 1976 Resource Conservation and Recovery Act (RCRA) will have significant impacts on current and future methods for disposal of municipal and industrial solid wastes. One of the key elements of this legislation is the designation of "hazardous" and "nonhazardous" solid wastes, with certain design procedures being required for each type of waste. At the present time, the EPA is proposing a toxicant extraction procedure (TEP) which will be used to indicate if a solid waste is to be classified as "hazardous." The TEP is a fixed leaching procedure which uses an acetic acid extractant with the resulting solution being subjected to further chemical, biological, carcinogenic, and mutagenetic testing. Failure to pass any of the proposed tests will designate the solid waste as being "hazardous" and will require the design of secure land disposal facilities to meet stringent environmental standards. At this time the details of the TEP and the requirements for land disposal have not been published in final form in the *Federal Register*. However, it appears that the implementation of RCRA will have significant economic impacts on coal conversion and combustion facilities, including DOE's current demonstration programs.

There are a large number of sources of solid waste in the coal utilization cycle. Solids are produced

---

*Environmental Sciences Division Publication No. 1337. Research sponsored by the U.S. Department of Energy under Contract W-7405-eng-26 with Union Carbide Corporation.

†Environmental Sciences Division, Oak Ridge National Laboratory, P.O. Box X, Oak Ridge, Tennessee.

during mining, coal cleaning, beneficiation, storage, and utilization. Figure 1 shows the types of solid waste that are generated from the mining to the utilization cycle. This paper deals only with the wastes produced during the utilization of coal for energy production, with particular emphasis on the ash and slag produced by the gasifier. Ultimately, the sulfur, fly ash, and the sludges produced during the treatment of liquid effluents will probably require land disposal.

Since coal conversion (either gasification or liquifaction) is still a developing technology, very few studies have been undertaken to design secure landfills for these materials. At the present time what little ash and slag is produced by process development units (PDU) or pilot plants is disposed of in existing municipal landfills. However, as the technology approaches the demonstration phase, this type of disposal may not be satisfactory. RCRA may also rule out the use of conventional landfills for ash/slag from coal conversion facilities. Due to the uncertainty over RCRA requirements, no landfills have yet been designed for these wastes.

A number of investigators have initiated studies related to the leachability of the ash and slag produced by various conversion processes. Essentially, most of these have been laboratory-scale studies because of the unavailability of relatively large quantities of waste materials. Only in the past year

have large enough quantities of waste become available to initiate larger scale experiments.

One of the earliest groups to leach ash and slag was at the University of Montana.[1] Leaching was performed in this study on CO_2 Acceptor, COGAS, Lurgi, HYGAS, and SRC wastes. Column leaching was carried out with solutions of various pH values. In general, the results reported do not indicate that this type of leaching produces leachates having high trace element concentrations. This project has ended and a final report is in preparation.

The University of North Dakota has performed leaching tests on Lurgi ash produced at Sasolburg, South Africa, using lignite from Mercer County, North Dakota.[2] They also planned to include Lurgi ash from Dunn County, North Dakota, in their studies. Their leaching procedures attempted to maximize ("worst case") the quantity of elements leached from the ash and slag. The material was ground to a fine powder and refluxed from 16 to 24 hours using boiling distilled water. Since Sasol ash was not available for their study from the Dunn County lignite, they attempted to produce ash using laboratory ashing procedures. Unfortunately, these ashing procedures did not produce ash with composition similar to the Mercer County ash from South Africa, even though the two feed lignites had somewhat similar chemical compositions. It was recommended that a better, more representative

ORNL—DWG 77-21322

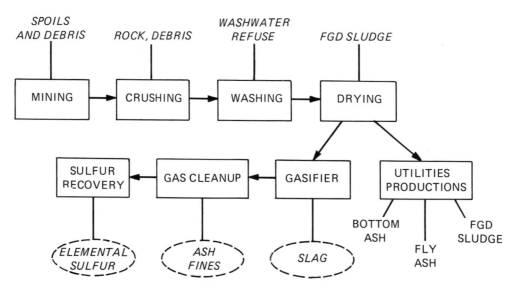

Fig. 1. Waste solids produced in a gasification process.

ashing procedure should be developed to produce ash in the laboratory from feed coals when pilot plant wastes are not available. They also concluded that the quantity and variability of their data precluded use of their analytical data in a mathematical model. This conclusion was drawn because the trace metal chemistry in the geochemical setting of the landfill was not investigated. However, they concluded that even under "worst case" conditions the trace elements should have a minor impact.

The Illinois State Geological Survey studied the solubility and toxicology of leachate pollutants from Lurgi and H-Coal processes.[3] Their leaching procedure was as follows: a 10% aqueous slurry of the ash was adjusted to 4 pH values over the range of 2 to 11, the pH's were monitored daily and readjusted to the specific value desired, and chemical equilibrium was assumed to occur when the pH remained constant. In general, this process took about 3 months. The Lurgi ash, however, reached complete equilibrium within one week.

Of the approximately 60 chemical constituents measured in the raw Lurgi ash and H-Coal residue, about 31 elements were found to be present at concentrations that could present a potential hazard. The remainder were present at such low levels that, even if completely soluble, they would pose no particular problem. Of the 31 elements that were a potential problem, 16 were found to be in forms soluble enough to exceed recommended water quality levels in some samples at pH values between 3 and 8. Seven of the elements—Al, C, Co, Cu, F, Fe, and Zn—exceeded the recommended levels in water only under certain strong acid pH conditions. The other nine constituents—B, Ca, Cd, K, Mn, NH_4, Pb, SO_4, and Sb—exceeded the recommended levels in all Lurgi ash solutions over the pH range 3 to 8. These nine constituents are thought to represent the highest potential pollution hazard (see Table 1). Discharges of the 16 constituents could cause some environmental degradation and require some form of wastewater treatment.[3]

Although literature data and reports are sparse, it would appear safe to assume that each of the pilot plant developers would have conducted some ash leaching on their residues. At the present time,

Table 1. Elements exceeding recommended water quality levels
in Lurgi and H-Coal residue leachates

Constituent	Lurgi ash solubility		H-Coal residue solubility		Recommended levels (mg/liter)
	pH 3 (mg/liter)	pH 8 (mg/liter)	pH 3 (mg/liter)	pH 8 (mg/liter)	
Al	132.	<0.5	5.5	<0.5	0.10
B[a]	5.5	4.0	13.6	13.0	0.75
Ca[a]	570.	290.	497.	175.	50.
Cd[a]	0.06	0.02			.01
Cr	0.12	<0.02			0.05
Co	0.19	<0.10			0.05
Cu	0.75	0.01			0.20
F			0.86	1.15	1.00
Fe	560.	0.06	31.50	<0.10	0.30
K[a]	26.	42.			5.00
Mn[a]	3.80	0.45	2.68	0.04	0.05
NH_4[a]	11.	17.	8.	6.	0.02
Pb[a]	0.20	0.10	0.25	<0.10	0.03
SO_4[a]	338.	820.			250.
Sb[a]	0.60	0.20			0.05
Zn	17.00	0.12	0.27	0.01	0.20

[a]Highest pollution potential, leachate concentration exceeds recommended levels.
Source: R. A. Griffin et al., "Solubility and Toxicity of Potential Pollutants in Solid Coal Wastes," pp. 506–518 in *Environmental Aspects of Fuel Conversion Technology, III,* EPA-600/7-78-063 (April 1978).

ORNL is attempting to locate and compile this information. Any assistance in locating additional ash or coal leaching data would be appreciated.

Some investigators are evaluating the economic potential of recovering certain trace elements from the solids resulting from coal gasification or combustion. Main elements of interest are iron and aluminum. Another reason for these kinds of studies is the selective removal of a selected trace element which might cause the waste to be classified as "hazardous" and whose removal would allow the waste to be disposed of as "nonhazardous" under RCRA criteria. If this were to be the case, certain of the treatment costs could be covered by reduced disposal costs.

To date, most of the leaching studies on ash and slag have been limited to chemical characteristics of leachates produced either by small laboratory columns or in batch systems. We have found no instance where soils were contacted with the waste and its leachate to determine potential attenuation by soil minerals. Attenuation of trace elements may well be the most important consideration in the design of landfills for hazardous waste, although RCRA addresses the permeability of the soil only in its draft landfill criteria. It seems apparent that soil-waste interactions should be performed on ash and slags as a vital consideration in landfill design. It now appears, however, that the RCRA regulations may be modified to include this important consideration.

In line with this philosophy, ORNL has proposed a landfill design program which involves laboratory characterization, batch and static leaching, field scale column studies with ash and soil, and a large-scale lysimeter using representative soils. This program is now funded by the Department of Energy/Fossil Energy Division and is called the Stored Solids Study. The reason for this designation is that the project is also evaluating the effects of rainfall and the attendant runoff on the environment adjacent to the coal storage pile.[4]

The objectives of the Stored Solids Study are listed in Table 2. The prime thrusts of this study are directed toward the demonstration programs of fossil energy in coal gasification. Table 3 lists the wastes being studied, along with the priorities, as of June 1, 1978. Not all of the wastes are currently available at ORNL. Most of the wastes are, however, either in storage at ORNL or in transit to the Laboratory. Some of the wastes are available in multithousand-pound quantities. Our collection of wastes and representative feed coals is probably the largest collection of these materials in the United States, in fact, maybe in the

Table 2. Objectives: Stored Solids Study

- To provide DOE/FE with technical data which can be used for designing and cost estimating landfills for coal gasification ash and slags that will meet RCRA criteria

- To provide DOE/FE with technical information on the potential environmental impacts resulting from rainfall on coal storage piles and the attendant runoff to surface streams and/or groundwater

- To assist DOE/FE in evaluating adequacy of solid waste disposal operations for coal gasification demonstration plants

Table 3. Priority for solid wastes

1. SRC (I AND II)
2. COGAS
3. Slagging Lurgi
4. HYGAS
5. BIGAS
6. Synthane

world. In addition to the studies being conducted, plans are under way to store extra waste for future use. The rapidly evolving technology will undergo changes with time, and samples of these early pilot plant wastes may provide valuable insight to changes in the waste, possible treatments of the waste, and alternative ultimate disposal methods.

The components making up the Stored Solids Study are illustrated in Fig. 2. Each level of data collection will support the next such that as information accumulates it should be possible to evaluate and predict not only the short-term but also the long-term behavior of buried wastes.[5] It is also likely that results from higher level efforts will feed back to lower level efforts and in so doing will provide more refinement of predictive capability for trace element migration in the soil.

LABORATORY STUDIES

Laboratory experiments are especially important when dealing with coal conversion wastes. Not only is the composition of the wastes variable and process related, but they also contain an abundance and wide variety of trace element and organic materials. Since little is known about their leaching behavior, controlled laboratory leaching studies are also needed to identify the effluents for the more comprehensive landfill studies.

Each residue will be characterized physically and chemically, and its properties will be compared with those of the feed coal from which it was derived. Properties such as solubility, density, porosity, compactability, and particles size distribution will be

ORNL—DWG 77-21321 RI

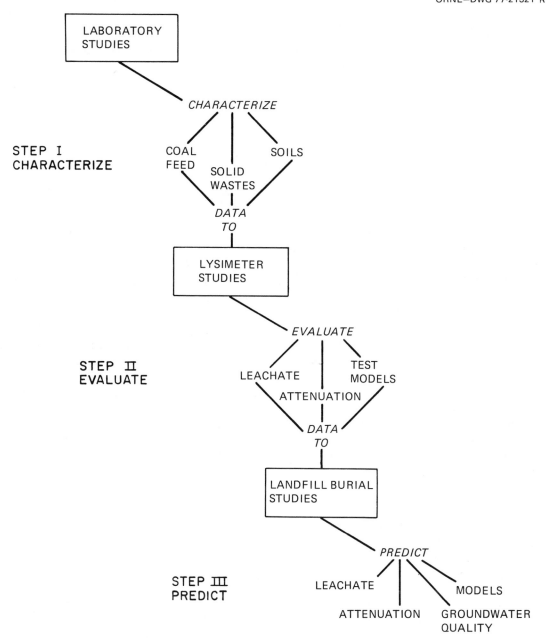

Fig. 2. Tiered approach of the ORNL Experimental Landfill Program.

determined. Chemical analyses of the solids will include major components, trace elements, and organic content, if any. Coals will be ashed under standard laboratory conditions to compare the physical and chemical properties with those of the process-derived residues. Certain of the physical property data developed to date are given in Table 4.

Soils can act as a large interactive medium for many toxic leachate effluents which can be either absorbed or precipitated by soil components. Like coal, the physical, chemical, and mineralogical composition of soil varies widely and, consequently, its capacity to retain toxic effluents varies also. Soils will be characterized for their pH, buffer capacity, ion-exchange capacity, lime equivalent, and mineral-

Table 4. Selected physical properties of solid coal conversion wastes

Property	Process A	Process B	Process C	Process D
Color	Tan	Gray	Shiny, dark gray	Shiny, dark green
Predominant size range	0.8–2.0 mm	Not available	2–3 mm	0.8–5 mm
Shape	Spherical	Spherical	Irregular, sharp edges	Irregular, sharp edges
Dry density (g/cm^3)	1.41		1.47	1.70
Particle density (g/cm^3)	2.82	2.69	2.81	2.81

ogical composition. The absorption/desorption behavior of soils will be examined with the aim of characterizing the persistence of leachate components in a landfill environment.

Leaching under laboratory conditions is being conducted both in the batch mode and by use of leaching columns. Emphasis will be on eluting trace elements. Batch testing is fast and simple. It simulates natural flooding conditions or submergence of landfill materials as when mounding of a shallow groundwater table occurs. Analyses of the batch effluents will indicate the inherent solubilities of trace substances in the wastes, and varying the pH of the effluent will provide data on waste behavior under neutral, dilute acidic, or dilute alkali conditions. Results of batch leaching at various pH values are shown in Table 5.

COLUMN LEACHING TESTS

Leaching in columns allows control of conditions to simulate the natural movement of soluble constituents more closely. The apparatus is simple, quickly assembled, and gives readily reproducible results. The leaching solution will be varied between weakly acidic and weakly alkaline, and fractions will be collected as functions of time and volume capacity. Samples, which will be compared with background-level blanks, will be analyzed by suitable methods to obtain accurate qualitative and quantitative characterization of the constituents. As with the batch tests, column leaching will be performed using representative soil and residue combinations (see Figs. 3 and 4).

Table 5. Typical analysis of batch leachate from various pilot processes

Element	Process A Extracting solution H$_2$O[a]	0.1N HCl[b]	Process B Extracting solution H$_2$O[c]	0.1N HCl[d]	Process C Extracting solution H$_2$O[c]	0.1N HCl[a]	Process D Extracting solution H$_2$O[a]	0.1N HCl[a]	0.1N NH$_4$OH[a]
	ppm		ppm		ppm		ppm		
Cadmium	<0.05	<0.05	<0.05	<0.05	<0.05	<0.05	<0.05	<0.05	<0.05
Copper	<0.05	<0.05	<0.05	0.55	<0.05	<0.05	<0.05	<0.33	<0.05
Lead	0.002	0.40	<0.001	0.09	<0.001	0.06	<0.001	0.05	<0.001
Nickel	<0.01	<0.03	<0.01	0.61	<0.01	5.6	<0.01	6.2	<0.01

[a]Results after double extraction with solid waste to solution ratio of 1:10.
[b]Results after double extraction with solid waste to solution ratio of 1:170.
[c]Results after double extraction with solid waste to solution ratio of 1:12.
[d]Results after double extraction with solid waste to solution ratio of 1:13.

Fig. 3. Laboratory leaching columns.

Fig. 4. Large ash/soil leaching columns.

FIELD-SCALE LYSIMETER STUDIES

Field studies will be initiated as soon as laboratory experiments are well under way. Field-size lysimeters have been used in conventional sanitary landfill studies as well as for examining burial ground behavior of radioactive solid wastes. Figure 5 shows the typical lysimeter which will be exposed to natural weather conditions during field studies.

The lysimeters will be loaded with soil and waste to simulate a vertical cross section of a landfill trench and will be designed to prevent escape of leachate into the environment. Both soil and leachate will be sampled at regular time intervals and analyzed to determine the extent of trace element and organic compound attenuation. Analytical methods, developed during the laboratory phase, will be applied to both the solids and leachates.

As a first step toward predicting long-term effects, data gathered during the lysimeter phase will be used to test and modify in-house transport models for substrate behavior of trace contaminants. These results will be used in designing and establishing the field burial studies.

FIELD VALIDATION

Currently, disposal of coal conversion solid waste is assumed to be accomplished using landfill methods similar to those used for conventional municipal or industrial solid wastes. Results of the laboratory- and field-scale lysimeter tests, when completed, will indicate whether this conventional method of solid waste disposal will be environmentally acceptable or whether unconventional designs will be required for disposal at conversion facilities.

Burial studies will be conducted at the demonstration site. Landfill designs developed during the lysimeter stage of the program and including any desirable amendments will be implemented in full-scale field studies after characterizing the subsurface hydrogeology of the proposed burial area. A monitoring plan will be established at the landfill site to sample both the landfill and subsurface soils and to monitor groundwater movement and quality. Data obtained from the preliminary modeling studies will help establish both the site and the monitoring plan.

Full-scale modeling studies will be conducted concurrently with the burial work. The rationale for modeling comes from the need to predict possible long-term, long-distance environmental impacts. Most incidents of groundwater contamination are

Fig. 5. Field-scale ash/soil lysimeters.

found to occur not only 20 to 30 years after burial but often long distances from the burial site. Mathematical models simulate both the spatial and temporal components of trace contaminant mobilization. The movement of contaminants will be described in terms of hydrogeologic parameters such as subsurface lithology, mineral constituents, texture and stratigraphic structure, and aquifer geometrical properties. Scenarios for transport will be developed and tested using data collected in the field monitoring operations.

CONCLUSIONS

1. Only limited research has been reported on leaching of coal conversion residues. What work that has been reported is mainly concerned with what amounts of trace elements are released at various pH's.

2. Engineering type studies using wastes and soils are needed to provide design parameters for coal gasification/liquifaction landfill designs.

3. Predictive models are needed to permit evaluation of the ultimate fate of the trace element pollutants both in space and time.

REFERENCES

1. W. P. Van Meter and R. E. Erickson, *Environmental Effects from Leaching of Coal Conversion By-products,* Quarterly Reports FE-2019-1 to FE-2019-6, 1975–1976.

2. M. H. Somerville and J. L. Elder, "A Comparison of Trace Element Analyses of North Dakota Lignite Laboratory Ash with Lurgi Gasifier Ash and Their Use in Environmental Analyses," pp. 292–311 in *Environmental Aspects of Fuel Conversion Technology, III,* EPA-600/7-78-063, April 1978.

3. R. A. Griffin et al., "Solubility and Toxicity of Potential Pollutants in Solid Coal Wastes," pp. 506–518 in *Environmental Aspects of Fuel Conversion Technology, III,* EPA-600/7-78-063, April 1978.

4. E. C. Davis and W. J. Boegly, Jr., *A Review of the Literature on Leachates from Coal Storage Piles,* ORNL/TM-6186, January 1978.

5. W. J. Boegly, Jr., et al., *Quarterly Report—Experimental Study of Leachate from Stored Solids, June 1, 1977, to January 1, 1978,* ORNL/TM-6304, January 1978.

DISCUSSION

Bill Rhodes, Environmental Protection Agency: Just as a point of additional information, the work that you refer to by Illinois State Geological Survey, which is being done under an EPA contract, is also looking at the attenuation of the material leachates through various clay materials, and that information will become available in the next few months.

Don Gardiner, Union Carbide Corporation, Nuclear Division, Mathematics and Statistics Research: Bill, you mentioned almost in passing a one-dimensional model and a two-dimensional model, and I wonder just how important these models are to the study and if any work has been done on validating or evaluating these models.

Bill Boegly: We plan to use the Reeves-Duguid model and will probably rely on some of the validation being done as part of our low-level waste disposal studies and using the lysimeters, will do further verification. Ward, Yeh, and Sri Rao are all working on the modeling aspect of this thing.

Don Gardiner: So it is an important concept? Would you mention again the name of the model?

Bill Boegly: Yes, the Reeves-Duguid.

Steve DeCicco, Oak Ridge National Laboratory: Bill, I am curious about whether we can rely on attenuation. I guess I'm thinking that after a while the zone of saturation will just move down to a point where, ultimately, it would be in contact with groundwater. Is that true?

Bill Boegly: It theoretically could be true, but as a part of this design, and again you see we are shooting for a moving target here, we don't know what RCRA is really going to tell us. If we are allowed to say that trace elements must not move off the property line or something, that is one thing. If we are being told we must ensure that trace elements get no further than ten feet from the landfill or something like that—it's a different game. If it becomes that stringent, then the concern you have is real, because this stuff is mobile, and I'm not convinced that even if you have a permeability of 10^{-7} in ten feet of it that you're really that well off, because I don't know of anybody that can guarantee—you remember, I talked about a trench 30 feet wide and 2 miles long a year—and to guarantee that kind of permeability along the whole length of it is pretty hard to believe, and maybe engineering-wise unfeasible or impossible.

Ron Neufeld, University of Pittsburgh: Bill, I share your concern about getting the samples of solid residuals. As you know from prior conversations, we, too, are doing some leaching and toxicity testing of coal conversion solid residuals. In your opinion, in the long run, which is the major problem, the ash material that you spent most of your paper talking about or the chemical process residuals? And secondly, have you been able to obtain any samples of the process residuals, the wastewater residuals, and other residuals for your work at Oak Ridge?

Bill Boegly: At the present time our charter is only on the gasification ash itself. I personally—this is only my reaction—am much more worried about the process waste sludges because of their organic nature and this type of thing, whereas the material I'm handling has been gasified at 2000 or 3000°. It's obviously a product that they wanted to remove as much of the organic from as possible because they want to make gas—that's where the economics are—so they have left me with a collection of essentially inert material. Now I might mention that there are studies, and Oak Ridge has them too, looking at resource recovery from this material. Whether this can be shown to be profitable I have no knowledge, but I would personally be more concerned with the liquid bio-sludges myself.

Bob Hightower, Oak Ridge National Laboratory: Can you comment on the adequacy of research into treatment for that type sludge?

Bill Boegly: I really can't be very authoritative about it because it strikes me as being quite fragmented and being done a little bit here, a little bit there, and I think we will be seeing that here as this meeting goes on, that there does seem to be a need in all the waste disposal conversion to have something pulling it all together and making sure that all parts of the program are looked at equally.

HOT GAS DESULFURIZATION WITH GASIFIER ASH SORBENTS*

J. Thomas Schrodt†

ABSTRACT

Four gasifier coal ashes containing 5–23 wt % iron oxide have been used in small, fixed beds to sorb the gaseous sulfur components selectively from hot, synthesized, low-Btu fuel gas. Results show sulfide removal efficiencies in excess of 99% and sorption capacities of up to 400 grains of sulfide per pound of ash. Spent ashes, regenerated up to 20 times, show cyclic use improves the ash capacities and the process efficiency. Desulfurization data have been used to evaluate rate parameters in two models developed to simulate the process steps dynamically. The kinetic gas-film and pore-diffusion gas-film models are both needed to describe the desulfurization process over the full range of operation. Computer simulation of an adiabatic fixed-bed process predicts the temperature and concentration of hydrogen sulfide and the availability of unreacted iron oxide as a function of time and position in the bed. For the regeneration phase of the process, the program will predict the temperature and concentrations of oxygen and unreacted ferrous sulfide. This process will offer a superior alternative to low-temperature desulfurization processes.

INTRODUCTION

A continuing interest in the technology of coal gasification to serve both the energy requirements and the environmental needs of the nation emphasizes the importance of deriving from coal a combustible gas which not only has an energy content adequate for power generation, but which is environmentally acceptable.

Today's air-blown gasifiers produce gases having heating values of 150–200 Btu/scf. As they leave these units the gases are at 1000°–1500° F and 1–20 atmospheres pressure and are laden with particulates, tars, and noxious sulfur compounds. At these temperatures and pressures, 5–10% of their energy is in the form of sensible heat. If the fuel gas is cleaned while still hot, this sensible heat is retained for power generation.

There are several processes now under development for desulfurizing hot fuel gases. For the most part these processes rely upon the sorption of the gaseous sulfides into solid metal oxides or molten metals or metal salts.[1,2]

A significant fraction of the sulfur in many U.S. coals is in the pyrite (FeS_2) and sulfate forms ($FeSO_4$). During gasification the iron is converted to various oxides, particularly hematite (Fe_2O_3) and magnetite (Fe_3O_4), and the sulfur is converted to hydrogen sulfide (H_2S), carbonyl sulfide (COS) and carbon disulfide (CS_2). The iron oxides in the ash residues can serve as sorbents and oxidizers for the sulfur compounds in the gases, even at high temperatures and pressures. These findings form the conceptual basis of a fixed-bed, hot gas desulfurization process that uses gasifier ashes as selective sorbents for the sulfur compounds in coal-derived fuel gases.

The basic chemistry of the process is described. Procurement and reduction of bench-scale data in the form of H_2S breakthrough curves and x-ray and SEM-EDAX solid phase analyses are discussed. Mathematical models, applicable to both the desulfurization and regeneration steps, are developed and their rate expressions for desulfurization are evaluated. Finally, results of a dynamic simulation of the desulfurization step of the process are presented and discussed.

*This work was supported by the U.S. Department of Energy under contract E-(40-1)-5076 and the Commonwealth of Kentucky through the Institute for Mining and Minerals Research.

†Department of Chemical Engineering, University of Kentucky, Lexington, Kentucky 40506.

BASIC CHEMISTRY

Upon exposure to hot reducing fuel gases, hematite is rapidly reduced to magnetite:

$$3Fe_2O_3 + H_2 \rightarrow 2Fe_3O_4 + H_2O$$

$$(K_{1000\ K} = 1.0 \times 10^5) , \quad (1)$$

$$3Fe_2O_3 + CO \rightarrow 2Fe_3O_4 + CO_2$$

$$(K_{1000\ K} = 1.1 \times 10^5) . \quad (2)$$

Further reduction is not thermodynamically favored at the prevailing temperatures and gas compositions.[1,3]

Secondary sulfur components, COS and CS_2, are hydrolyzed:

$$COS + H_2O \rightarrow CO_2 + H_2S$$

$$(K_{1000\ K} = 3.0 \times 10^1), \quad (3)$$

$$CS_2 + H_2O = COS + H_2S$$

$$(K_{1000\ K} = 2.7 \times 10^2) . \quad (4)$$

Desulfurization leads to the formation of pyrrhotite, FeS:

$$Fe_3O_4 + 3H_2S + H_2 \rightarrow 3FeS + 4H_2O . \quad (5)$$

The equilibrium constant ($K_{1000\ K} = 2.7 \times 10^7$) favors a nearly complete removal of H_2S, such that the residual sulfide concentration is considerably below the National Source Performance Standards (NSPS) set by the government.

Regeneration of spent sorbents containing primarily FeS with low concentrations of O_2 releases sulfur in the pure and oxide forms:

$$4FeS + 5O_2 \rightarrow 2Fe_2O_3 + 2SO_2 + S_2 . \quad (6)$$

Although the latter is a highly exothermic reaction ($K_{1000\ K} = 1.3 \times 10^{69}$), control of its rate of use of low O_2 levels will prevent thermal damage to the ash sorbents and result in significant amounts of pure sulfur. High levels of O_2 lead to complete oxidation of the sulfur to SO_2.

Other reactions of importance to the process include the water-gas-shift reaction and reactions that lead to carbon deposition.

EXPERIMENTAL

Four gasifier bottoms ashes, containing 5–23 wt. % Fe_2O_3, were selected for testing as desulfurization sorbents. These were crushed, and three Tyler-mesh sized fractions (4/8, 10/20 and 20/35) were prepared (Table 1).

Bench-scale, fixed-bed tests were carried out at near atmospheric pressure in a 35-mm ID quartz reactor tube located in a high temperature controlled furnace. The tube was packed with 5–10 cm of ash over a 25-cm bed of inert ceramic chips. Through these chips was passed a gas of approximately 17% CO, 15% H_2, 10% CO_2, 4% CH_4, 3% H_2O, 50% N_2 and the remainder H_2S (0.5–1.6%). For selected tests COS and/or CS_2 were added. Temperatures were monitored at selected points with corrosion resistant 442-SS-sheathed chromel-alumel thermocouples. Inlet and outlet gas compositions were determined by a dual column GC equipped with a thermal conductivity detector and an on-line control data system. The test program measures ten variables in desulfurization and regeneration experiments (Table 2).

The desulfurization experiments were to measure the effect of operating conditions on:

1. Sorption capacities of different ashes and ash sizes.

Table 1. Properties of ash sorbents

Property	Ash			
	W. Ky. No. 9	Virginia Sprint	Rosebud	Elkhorn No. 3
Fe_2O_3 (wt %)	23.1	5.02	8.19	9.51
Mesh size	10/20	10/20	10/20	10/20
Bed porosity, ϵ	0.51	0.63	0.49	0.57
Bulk density (lb/ft^3)	72.3	34.3	60.5	51.3
Sphericity, ϕ_s	0.66	0.48	0.69	0.56
Particle radius (10^{-3} ft)	2.1	1.7	1.9	1.8
Bulk surface area (ft^2/ft^3)	1073	1340	1125	1157

Table 2. Scope of test program

Variable	Range of test	
	Desulfurization	Regeneration
Temperature (°F)	800–1400	800–1300
Space velocity (h^{-1})	300–4000	500–4000
Pressure (psig)	0–400	0–200
Linear velocity (ft/sec)	0.01–2.5	0.005–3.0
Gas concentrations (%)		
H$_2$S	0.5–1.6	
COS	0.17	
CS$_2$	0.15	
O$_2$		2–21
Particle size		
(Tyler mesh)	4/8, 10/20, 20/35	10/20, 20/35
Number of ashes	4	2

2. The effectiveness of the ashes to remove the gaseous sulfides.

3. The efficiency, τ, of each test, measured as the ratio of the amount of sulfide gases introduced to the bed to the ultimate sulfide sorption capacity of the ash bed at the 15% breakthrough point.

Formulation and solution of a dynamic simulation model depended greatly upon a proper interpretation of the sulfide breakthrough curves. In dealing with ashes of different physical, chemical, and geometric character, it did not appear practical to attempt an evaluation of the mechanical kinetic steps of the desulfurization process. Rather, an approach was taken that relied on differential forms of the material and energy balances of the system, which could be solved with rate parameters and boundary conditions extracted from the test data.

RESULTS

Hydrocarbon gases remain at nearly constant concentrations (Fig. 1). Initial decreases in the concentrations of CO and H$_2$ result as these gases react with the hematite in the ash. Reduction occurs rapidly, forming magnetite and releasing H$_2$O and CO$_2$. Following this, CO, CO$_2$, H$_2$, and H$_2$O attain equilibrium concentrations relative to the water-gas shift reaction (Figs. 1–4).

As the make-up gases penetrate the sorbent bed, reaction between H$_2$S and Fe$_3$O$_4$ occurs [Eq. (5)]. For long on-stream times there is no GC-detectalbe concentration of any sulfide in the gaseous effluents. Concentration waves of H$_2$S and COS form and move through the bed slowly, according to the bed's

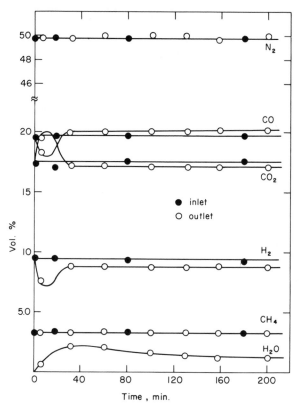

Fig. 1. Fuel gas concentrations during desulfurization at $T = 1000°F$, $SV = 1400$ h^{-1}, and $P = 0$ psig using 10/20 W. Ky. No. 9 ash.

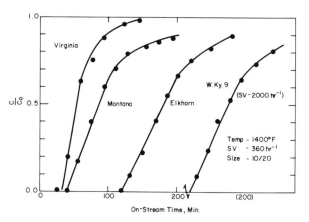

Fig. 2. Sulfide breakthrough curves for four ashes.

sulfur capacity and the operating space velocity. When the iron oxide at the bed entrance is spent, the shape of the reaction wave is set. If axial diffusion is assumed to be negligible relative to convective flow, the wave penetrates the bed in a constant pattern form. The breakthrough curves, as defined by the GC-detected concentrations of H$_2$S and COS, are

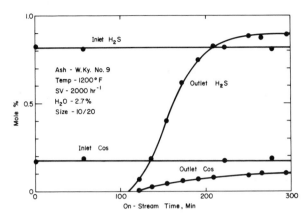

Fig. 3. Effect of COS on desulfurization.

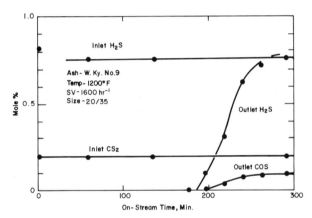

Fig. 4. Effect of CS₂ on desulfurization.

mirror images of the reaction waves and therefore describe the reaction rates within the bed (Fig. 2). Even when no COS is introduced to the bed, it is formed at an equilibrium concentration [Eq. (3)]. No CS_2 is formed. Figures 3 and 4 show the effect of introducing COS and CS_2 to the bed.

Total ash sulfur capacities were evaluated by the equation:

$$M_0 = SV \cdot C_0 \int_0^t (1 - \bar{Y}) \, dt \,, \quad \bar{Y} \to 1 \,. \quad (7)$$

Sorption bed efficiencies, at specified breakthrough concentrations of H_2S, were evaluated from the equation:

$$\tau = t / \int_0^t (1 - \bar{Y}) \, dt \,, \quad \bar{Y} \to 1 \,. \quad (8)$$

In this work the efficiencies were evaluated at the time when $\bar{Y} = 0.15$. Total sulfur capacity was expected to be a function of the inherent iron content of the ashes

only; however, it was found that fresh ashes—those which had not been subjected to a series of sorption-regeneration cycles—had sulfur capacities that were only a fraction of their theoretical capacity, typically representing 10–20% of their iron contents. As the number of cyclic uses was increased, the sulfur capacity increased to a constant value beyond 10–12 cycles. With the Western Ky. No. 9 ash, at 1400°F, after 10 cycles the capacity was a relatively stable 410 grains of sulfur per pound of ash. This represents 60% use of the total Fe_2O_3.

Analysis of random particles of ash by a scanning electron microscope with an EDAX attachment indicated that cyclic use of the ash results in a diffusion of the iron from deep within to the surface of the particle and its pores (Fig. 5). This finding suggested that ash beds subjected to cyclic use would show increased capacity and improved bed efficiencies, and such seems to be the case (Table 3).

Selected specimens of ash were also examined by x-ray analysis to identify the major iron compounds (Fig. 6). Magnetite and hematite were the prevalent iron oxides in the fresh, and pretreated and regenerated samples respectively. Pyrrhotite (FeS) was the only iron-sulfur compound identified in the spent samples. These findings support the basic chemistry given earlier.

Further analysis of results (Table 3) indicates that ash capacities and bed efficiencies increase with increasing cycles (to a point), temperature, Fe_2O_3 concentration and decreasing particle sizes. At constant linear velocity, space velocity has no significant effect on capacity or τ.

Mathematical analyses of the results, for the purpose of model development, were confined to those tests where ashes had experienced ten or more cycles, because only then were results reproducible. To define the effects of various physical and operating parameters on the dynamics of the process better, two models were formulated, solved in exact fashion subject to known boundary conditions, solved in constant pattern fashion, and then evaluated following application of the rate parameters.

MATHEMATICAL MODELS

The reaction of H_2S with Fe_3O_4 within ash particles involves the following transport and kinetic steps:

- Diffusion of H_2S from the bulk gas phase to the solid particle.

36   J. T. Schrodt

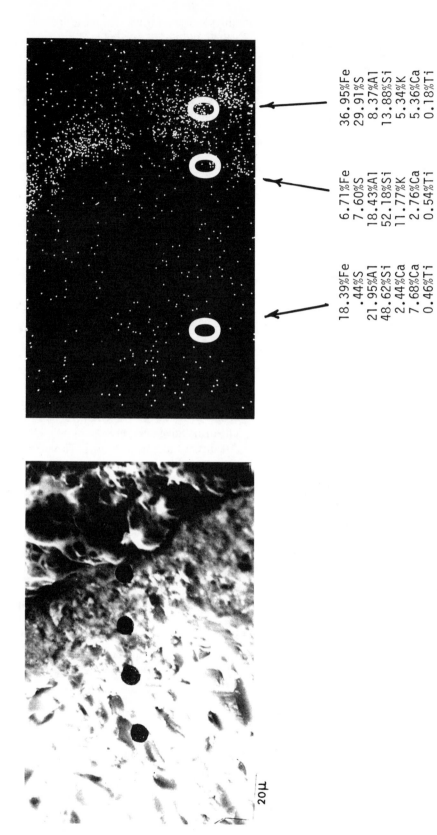

Fig. 5. Scanning electron micrograph of cross section of ash particle after H₂S adsorption. Right hand picture shows distribution of sulfur along particle edge. Normalized EDAX results indicate high sulfur concentrations along edge.

Table 3. Selected desulfurization results

Run[a]	Ash type	Fe$_2$O$_3$ (wt %)	Particle size (Tyler mesh)	Temperature (°F)	GHSV (h^{-1})	Number of cycles	Sulfur capacity of ash (grains/lb)	Efficiency, τ (at 15% S)
4-2	W. Ky 9	23.1	10/20	1200	2000	2	84.0	0.43
4-25	W. Ky 9	23.1	10/20	1200	2000	25	353.0	0.89
4-21	W. Ky 9	23.1	10/20	1400	2000	21	414.0	0.92
4-17	W. Ky 9	23.1	10/20	800	2000	17	171.0	0.67
4-23	W. Ky 9	23.1	10/20	1200	4000	23	363.0	0.90
6-17	W. Ky 9	23.1	4/8	1200	1500	17	282.0	0.67
7-10	W. Ky 9	23.1	20/35	1200	2000	10	431.0	0.86
8-18	Virginia	5.02	10/20	1400	363	18	29.2	0.72
10-11	Elkhorn 3	9.51	10/20	1400	363	11	121.0	0.89
11-11	Montana Rosebud	8.19	10/20	1400	363	11	52.6	0.65

[a]All runs were at one atmosphere of pressure.

- Diffusion of H$_2$S through interstices of the particle to reaction sites.

- Chemical reaction of H$_2$S with unreacted Fe$_3$O$_4$.

- Counter-diffusion of the product gas, H$_2$O, out of the solid.

SEM analyses showed the oxide is concentrated after cyclic use in a thin layer near the particle surface. The layers approximate flat plates of thickness 10–50 μ fixed by the parent iron oxide concentrations. Two asymptotic regimes of reaction control are suggested: a region of chemical control where Fe$_3$O$_4$ grains react in a spatially uniform way with a uniform concentration of H$_2$S, and a region of diffusion control where reactant gas diffuses through interstices and reacts rapidly at a sharp solid product-reactant interface.

The uniform chemical reaction rate model describes the behavior of the bed when chemical reaction controls the process. Basic assumptions include isothermal plug flow, irreversible reaction, pseudosteady-state relative to bed and particles, and uniform dispersal of H$_2$S in the particles. When material balances were developed for the sulfur components in both the gas and solid phases in a differential element of the reactor bed, the following two coupled differential equations expressed in dimensionless form were obtained:

$$\frac{\partial \bar{Y}}{\partial Z} = -\frac{\bar{Y}(1 - \bar{X})}{\dfrac{1}{N_K} + \dfrac{(1 - \bar{X})}{N_F}} , \tag{9}$$

$$\frac{\partial \bar{X}}{\partial \tau} = \frac{\bar{Y}(1 - \bar{X})}{\dfrac{1}{N_K} + \dfrac{(1 - \bar{X})}{N_F}} . \tag{10}$$

The dependent variables $\bar{Y}$ and $\bar{X}$ are the dimensionless gas and solid phase sulfur concentrations respectively, and the independent variables $\bar{Z}$ and τ are the dimensionless bed depth and time respectively. The latter term is defined as:

$$\tau \equiv \frac{\rho_m y_0 G/Vt}{M_0(1 - \epsilon)} . \tag{11}$$

Equations (9) and (10) were solved in exact form[4] with the boundary conditions:

$$\bar{Y} = 0 \text{ for } \bar{Z} > 0, \tau = 0 ,$$
$$\bar{Y} = 1 \text{ for } \bar{Z} = 0, \tau \geqslant 0 ,$$
$$\bar{X} = 0 \text{ for } \bar{Z} \geqslant 0, \tau = 0 .$$

The solution is:

$$\tau = \frac{1}{N_F} \cdot \frac{1 - \lambda}{1 - \lambda \bar{Y}} - \frac{1}{N_K} \ln \frac{\lambda (1 - \bar{Y})}{1 - \lambda \bar{Y}} , \tag{12}$$

where

$$\lambda = \frac{\exp(-N_K)}{\bar{Y}^{[1+(N_K/N_f)]}} . \tag{13}$$

The numbers N_K and N_F are the dimensionless kinetic and bulk diffusion rate parameters defined as:

$$N_K \equiv k_k \frac{(1 - \epsilon)\alpha M_0}{G/V} , \tag{14}$$

$$N_F \equiv k_m \frac{a_p}{G/V} . \tag{15}$$

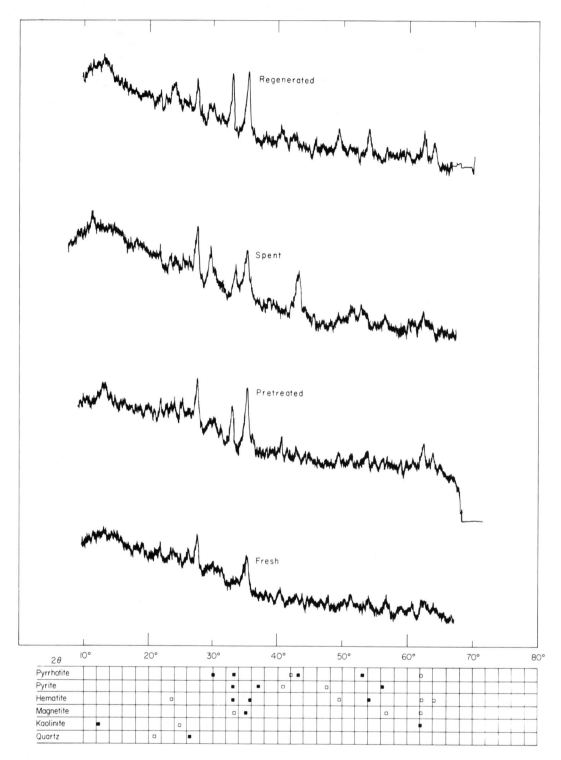

Fig. 6. X-ray analyses of fresh, pretreated, spent, and regenerated W. Ky. No. 9 coal ash (■—primary peak; □—secondary peak).

Equation (12) was used to predict the bed efficiency, τ, for values of $1.0 > \bar{Y} > 0$ at $\bar{Z} = 1.0$ and $\bar{X}(0,\tau) < 1.0$.

Equations (9) and (10) were also solved in constant pattern fashion[5] in which $\bar{X} = \bar{Y}$. A general integration gives the equation:

$$\tau = 1 + \frac{(1 + \ln \bar{Y})}{N_F} + \frac{\ln[\bar{Y}/(1-\bar{Y})]}{N_K}. \qquad (16)$$

If a bed of sorbent is long enough for the sorbate concentration wave to be fully developed before breakthrough, the bed efficiency can be approximated by Eq. (16) if the model is correct and N_K and N_F are known.

The second model describes the behavior of the bed when pore diffusion controls the process rate. In addition to the previous assumptions set forth, there is the added condition of instant chemical reaction and a thin flat plate geometry for the solid reactant. Gas and solid phase material balance equations developed previously apply to this model also. Thus in dimensionless form:

$$\frac{\partial \bar{Y}}{\partial Z} = -\frac{\bar{Y}}{\dfrac{1}{N_F} + \dfrac{\bar{X}}{N_P}}, \qquad (17)$$

$$\frac{\partial \bar{X}}{\partial \tau} = \frac{\bar{Y}}{\dfrac{1}{N_F} + \dfrac{\bar{X}}{N_P}}, \qquad (18)$$

where

$$N_P \equiv \frac{3a_p D_p(1 - \epsilon)}{\phi_s(r_p - r_o)r_p G/V}. \qquad (19)$$

When Eqs. (17) and (18) are solved by the previously cited method of Bischoff[4] with the cited boundary conditions, they yield

$$\tau = \frac{[N_P + (N_P/N_F)\ln \bar{Y}]^2}{2N_P(1 - \bar{Y})^2} + \frac{[N_P + (N_P/N_F)\ln \bar{Y}]}{N_F(1 - \bar{Y})}. \qquad (20)$$

Equation (20) was used to predict the bed efficiency, τ, for specified concentrations of sorbate, $\bar{Y}$, at $\bar{Z} = 1.0$ and $\bar{X}(0,\tau) < 1.0$. The constant pattern solution to Eqs. (17) and (18) when $\bar{X} = \bar{Y}$ is:

$$\tau = 1 + \frac{(1 - \ln \bar{Y})}{N_F} + \frac{(\bar{Y} - 1/2)}{N_P}. \qquad (21)$$

Graphic representations of the exact and constant pattern solutions for the two models (Figs. 7 and 8) were prepared for selected values of the dimensionless rate parameters N_F, N_P, and NK; however, as indicated, the results are not affected by variations in N_F above a value of 50. For all the test runs N_F was above 500; this clearly shows that bulk gas phase diffusion was not a controlling factor. It is noteworthy that the exact analytical solutions are valid only for $\bar{X}(0,\tau) \leqslant 1.0$; this is the maximum concentration of FeS. At $\bar{X}(0,\tau) = 1.0$ the constant pattern solutions are valid, because then $\bar{X} = \bar{Y}(0,\tau) = 1.0$. The curves also show that as N_P and N_K increase, the exact and constant pattern solutions merge at lower values of $\bar{Y}$ and τ. As expected, when N_P and

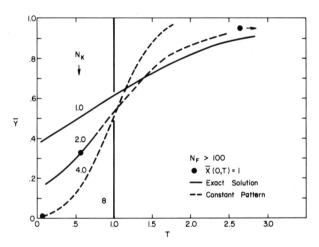

Fig. 7. Exact and constant pattern solutions for the uniform chemical reaction rate model.

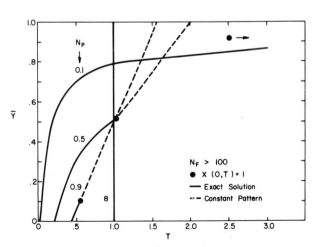

Fig. 8. Exact and constant pattern solutions for the pore-diffusion model.

N_K become infinite, a condition corresponding to instantaneous reaction rate, perfect sorption is predicted (i.e., all sorbate introduced to the reactor bed reacts with all available solid reactant before breakthrough occurs).

Neither of the models adequately describes the shape of the experimental breakthrough curves over the full concentration range $1.0 \geq \overline{Y} \geq 0$ nor the correct efficiency at $\overline{Y} = 0.15$. However, a combination of the two models does this. As the SEM-EDAX analyses showed, the iron oxide was concentrated near the surface of the particles in a highly porous or spongy structure, while some iron oxide remained in the deep core of the particles. For $\overline{Y} \leq 0.7$–0.8 the uniform chemical reaction rate model fits the data well, and for $Y \geq 0.7$–0.8 the pore diffusion model fits the data. Combining the models provided a good overall solution.

Each of the dimensionless rate numbers contains a rate coefficient, either k_m, k_k, or D_p. For each of the test runs k_m was evaluated from an established packed-bed, mass transfer correlation:[6]

$$k_m = 1.15 \frac{U_0}{\epsilon} \left(\frac{2r_p U_0}{v\epsilon} \right)^{-0.5} \left(\frac{v}{D_m} \right)^{-0.667} . \quad (22)$$

The reaction rate constant, k_k, was correlated using the test data and the Arrehenius expression:

$$k_k = S \exp \left(-E/RT \right). \quad (23)$$

For the four ash sorbents tested,

$$E = 4940 \pm 55 \text{ Btu/1b} \cdot \text{mol} . \quad (24)$$

The kinetic pre-exponential, S, varied from 0.68 to 2.1×10^5 for the four ash sorbents. No specific explanation can be given for the latter value. The upper portions of the experimental breakthrough curves were fitted to the pore diffusion model, and values of D_p in the range 10^{-3} to 10^{-4} ft²/h were obtained. These values are in the Knudsen diffusivity range and vary directly with the square root of the temperature. Specific details of the correlational procedures and applied data are given elsewhere.[7]

DYNAMIC SIMULATION

A finite-element computer program was prepared for solution of the material balance differential equations for the purpose of dynamically simulating

the desulfurization phase of the process and confirming the applicability of the models. Some curves (Fig. 9) show the development and progression of the H_2S concentration wave within the ash-sorbent bed; one curve (Fig. 10) shows the predicted breakthrough of H_2S at $\overline{Z} = 1.0$. The pore-diffusion model was activated at $\overline{Y} = 0.7$. Similar curves may be prepared for the solid phase concentrations. The computer program, incorporating the two models, probably can predict the capacity and efficiency of a fixed-bed fuel gas desulfurization process adequately.

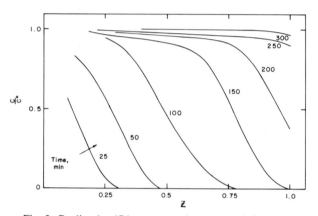

Fig. 9. Predicted sulfide concentration waves within a hypothetical reactor.

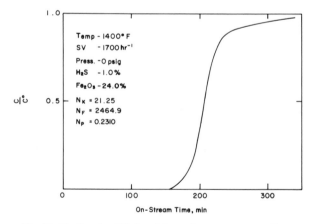

Fig. 10. Predicted sulfide breakthrough curve corresponding to predicted sulfide concentration waves within a hypothetical reactor.

SUMMARY COMMENTS

These results should not be regarded as independently conclusive, because they represent an incomplete phase of a broad program of research and development. No results on the heat effects or the regeneration phase of the process were presented here. When completed, the work probably will support a fuel gas desulfurization process that is superior to the low-temperature processes.

NOMENCLATURE

a_p	surface area of particles per unit volume of packed reactor, $\mathrm{ft^2/ft^3}$
C_0	inlet H_2S concentration $\mathrm{lb \cdot mol/scf}$
D_m	gas phase diffusivity, $\mathrm{ft^2/h}$
D_p	pore diffusivity, $\mathrm{ft^2/h}$
E	reaction activation energy, $\mathrm{Btu/lb \cdot mol}$
G	volumetric gas flow rate, $\mathrm{ft^3/h}$
k_k	specific reaction rate constant, $\mathrm{ft^3/lb \cdot mol \cdot h}$
k_m	gas phase mass transfer coefficient, $\mathrm{ft/h}$
M_0	ash sulfur capacity, $\mathrm{lb \cdot mol/ft^3}$ solid
N_F	dimensionless film diffusion number
N_K	dimensionless kinetic number
N_P	dimensionless pore diffusion number
r_p	particle radius, ft
S	kinetic pre-exponential
SV	space velocity, $\mathrm{h^{-1}}$
R	gas law constant, $1.987\ \mathrm{Btu/lb \cdot mol \cdot {}^\circ R}$
t	on-stream time, h
T	temperature, $^\circ R$
U_0	superficial velocity, $\mathrm{ft/h}$
V	packed bed volume, $\mathrm{ft^3}$
$\bar{X}$	dimensionless fraction of reacted solid
y_0	inlet bulk gas phase sorbate mole fraction
$\bar{Y}$	dimensionless sorbate concentration
$\bar{Z}$	dimensionless position in packed bed
α	moles of reactant solid per mole of sorbate
α_p	particle porosity in particle shell, $\mathrm{ft^3/ft^3}$
ϵ	bed porosity, $\mathrm{ft^3\ void/ft^3\ bed}$
ν	viscosity, $\mathrm{ft^2/h}$
ρ_m	bulk gas phase molar density, $\mathrm{lb \cdot mole/ft^3}$
τ	dimensionless time defined by Eq. (11)
ϕ_s	sphericity factor
λ	group defined by Eq. (13).

REFERENCES

1. J. T. Schrodt, and O. J. Hahn, *Hot Fuel Gas Desulfurization,* Report IMMR 15-PD11-76, University of Kentucky, May 1976.

2. B. N. Murthy et al., *Fuel Gas Cleanup Technology for Coal Gasification,* U.S. Department of Energy Report FE-2220-15, March 1977.

3. J. T. Schrodt, G. B. Hilton, and C. A. Rogge, "High-Temperature Desulfurization of Low-CV Fuel Gas," *Fuel,* **54,** 269 (1975).

4. K. B. Bischoff, "General Solution of Equations Representing Effects of Catalysts Deactivation in Fixed-Bed Reactors," *Ind. Eng. Chem. Fundam.* **8,** 665 (1969).

5. K. R. Hall et al., "Pore- and Solid-Diffusion Kinetics on Fixed-Bed Adsorption under Constant-Pattern Conditions," *Ind. Eng. Chem. Fundam.* **5,** 212 (1966).

6. J. J. Carberry, "A Boundary-Layer Model of Fluid-Particle Mass Transfer in Fixed Beds," *A.I. Ch.E. J.* **6,** 460 (1960).

7. J. T. Schrodt, *Hot Gas Desulfurization,* U.S. Department of Energy Report FE-5076-5, September 1978.

DISCUSSION

M. A. Shapiro, University of Pittsburgh: I was just wondering whether your capacity measurement was that per regeneration or for the total capacity, and I'm not sure I understood the relationship to efficiency.

Tom Schrodt: Okay, are you talking about the capacity or the efficiency or both of them? The capacity of the ash measured in terms of grains of salt per pounds of ash is essentially what you saw there—typically maybe 400 or 500 grains per pound. That, by the way, is a utilization factor of about 60% of the inherent iron within the ash particle, and that's pretty much irrespective of what ash I use, whether it's sub-bituminous or whatever. To reach that capacity, however, we have to regenerate a minimum of about 8 to 9 times to bring that iron to the surface. Now, if we were to leave the gas passing through the bed for three or four days as opposed to 200 or 300 min, I am sure that some of the hydrogen sulfide could migrate way down into the interstices of the particle, and that is why the breakthrough curves, as you approach one, tend to flatten out and continue on. So we still have some capacity left in there, that's true. The efficiency, the tau value that you see there, is a measure of the fraction of the 60% that was utilized at 15% breakthrough of the hydrogen sulfide. That is, when the H_2S concentration reaches 15% of the effluent, then a tau, say 0.85, means that we have passed 85% of the sulfur into the bed, and it has been adsorbed of the utilizable fraction. That sounds like a lot of double-talk but that's essentially what it means, Dr. Shapiro.

INTERPRETATION OF THE CO₂-ACCEPTOR ENVIRONMENTAL DATA BASE: INFLUENCE OF PROCESS AND SCALE-UP FACTORS

J. C. Craun*   M. J. Massey*

ABSTRACT

A comprehensive analysis of the effluent production characteristics and related process conditions for the CO_2-Acceptor coal gasification process has recently been completed. Methodologies and data needed to perform this type of scalable environmental data analysis are discussed, and examples from the CO_2-Acceptor data base are used to demonstrate these needs. Examples discussed include aspects of characterizing the behavior of sulfur, nitrogen, and trace elements in the CO_2-Acceptor process.

INTRODUCTION

A comprehensive analysis of the effluent production characteristics and related process operating conditions for a high-BTU coal gasification pilot plant has recently been completed for Conoco Coal Development Company's (CCDC) CO_2-Acceptor process.[1] This work was done under subcontract to Carnegie-Mellon University (C-MU), the U.S. Department of Energy's Fossil Energy environmental coordination contractor.[2]

This analysis of effluent characteristics in conjunction with process operating conditions, or process environmental analysis (PEA) (Fig. 1), is required for reliably predicting commercial-scale effluent production from pilot plant data. These predictions provide appropriate data for assessing effluent treatability and for predicting final effluent discharges and environmental impacts[3] (Fig. 2).

This paper presents the basic methodology for performing a process environmental analysis by utilizing the CCDC CO_2-Acceptor pilot plant data and the Stearns-Roger conceptual design of a commercial-scale CO_2-Acceptor coal gasification plant.

DESCRIPTION OF METHODOLOGY

Major Factors in PEA

The major factors involved in performing a process environmental analysis on pilot plant (or sub-commercial-scale) data are summarized in Figure 1. The key steps involved are characterization and data analysis for scaling. The characterization step consists of chemical and bio-assay measurements on samples from various process and effluent streams for a variety of components. This effort must be supplemented and linked with relevant process data such as operating temperatures and pressures, flow rates, and major component analyses to allow proper analysis for scaling.

In the case of the CO_2-Acceptor process, the primary effluent data was collected by Radian Corporation,[4] with support efforts by C-MU[5-7] and the South Dakota School of Mines and Technology.[8] The process data was provided by CCDC.[9]

---

*Environmental Research & Technology, Inc., Pittsburgh National Building, Fifth and Wood, Pittsburgh, Pennsylvania 15222.

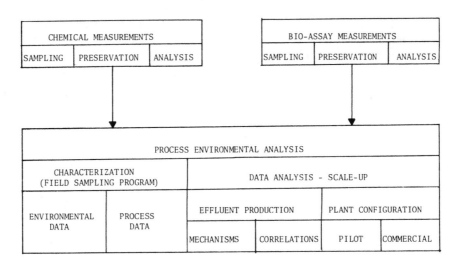

Fig. 1. Major factors associated with subcommercial-scale process environmental analysis.

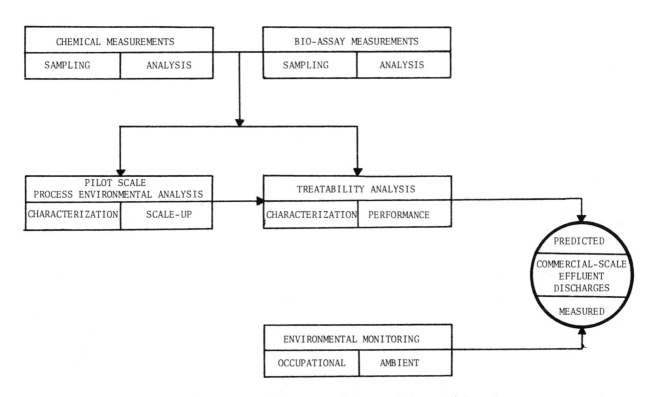

Fig. 2. Steps in the projection of commercial-scale environmental discharge data.

Data analysis for scale-up requires insight regarding basic effluent production mechanisms and knowledge of specific similarities and differences between pilot- and commercial-plant configurations. Some knowledge of effluent production mechanisms is needed to identify and interpret properly the relationship(s) between process conditions and effluent production. This knowledge can either be fundamentally based, e.g., reaction kinetics, phase equilibrium, or transport phenomena, or empirically derived from correlations which use gross process parameters such as carbon conversion, temperature, residence time, or partial pressure. Although fundamentally based understanding is preferred, often empirically based relationships must be used because of the lack of relevant data to conduct a fundamentally based analysis.

An understanding of the similarities and differences between pilot- and commercial-plant configurations is necessary to assess the scalability of the streams sampled and the need for data manipulation to obtain scalable results. The comparison between pilot and commercial plants is essentially concerned with two basic areas: scale-up considerations in the "heart" of the pilot plant (usually the coal gasifier or oil shale retort or coal liquefaction reactor and operation itself) and the effects of possibly non-scalable downstream equipment on characterizing scalable streams. These areas of concern arise from the fact that most pilot plants have been built and operated to produce scalable data for only the principal process operations, such as the reactor, with all ancillary equipment operated with little concern for scale-up. This mode of operation has essentially resulted from economic concerns being focused primarily on straight process development and demonstration. However, because of increasing concerns for environmental discharges and the high cost of pollution control equipment, it may become important for pilot plant operations to expand their focus to include development of scalable effluent production data.

In the case of CO_2-Acceptor, the pilot and commercial plant comparisons were made using the CCDC pilot plant[4] and Stearns-Roger conceptual commercial designs.[10]

Basic Strategies and Techniques for PEA

The basic strategy used in analyzing effluent data from a subcommercial-scale process is to characterize the behavior of environmentally significant species in the process. This means that production, speciation, and distribution results and mechanisms are investigated for individual generic species such as sulfur, nitrogen, hydrocarbons, and trace elements. Such a strategy clarifies from the outset the necessary linkages among process data, environmental data, behavior mechanisms data, and scalability considerations.

As Table 1 shows, there are three basic sequential data analysis steps required to implement this strategy. First, the intrinsic scalability of the data must be assessed. Then the reliability of the data must be assessed. Finally, the scalable and reliable data

Table 1. Basic strategy and techniques for process environmental analysis of sub-commercial effluent data

Basic Strategy

Characterize the behavior of environmentally significant species in the process.

- Behavior is production, speciation, and distribution results and mechanisms.
- Generic species include sulfur, nitrogen, hydrocarbons, and trace elements.

Implementation

1. Assess data scalability by comparing pilot- and commercial-plant configurations.

   - Identify scalable process streams and units.
   - Identify critical scale-up parameters in scalable process units.
   - Identify effects of non-scalable downstream units.

2. Assess data reliability.

   - Compare replicate sampling results.
   - Analyze results of sample spiking recovery tests.
   - Check mass balance closure.

3. Analyze scalable, reliable data for effluent production mechanisms and correlations.

   - Check results of time-series sampling.
   - Compare data with predictions from reaction kinetics, phase equilibria, etc.
   - Correlate data with relevant process parameters.

must be used to investigate effluent-production dynamics (mechanisms and correlations).

Table 1 shows a number of the specific techniques which can be used for each of these steps. The problem of assessing scalability was discussed in the previous section and most of the techniques for the following steps are self-explanatory. However, some emphasis is in order. For example, checking mass balance closure gives the definitive measure of the accuracy and reliability of the data, as well as important initial information on distribution, and should be done whenever possible. Time-series sampling is valuable for determining the sensitivity of measured results to process variations but requires intimate linkage of process and effluent data collection efforts. The correlation of observed results with process parameters is crucial and often requires considerable judgment and familiarity with the process involved.

In summary: Table 1 shows a number of techniques available for implementing the strategy of characterizing the behavior of individual, environmentally significant species. When they are applied to an effluent data base as part of the process environmental analysis procedure, each environmentally significant species can be categorized as being:

1. well characterized by fundamental behavior mechanisms

2. well characterized by empirical process correlation

3. poorly characterized because of conflicting and/or unreliable results

4. uncertainly characterized because of insufficient data

In this paper, examples from the CO₂-Acceptor effluent data base will be given for each of these categories to illustrate the use of process environmental analysis techniques and the kinds of problems that can arise in this effort.

CO₂-ACCEPTOR PROCESS DESCRIPTION

Basic Process Concepts

Figure 3 shows a schematic flow diagram of the two-reactor "heart" of the CO₂-Acceptor process. In the gasifier, lignite or sub-bituminous coal is gasified with steam at 1500–1550°F. The heat for the endothermic steam gasification reactions is supplied by the exothermic reaction of the CaO acceptor with CO_2 produced by the gasification reactions. The resulting $CaCO_3$ is transferred to the regenerator where it is calcined back to CaO at about 1850°F to close the acceptor recirculation loop. The heat for calcination is supplied by burning char residue from the gasifier. Purge and make-up acceptor streams are part of the recirculation loop because of the gradual loss of acceptor activity with repeated reaction. Ash is collected in a cyclone at the regenerator outlet.

The CO₂-Acceptor process has a number of advantages because of its use of two separate reactors. Because combustion occurs in the regenerator, the product gas is undiluted with nitrogen without the use of an oxygen plant. The product gas is rich in hydrogen (50–60%) so that a shift converter is not required before methanation. Product gas clean-up is minimized since the acceptor adsorbs both H₂S and CO₂. Only lignite and sub-bituminous coals can be gasified, however. At the higher temperatures needed to gasify bituminous coals, thermodynamic constraints imposed on the CO₂-

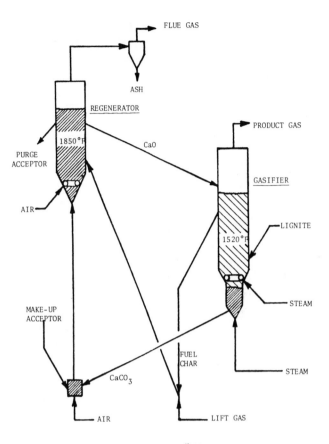

System pressure ≅ 11 atm.

Fig. 3. CO₂-Acceptor gasifier/regenerator schematic flow diagram.

Acceptor reactions in the gasifier and regenerator cannot be satisfied. The higher temperatures required to gasify bituminous coals would also result in ash fusion.

Scale-Up Concerns

The gasifier and regenerator operations for both the pilot and conceptual commercial-scale plants are essentially identical to those shown in Fig. 3. However, the mechanics of operating the pilot plant required using an auxiliary fuel in the regenerator, which was usually lignite. In the commercial-scale operation, it is anticipated that through proper materials handling the char residue from the gasifier will completely satisfy the regenerator heat requirements. With only this and some other minor differences in gasifier–regenerator operation, the major areas of difference between the pilot- and commercial-scale unit are in the downstream processing of the raw product and flue gases.

In the pilot plant the raw product and flue gases are both immediately cooled, passed through cyclones for particulate removal, and quenched with once-through city water. In the commercial plant the gas handling trains are significantly different, as illustrated in Fig. 4. From this figure it is obvious that pilot- and commercial-scale gas streams are directly comparable only at the immediate exit points from the reactors. The differences in gas handling beyond these points mean that downstream pilot plant units produce non-scalable effects to various extents, depending on the species and stream considered. Hence, significant data manipulation and judgment are often required to predict commercial-scale effluent production and behavior reliably from measurements of pilot plant streams.

The preceding discussion illustrates a number of typical concerns in scaling pilot plant data when the pilot operation only addresses the problems of the principal process unit. While this discussion is by no means exhaustive for the CO_2-Acceptor pilot plant, let alone for other processes, it does exemplify typical concerns in selecting and interpreting scalable effluent data from pilot plant operations.

PEA EXAMPLES FROM CO_2-ACCEPTOR

Process environmental analysis of the CO_2-Acceptor pilot plant effluent data gives examples of each of the categories of species characterization discussed previously. These are: (1) H_2O production in the gasifier to illustrate species behavior well characterized by fundamental mechanisms, (2) nitrogen conversion in the gasifier and regenerator to illustrate species behavior well characterized by empirical process correlations, (3) sulfur distribution in the gasifier and regenerator as an example of extensive data giving conflicting and/or unreliable results, and (4) trace element behavior in the gasifier and regenerator to illustrate the case of insufficient data giving uncertain results on species behavior.

H₂S Production in the Gasifier

Characterizing H_2S production in the CO_2-Acceptor gasifier is a good example of characterizing the behavior of an environmentally significant species by fundamental mechanisms. The method used was to compare observed H_2S concentration in the raw product gas with the levels predicted by reaction equilibrium and thermodynamics.

The primary sulfur reaction in the gasifier is

$$CaCO_{3(s)} + H_2S_{(g)} \leftrightarrow CaS_{(s)} + CO_{2(g)} + H_2O_{(g)} \ .$$

This is the means by which coal sulfur liberated as H_2S is transferred to the regenerator as CaS in the recirculating acceptor. Laboratory work by CCDC[9] and others[11] indicates that this reaction achieves equilibrium quickly and reaction equilibrium constants at different temperatures are available. Figure 5 compares observed H_2S levels with levels predicted from reaction equilibrium considerations. It is apparent from the data that measured H_2S levels in the pilot plant raw product gas correspond closely to the predicted equilibrium levels. The pilot plant data represents H_2S measurements from a number of different runs using different sampling and analytical methods, which lends confidence to the equilibrium conclusion.

The result indicated by Fig. 5 is that achieving equilibrium in the $CaCO_3/H_2S$ reaction is the proper basis for projecting raw product gas H_2S concentrations. This result is significant because factors

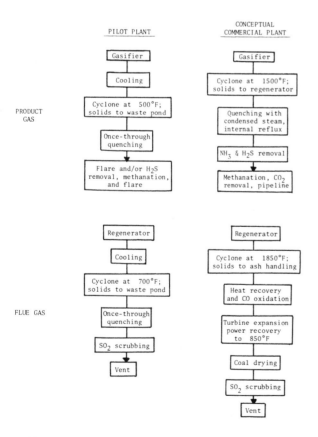

PILOT PLANT

PRODUCT GAS

Gasifier
↓
Cooling
↓
Cyclone at 500°F; solids to waste pond
↓
Once-through quenching
↓
Flare and/or H_2S removal, methanation, and flare

FLUE GAS

Regenerator
↓
Cooling
↓
Cyclone at 700°F; solids to waste pond
↓
Once-through quenching
↓
SO_2 scrubbing
↓
Vent

CONCEPTUAL COMMERCIAL PLANT

Gasifier
↓
Cyclone at 1500°F; solids to regenerator
↓
Quenching with condensed steam, internal reflux
↓
NH_3 & H_2S removal
↓
Methanation, CO_2 removal, pipeline

Regenerator
↓
Cyclone at 1850°F; solids to ash handling
↓
Heat recovery and CO oxidation
↓
Turbine expansion power recovery to 850°F
↓
Coal drying
↓
SO_2 scrubbing
↓
Vent

Fig. 4. Comparison of downstream gas handling in pilot- and commercial-scale CO_2-Acceptor plants.

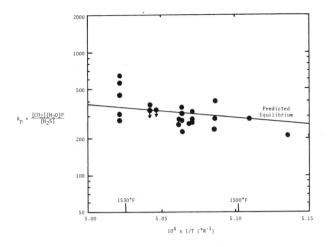

$$k_p = \frac{[CO_2][H_2O]P}{[H_2S]}$$

Fig. 5. Comparison of pilot plant data with thermodynamic equilibrium for the reaction $CaCO_3 + H_2S \leftrightarrow CaS + CO_2 + H_2O$.

influencing the equilibrium H_2S level, such as steam and CO_2 concentrations, will change in process scale-up and the effect of these changes can only be predicted with this fundamental understanding of gasifier H_2S behavior.

It is instructive to examine in some detail the effort required to achieve fundamentally based characterization of an environmental species such as gasifier H_2S, given the importance of such fundamental understanding in scale-up projections. Three distinct types of data are required: (1) environmental data such as H_2S measurements; (2) process data such as temperature, pressure, and H_2O and CO_2 concentrations; and (3) external fundamental behavior data such as kinetics for the $CaCO_3/H_2S$ reaction. The process and environmental measurements need to be closely linked in timing and location. Part of the data scatter in Fig. 5 is due to the fact that this required linkage was not always achieved at CO_2-Acceptor. The effluent sampling of raw gas streams at CO_2-Acceptor and almost any gasification pilot plant is very difficult because of high temperatures and pressures, heavy particulate loadings, and the presence of considerable amounts of steam. However, in many cases the raw gas must be sampled if scalable data is to be obtained.

Nitrogen Conversion in the Gasifier and Regenerator

Nitrogen conversion in the gasifier and regenerator of the CO_2-Acceptor pilot plant provides a good example of characterizing the behavior of an environmentally significant species with empirical

process correlations. This is achieved by correlating the nitrogen species NH_3 and HCN in the product and flue gases with relevant process operation parameters. These gaseous nitrogen species result from the thermal conversion of solid-phase nitrogenous compounds in the gasifier coal feed and the regenerator fuel.

Figure 6 shows that coal nitrogen conversion in the gasifier is correlated with coal carbon conversion and that nitrogen conversion levels are in the range of 85–90%. Each data point is based on feed coal and residual char nitrogen analyses for a separate pilot plant run. The correlation of high nitrogen

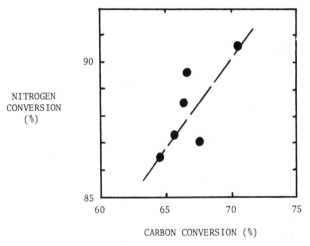

Fig. 6. Nitrogen conversion in the gasifier.

conversion with carbon conversion is also seen in other high-Btu coal gasification plants* (Refs. 2,12). Although the fundamental mechanisms behind this correlation are at present unknown, the evidence of correlation for different pilot plants and the fact that carbon conversion reflects the interaction of critical gasification parameters such as temperature and residence time clearly supports the correlation as a means of confidently predicting commercial-scale nitrogen conversion levels.

Even though the data in Fig. 6 were not derived from raw product gas analyses, the limited product gas measurements available indicate that essentially all of the converted nitrogen is released as NH_3 with a trace amount, less than 0.5%, released as HCN.

*Preliminary data from low-Btu gasification plants indicates that coal nitrogen conversion in air-blown gasifiers may be considerably lower than 80–90%.

48   J. C. Craun and M. J. Massey

Release as N₂ cannot be determined accurately but appears to be negligible. At this time there is insufficient data to understand the cause(s) of speciation between NH_3 and HCN.

Figure 7 shows that converted nitrogen measured as NH_3 and HCN in the regenerator flue gas appears to come entirely from auxiliary fuel nitrogen and not from char nitrogen. Each point on this plot represents raw flue gas measurements by Radian Corp. and solids analyses and flow rates from CCDC for a separate pilot plant run. The indication that essentially all gaseous nitrogen is released by the auxiliary fuel is significant because at the commercial scale, all regenerator heat requirements should be met by gasifier char and the nitrogen in this char is

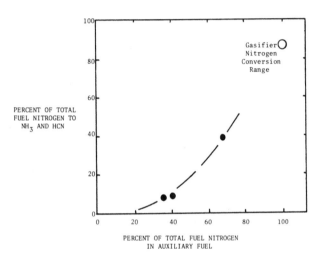

Fig. 7. Nitrogen conversion in the regenerator.

apparently refractory and unreactive. Hence, a simple extrapolation of commercial-scale flue gas NH_3 and HCN levels from pilot plant measurements where auxiliary fuel was used in the regenerator would give erroneous results.

The available data do not permit a complete material balance for nitrogen in the regenerator. The residual nitrogen, that which is not converted to NH_3 and HCN, may remain in the ash or may be driven off as N₂ gas. Spot sampling of the gas streams for NO_x showed these species were absent.

Sulfur Distribution in the Gasifier and Regenerator

Sulfur distribution between gasifier and regenerator solid and gaseous streams is an example of the case where extensive data gives conflicting and/or

unreliable results. The distribution problem is important because sulfur is a major environmentally significant species in the CO_2-Acceptor process, and treatment design for sulfur-bearing streams will obviously depend on the sulfur distribution.

The conflicting sulfur-distribution results arise from comparing two independent and ostensibly reliable data bases. One data base consists of gasifier–regenerator sulfur balance measurements by Radian Corporation, and the second consists of sulfur balance results from CCDC's heat and material balance model where only solid streams were actually analyzed for sulfur. This data comparison, presented in Table 2, shows considerable disagreement with the exception of Run 39.

From Table 2, it is difficult to choose which of the two distribution results is correct. The Radian distribution is based on measured data which did give closing sulfur balances within the limits of experimental error. CCDC's heat- and material-balance model used measured sulfur contents for all solid streams, assumed $CaCO_3/H_2S$ equilibrium to calculate sulfur in the raw product gas,* and then fixed the flue gas sulfur as necessary to force balance closure. The model was developed as a process analysis tool by CCDC because of difficulties in measuring certain pilot plant streams and was used to close material balances for carbon, hydrogen, oxygen, and nitrogen in addition to sulfur.

Detailed comparisons of Radian and CCDC results for individual balances are of uncertain value since the balances do not match in timing and analyses were not performed on the same samples. However, for certain solids samples, the Radian measurements appear to be consistently lower than those reported by CCDC, suggesting possible problems with the sampling and/or analytical procedures used by the different groups. Furthermore, discrepancies can result from fluctuations in the pilot plant operation during the sampling period.

The net result of this discussion is that there is no definitive way to determine from this limited data base which sulfur distributions are correct. Both may be correct and simply reflect fluctuations in operating conditions. With the limited and conflicting data available the choice is essentially a matter of judgment for the design engineer, and it is not in the province of this work to make this judgment. However, it is significant that such difficulty exists

---

*H₂S accounts for over 90% of the product gas sulfur.

Table 2. Comparison of measured and model overall sulfur distributions

Run	Date and time of sampling		Total outlet sulfur distribution, percent Radian measured/Conoco model		
	Radian[a]	Conoco[b]	Product gas[c]	Flue gas[d]	Solids[e]
47B	2015–0830 (9/28–30/77)	1230–0700 (9/29–10/1/77)	43/23	31/31	36/46
45[f]	0945–0600 (6/5–6/77)	0700–0700 (6/6–7/77)	38/20	39/30	23/50
39	1200–0400 (9/29–30/76)	1900–2300 (9/29/76)	23/20	14/20	63/60

[a]Radian time period for composite solids sampling with gas grab samples at various times during the period.
[b]Conoco time period for composite solids sampling during defined steady-state conditions.
[c]Conoco values derived from $CaCO_3/H_2S$ reaction equilibrium calculation.
[d]Conoco values derived by setting flue gas sulfur content to force balance closure.
[e]Solids are purged acceptor, gasifier and regenerator cyclone solids, and entrained particulates in the product and flue gases.
[f]Radian values given as maximums and minimums because the product gas H_2S measurement was beyond the analytical range.

for a major, fairly well understood and characterized species like sulfur. The lesson is clear that considerable effort is required in linking process and effluent data and in validating sampling and analytical procedures if such conflicts are to be avoided.

Trace Element Behavior in the Gasifier and Regenerator

Trace element behavior in the CO₂-Acceptor process is a good example of a situation where insufficient data exist to give a reliable data base and to establish behavior mechanisms or correlations confidently. Radian Corp. conducted a sampling program during Run 45 to develop material balance data around the gasifier–regenerator, gasifier quench, and regenerator quench for 29 elements. One indication of the reliability of this data is presented in Fig. 8, where the relative volatility of the different elements is correlated with their material-balance closure. This correlation shows a distinct trend of poor material-balance closure for the highly volatile elements. However, not all highly volatile elements have poor balance closure, as shown by chlorine and cadmium.

The data were further examined for reliability by comparing the concentrations of the different elements in samples collected from the same stream during two separate balance periods (Table 3). These comparisons clearly show much poorer agreement for two separate gas samples than for two separate particulate samples. Changing pilot plant operating conditions between the two samples does not appear to account for the disagreement since normalizing concentrations for flow rate and feed coal trace element content changes has little effect on the

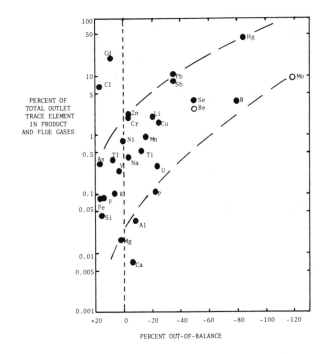

Fig. 8. Balance closure vs relative volatility for Run 45 trace elements. Values for Be and Mo are uncertain since all stream concentrations were below detection limits. Gas-phase measurements for Al, As, Li, Si, and Tl were below detection limits so relative volatility plotted is maximum possible. Percent out-of-balance = (100) (out − in)/(out + in/2).

agreement and individual comparisons show that vapor-phase concentrations appear both to increase and to decrease randomly from one balance period to the next.

The poor material balance closure correlation in Fig. 8 and the wide discrepancies between "duplicate" gas samples shown in Table 3 could result from

Table 3. Comparison of vapor-phase and entrained particulate trace element concentrations for two separate samples taken during Run 45

Stream	Number of elements whose concentration in two different samples differed by factor of:		
	>10	2–10	<2
Unquenched product gas[a]	8	8	13
Unquenched flue gas[b]	6	14	9
Unquenched product gas particulates[a]	2	7	20
Unquenched flue gas particulates[b]	0	7	22

[a]Sampled in the periods 0945–0600 (6/5–6/77) and 2245–1520 (6/6–7/77).

[b]Sampled in the periods 0945–0600 (6/5–6/77) and 0930–2000 (6/8/77).

problems in sampling and/or analytical procedures for gas streams. However, as shown in Fig. 8, such difficulties have a significant effect on balance closure only when the element is highly volatile.

The results of a number of data-reliability checks in addition to those shown in Fig. 8 and Table 3 are summarized in Table 4. This table categorizes the 29 elements as being well or poorly characterized on the basis of material balance closure and data reliability. The importance of this categorization is indicated by classifying the elements as environmentally significant or relatively innocuous according to the toxic pollutant list of the Water Pollution Control Act,[13] the tentative OSHA Class I carcinogens list,[14] and EPA wastewater effluent guidelines.[15] The significant result of Table 4 is that 12 of the 19 environmentally significant trace elements were poorly characterized by the Run 45 work. More importantly, none of the trace elements are well characterized in terms of process correlations or behavior mechanisms since the data was from only one pilot plant run. At best, the trace element data provides screening information for future sampling and analytical methods development.

COMMERCIAL-SCALE PROJECTIONS

The impact of the process environmental analysis for the CO_2-Acceptor process on the prediction of commercial-scale effluent production is shown in Table 5. The table compares the scale-up projection

Table 4. Summary of trace element characterization for Run 45

Environmentally significant[a]				Relatively innocuous	
Well characterized	Poorly characterized[b]			Well characterized	Poorly characterized
Al	Sb	Pb		Ca	B
F	As	Mn		Mg	Li
Fe	Be	Hg		K	Mo
Ni	Cd	Se		Si	Ti
P	Cl	Tl		Na	V
U	Cr				
Zn	Cu				

[a]Underlined elements are on toxic pollutants list of the Federal Water Pollution Control Act. In addition, As, Be, Cd, Cr, Pb, and Ni are on the tentative OSHA category I carcinogens list. All toxic substance list elements except Sb, Be, and Tl are regulated by EPA Effluent Guidelines for the inorganic chemicals, nonferrous metals, iron and steel, steam electric power, electroplating, ore mining, and/or coal mining industrial categories, as are Al, Cl, F, Fe, Mn, P, and U.

[b]Poorly characterized elements were over 25% out-of-balance and/or had significant variability on the basis of the concentration comparisons summarized in Table 3. Variability was considered to be significant if it was over a factor of 2 for vapor-phase streams for relatively volatile elements (over 1% of the total outlet element in the product and flue gases).

Table 5. Impact of PEA of pilot plant effluent data on prediction of commercial-scale effluent production

Species	Basis for predicting raw product gas effluent production	
	Conceptual commercial design	PEA effluent data base
H_2S	◀—Equilibrium in $CaCO_3$ + H_2S reaction—▶	
NH_3	60% of coal N	~90% of coal N
HCN	None	<0.5% of coal N
N_2	23% of coal N	Negligible
Trace elements	Not considered	Variable and uncertain but significant

bases used in the Stearns-Roger conceptual commercial design with those resulting from this PEA for the environmentally significant species discussed in the previous examples. The conceptual commercial design was based on the limited effluent production data available at the time, so the impact of a comprehensive pilot-plant effluent data base on predicting commercial-scale effluents is clear.

Table 5 shows significant differences between the scale-up bases for all species except H_2S. For the other species the conceptual commercial-design basis either disagrees with this analysis or else did not consider the species because of a lack of data or a lack of significance in preparing a first-order cost estimate design. The H_2S exception is significant because it is the only species whose behavior is well characterized in a fundamental manner, which is by equilibrium in the $CaCO_3 + H_2S$ reaction. This fundamental characterization is important because the conceptual commercial plant raw product gas differs considerably from the pilot plant gas in major component concentrations. This difference shifts equilibrium H_2S concentrations in the commercial raw product gas to about 30% of measured levels in the pilot plant. Hence, understanding the mechanisms of H_2S behavior clearly has an impact on projecting commercial-scale production.

Table 5 shows considerable differences in nitrogen species projection. This analysis indicates 50% higher NH_3 production than the conceptual design (90% of coal nitrogen vs 60% of coal nitrogen, respectively) which has obvious importance in designing ammonia scrubbing and recovery equipment. For HCN, this analysis indicates that significant production occurs —although a clear understanding of the production mechanism is not available—while the conceptual design did not consider this species.

While trace element production was not considered in the conceptual commercial design, this analysis indicates that significant amounts of hazardous volatile elements such as Hg, Pb, and Sb exist in the raw product gas. However, a reliable basis for accurately predicting trace element behavior is not available.

CONCLUSIONS

The following conclusions regarding needs for comprehensive environmental data bases can be reached on the basis of the comprehensive effluent data base analysis for the CO_2-Acceptor process pilot plant and the methodologies presented here for performing such an analysis.

1. Whenever possible, environmentally significant species should be characterized in the context of fundamental behavior mechanisms. This requires intimate linkages between process and effluent data and fundamental data external to the process which addresses species formation–decomposition behavior.

2. At a minimum, the behavior of environmentally significant species should be characterized by process-specific empirical correlations. This minimal requirement is needed to allow confident extrapolation when fundamental behavior mechanisms are unknown and process conditions are subject to change. To perform this kind of characterization requires intimate linkage, again, between process and effluent data and a technical basis for formulating species behavior hypotheses for testing with empirical process correlations.

3. Major constraints in performing fundamental or empirical correlation characterization of environmentally significant species are:

- Sampling and/or analytical problems which affect the validity of available data.

- Intrinsic lack of scalability of the data—usually due to a poor choice of sampling location.

- Insufficient linkage between process and effluent data to allow fundamental or empirical species behavior characterization.

- Insufficient data on fundamental behavior mechanisms to perform this type of characterization analysis for pilot plant data.

The efforts of many researchers throughout the synthetic fossil fuel industry will be required to meet the goal of developing scalable effluent data bases for synthetic fossil fuel plants. It is hoped that this paper has illustrated data analysis requirements and methodologies that will help in achieving this goal.

REFERENCES

1. J. C. Craun and M. J. Massey, *Final Analysis of the Environmental Data Base for the CO₂-Acceptor Process,* FE-2496-35, November 1978.

2. M. J. Massey, R. G. Luthy, and M. J. Pochan, *A Synopsis of the C-MU Technical Contributions to the DOE Coal Gasification Environmental Assessment Program: July 1976–July 1978,* FE-2496-39, August 1978.

3. *Report of Findings: Source, Occupational and Ambient Characterization, Measurement and Monitoring in Coal Gasification and Liquefaction Processing,* Workshop on the Health and Environmental Effects of Coal Gasification and Liquefaction Technologies, Leesburg, Va., September 6, 1978 (in preparation).

4. *Environmental Characterization of the CO₂-Acceptor Process Gasification Pilot Plant, Final Report Volume 11, Period July 1976–March 1978,* Conoco Coal Development Co. and Radian Corp., FE-1734-44 (in preparation).

5. M. J. Massey, R. W. Dunlap, F. C. McMichael, and D. V. Nakles, *Characterization of Effluents from the Hygas and CO₂-Acceptor Pilot Plants,* FE-2496-1, November 1976.

6. R. W. Dunlap, R. G. Luthy, D. V. Nakles, and M. J. Massey, *Characterization of Effluents from the CO₂-Acceptor Coal Gasification Process,* FE-2496-5, February 1977.

7. J. P. Fillo, G. P. Curran, and M. J. Massey, *Probe Studies and Analysis of Chemical Composition within the CO₂-Acceptor Gasifier,* FE-2497-37 (in preparation).

8. *Support Studies by South Dakota School of Mines, CO₂-Acceptor Process Gasification Pilot Plant, Final Report Volume 9, Period March 1971–June 1977,* Conoco Coal Development Co., FE-1734-42 (in preparation).

9. *Plant Operations Final Report, CO₂-Acceptor Process Gasification Pilot Plant, Final Report Volume 12, Period January 1972–December 1977,* Conoco Coal Development Co., FE-1734-45 (in preparation).

10. *Commercial Plant Conceptual Design and Cost Estimate, CO₂-Acceptor Process Gasification Pilot Plant, Final Report Volume 10, Period August 1976–December 1977,* Conoco Coal Development Co. and Stearns-Roger Engineering Co., FE-1734-43 (in preparation).

11. L. A. Ruth, A. M. Squires, and R. A. Graff, "Desulfurization of Fuels with Half-Calcined Dolomite: First Kinetic Data," *Environ. Sci. Technol.,* **6**(12), 1009 (1972).

12. M. J. Pochan and M. J. Massey, *Computer-Aided Analysis of Coal Gasification Pilot Plant Data Bases,* FE-2496-26, August 1978.

13. *Federal Register* **43**(21), 4109 (January 31, 1978).

14. "OSHA Issues Tentative Carcinogen List," *Chem. Eng. News* **56**(31), 20 (July 31, 1978).

15. Title 40, *Code of Federal Regulations,* Parts 413, 415, 420, 421, 423, 434, 440, revised as of July 1, 1977.

DISCUSSION

Jim Evans, Enviro Control: I'm interested in the side issue in just where you found the hydrocarbons, in which of the streams from the CO₂-Acceptor process?

John Craun: The hydrocarbons were measured in the raw gas streams. Samples of the raw gas were collected, the vapor was analyzed for hydrocarbons, and the hydrocarbons were also analyzed on the particulates collected by extraction. Hydrocarbons are found in the quench water downstream, as well, but those come from the raw product gas in them. The raw gas is the place to measure them.

Ron Neufeld, University of Pittsburgh: Your problems of characterization of materials, understanding their formation, is a pertinent problem, of course, but has there been any industry that has solved that problem for their specific wastes?

John Craun: I don't know if I can answer that or not. I like to use the analogy of looking at environmental characterization much as an engineer looks at developing a process. If you are developing many of the petroleum technologies, for example, where you start at pilot scale, you have some idea of the fundamental catalytic reactions that may be taking place, but there is still a lot of mystery, but at least you want to get enough understanding that you can continue extrapolating with some real confidence in predicting what the next step is going to behave like. I think the same holds true at the environmental level; if it turns out that you don't really need to understand fundamental mechanisms, fine. As an example—the nitrogen in the regenerator—we don't really know how it's being liberated, and yet the important fact is that it apparently is coming strictly from the auxiliary fuel, pilot plant stream only. And yet in the case of H₂S, you have to have that fundamental understanding, otherwise your predictions for commercial scale, just from blind extrapolation of concentrations, would be wrong because of changes in the gas composition from pilot to commercial scale and how that affects that particular reaction. So, I guess the answer is that it depends on what you are looking at, and it's a relative answer. It depends, as you do the work and start to see how things seem to behave and how much they are affected by changes in process conditions, then you can start to get a handle on whether or not it really is important to go back and try to get some more fundamental understanding of what is going on.

SESSION II: CHEMICAL CHARACTERIZATION IN PROBLEM IDENTIFICATION

Chairman: M. R. Guerin
Oak Ridge National Laboratory

SESSION II: CHEMICAL CHARACTERIZATION IN PROBLEM IDENTIFICATION

SUMMARY

M. R. Guerin*

The physical scientist has become the interface between the production-oriented engineer and the environmental-health-oriented ecologist and biologist. Working in conjunction with the engineer, the physical scientist is charged with sampling and delivering meaningful materials for study to the health scientist and ecologist. Furthermore, he is charged with providing data and information on the physical nature and general chemical composition of the materials. In conjunction with the biologists and ecologists, the physical scientist provides guidance in experimental design and assists in developing systems to deliver the test material to the bioassay media in an interpretable manner.

Sampling and both physical and chemical measurement methodologies require further development before completely interpretable physical support is possible. The most urgent need, however, is for the systematic application of existing capabilities to the collection and characterization of materials of importance to the life scientists. This session of the Life Sciences Symposium was designed to address this basic need. Information, no matter how 'qualified', on the nature of potential environmental and occupational releases is required by life scientists to determine priorities for materials and issues to be investigated.

Edo Pellizzari reports on the identification and quantitation of organic contaminants of waters from low-Btu coal gasification and from in situ coal gasification and shale retorting. Many heteroatom-containing organics were visualized, and both qualitative and quantitative differences between processes were observed. Volatile and other water-soluble organics produced during in situ conversion were found to migrate from the source of reaction. Most importantly, data are tabulated to provide guidance for ecological and human health-oriented research.

George Newton presents data illustrating the complexity of acquiring samples representative of inhalation exposure to materials issuing from low Btu coal gasification and fluidized-bed coal combustion. Experiences to date suggest that respirable-sized particles associated with these new technologies are significantly different from those associated with traditional coal combustion. Inorganic elemental enrichment and organic chemical composition data are presented to alert life scientists to potentially unique occupational or general environmental hazards.

Ron Filby has addressed the distribution and fate of trace elements through the Solvent Refined Coal (SRC) processes and relates his observations to the trace element content of native coals. SRC I and SRC II products are compared with other fossil fuels and the fate of trace elements during processing is addressed. Environmental control technologies currently being used are found generally effective for the removal of trace elements from process-derived waters. Quantitative data for 27 elements in SRC I process streams are reported, and the forms in which selected elements may be found in coals or released upon processing are discussed.

Jim Dooley reports on the extensive, systematic evaluation of the nature of coal-derived crude oils as compared with petroleum crudes. The work points to a higher polyaromatic-polar content of coal-derived oils and to a higher nitrogen content of the coal oils.

---

*Oak Ridge National Laboratory.

55

These factors are suggested as likely to result in refining and environmental difficulties not currently associated with petroleum refining. The data presented may be used by life scientists to estimate the relative importance of synthetic fuels to petroleum as regards transportation and subsequent handling or processing.

Bruce Clark summarizes experiences on the combustion of a shale-derived crude oil. The ease of handling and the combustibility of the oil make it excellent fuel. Emissions of nitrogen oxides upon combustion are found to be greater than for a petroleum-derived oil, as would be predicted from the higher nitrogen content of the shale oil. Of special interest is the finding of halocarbons in the stack effluent. The relatively high halogen content of solid fossil fuels may result in halogens being more important than nitrogen- or sulfur-bearing constituents as biological concerns upon combustion or other utilization of syncrudes.

Larry Johnson summarizes the methodology with its application of the current Environmental Protection Agencies Level I Environmental Assessment. The method is designed to prioritize technologies and technology modules that might require environmental controls development of a significant environmental hazard. The method is shown to be both cost effective and capable of identifying unsuspected environmental concerns.

ORGANIC CONSTITUENTS OF WATERS ASSOCIATED WITH COAL AND OIL-SHALE CONVERSION*

E. D. Pellizzari†
N. P. Castillo†

ABSTRACT

Volatile (purgeable) and semivolatile (extractable) organic constituents were analyzed and compared in aqueous samples from *in situ* and low-Btu coal gasification processes and from *in situ* oil-shale retorting experiments. Qualitative and quantitative information was obtained using glass capillary gas chromatography combined with computerized mass spectrometry. Many oxygen-, nitrogen-, and sulfur-containing compounds were detected and qualitative and quantitative differences were observed among the processes.

The transport of volatile and water-soluble organics away from the burn zone after *in situ* coal gasification was demonstrated for the Felix II coal seam in Gillette, Wyo.

INTRODUCTION

The growing divergence between the demand and the domestic supply for liquid and gaseous fuel necessitates a strong, balanced energy program to develop alternate fuel sources. The domestic coal reserves are estimated to be adequate for several hundred years usage, and large energy resources are also available from oil shale and tar sands. Over 230 billion tons of coal are considered mineable by existing methods. This coal could supply the country's estimated requirements for solid, liquid, and gaseous fuels for decades.

Conversion of coal to liquid or gaseous fuel consists essentially of two basic steps, the cracking of heavy hydrocarbons into lighter ones and the enrichment of the resultant molecules with hydrogen. This may be accomplished either by underground, i.e., *in situ* methods, or by mining the coal and transferring it to the surface conversion plants. The energy form may then be produced as a liquid, by liquefaction, or as a gas, by gasification, along with the production of by-products.

With the increased emphasis on energy production, there is a potential for large-scale environmental degradation. Effluents and wastes produced from energy-producing activity can have a profound effect on water quality.

This report addresses the problem of water contamination arising from energy-related activities, with a major emphasis on the identification and quantification of organics in water. The potential contamination of water from an *in situ* and low-Btu coal gasification and the composition of organics associated with the waters from oil-shale retorting systems are discussed.

EXPERIMENTAL PROCEDURES

Sample Collection and Techniques

Aqueous samples were collected in glass containers which had been cleaned with dilute HCl, rinsed with deionized-distilled water and then heated to 450°C for 2 h to remove any residual organic material.[1-3] After sample collection they were immediately shipped to the laboratory by Federal Express to ensure a 24-h delivery and allow for their immediate processing for organic analysis. Sampling

---

*This research was supported by EPA Contract No. 68-03-2368 and monitored by the Analytical Chemistry Branch of the Environmental Research Laboratory, Environmental Protection Agency, Athens, Georgia.

†Research Triangle Institue, Research Triangle Park, North Carolina 27711

containers were checked for contamination before shipment to the field site.

Samples were collected from the low-Btu coal gasification pilot plant in Morgantown, W. Va. (Fig. 1). Grab samples were taken representing approximately a 30-min time span prior to the effluent's leaving for the water treatment facility (Fig. 2).

Aqueous samples were also obtained from *in situ* gasification processes in Gillette and Hanna, Wyo.[1] To demonstrate transport of organics in water away from the gasification zone, samples were obtained from dewatering, environmental monitoring, and production wells at the Hoe Creek I gasification site in Gillette (Fig. 3).

Boiler blowdown and OMEGA-9 retort water samples were from *in situ* oil-shale retorting processes in DeBeque, Colo., and Rocky Springs, Wyo., respectively.[1]

Phases were separated when necessary and aqueous and tar samples were stored at 4°C until analysed.

Analytical Methods

Volatile (purgeable) and semivolatile (extractable) organic constituents were analyzed by a glass capillary gas chromatograph followed by a computerized mass spectrometry system (Fig. 4).[1] Volatile organics were recovered from the aqueous effluents by a purge and trap technique using Tenax GC sorbent.[1,2] The less volatile organics not recovered by the purge and trap method were extracted from water using a solvent partitioning scheme yielding three fractions—neutrals, acids, and bases.[1,2]

Qualitative identifications and quantification techniques were performed as previously reported.[1-3]

Strict quality control measures were followed at all stages of the analytical process. All reagents were repeatedly analyzed for contamination, and all glassware was carefully cleaned and subsequently checked on a frequent basis for contamination. The condition and calibration of the mass spectrometer system were also monitored on a daily basis by using standard reference mixtures.[1-3]

RESULTS AND DISCUSSION

Transport of Organics Away from *In Situ* Gasification Zone

The environmental monitoring wells, EM-1 through EM-5, were located approximately on the downstream side of the gasification zone (Fig. 3).[4] The curve around the injection and production wells indicates the estimated extent of gasification which was deduced from thermocouple data.[4] The dewatering wells 1–6, in a nearly circular pattern surrounding the zone prior to gasification, were used to keep excessive water from invading the burn zone during the gasification phase of the experiment.

Figure 5 depicts three high resolution gas chromatograms obtained by the analytical system described above of organics in water from wells DW-5 (5 ft from burn) and EM-4 (45 ft from burn), and a tar sample from the product stream during gasification. It is apparent from the upper chromatogram that the more volatile constituents were transported a greater distance through the coal seam than the less volatile compounds represented by those chromatographic peaks with longer retention times. For example, many of the volatile constituents which eluted before xylene had been carried to the environmental monitoring well 45 ft from the burn, while those heavier and less volatile substances which eluted after xylene were found primarily only in the well 5 ft from the burn. Although various volatile contaminants were detected at above-background levels in the most distant environmental wells as far as 100 ft from the burn zone, their transport cannot be attributed solely to the natural groundwater movement.[4] The groundwater flow velocity is believed to be only a few meters per year in the Felix II coal seam. Consequently, some of this contamination of the outermost wells probably occurred by vapor-phase transport during gasification.

Table 1 presents quantitative data on individual compounds as a function of transport through the coal seam after gasification. The two major groups of volatile and semivolatile organics are listed essentially according to the method of work-up. In general, the concentrations of the organics were higher 6 days after gasification in DW-4 (2 ft from burn) than in EM-4 (45 ft from burn). In contrast, the concentrations of the more volatile organics, e.g., benzene, increased considerably in EM-4 but decreased in DW-4 at 83 days after gasification. The levels of the more water soluble compounds such as phenols and cresols (Table 1) have increased in the distant well EM-4 and are higher than those for DW-4. These data indicate movement of organics, particularly those which are volatile and water soluble, through the coal seam.

Organics in Water from Coal Gasification

A comparison of the organic composition in water from the Hoe Creek I, Hanna II (*in situ* methods), and

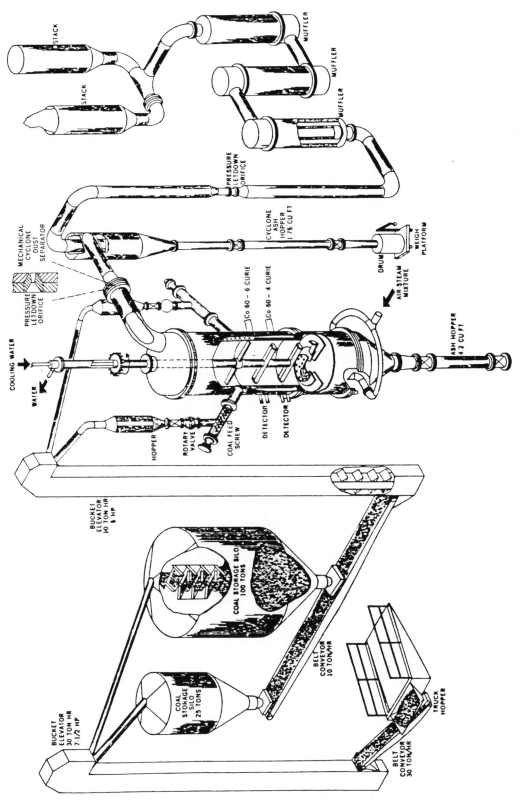

Fig. 1. Schematic of Low-Btu Gasifier at Morgantown, W. Va.

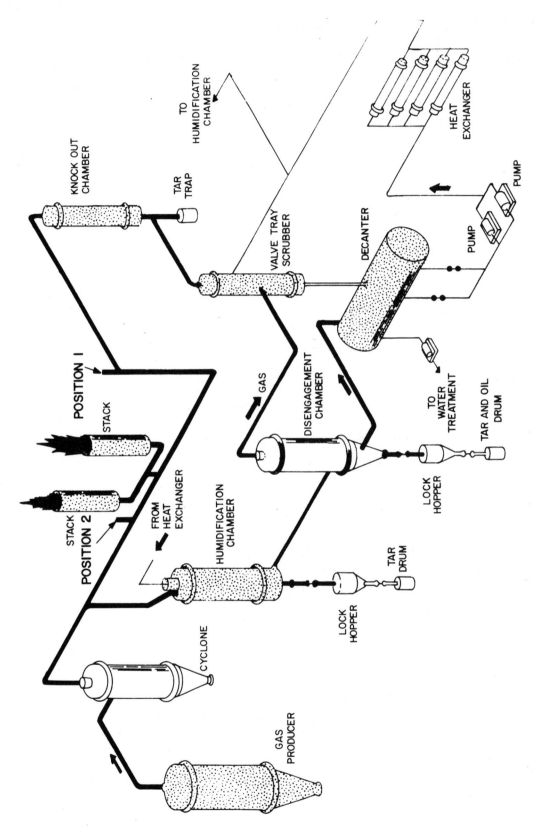

Fig. 2. Partial flow gas cleanup system of Low-Btu Gasifier at Morgantown, W. Va.

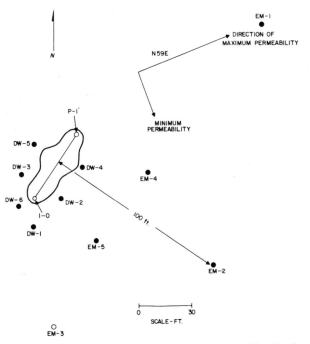

Fig. 3. Principal wells used for water sampling at the Hoe Creek I gasification site near Gillette, Wyo.

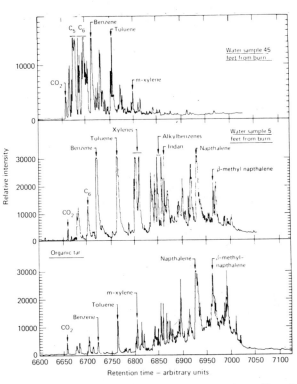

Fig. 5. Gas chromatographic/mass spectrometric profiles of volatile organics in water from EM-4, DW-4, and production wells at Hoe Creek I site. EM-4 and DW-4 were 45 and 4 ft, respectively from the burn zone.

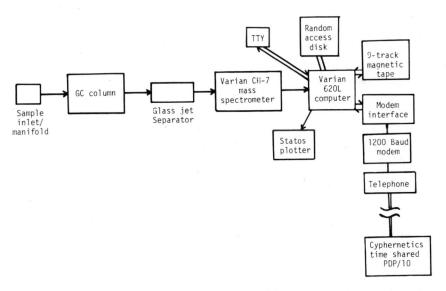

Fig. 4. Schematic diagram of Varian MAT gas chromatograph/mass spectrometry computer system.

Table 1. Transport of organics through coal seam at two time periods after gasification[a]

Group/Compound	6 days			83 days		
	DW-4 (2 ft)	EM-4 (45 ft)	DW-5 (5.5 ft)	DW-4 (2 ft)	EM-4 (45 ft)	DW-5 (5.5 ft)
Volatile						
Benzene	363	180	159	98	290	12
Thiophene	28	0.7	41	6	0.9	8
Toluene	240	0.8	219	5	T	58
Ethylbenzene	12	0.4	34	33	0.2	6
Xylenes	252	1.4	628	128	0.5	106
Naphthalene	74	T	133	0.7	0.5	0.2
Total volatile compounds	3280	NC	2427	457	NC	381
Semivolatile						
n-heptanal	1.8	ND	ND	6	ND	ND
n-octanal	13	3	9	3	ND	T
n-nonanal		34	41		66	ND
Phenol	120	138	141	163	93	59
Dimethyl phenol isomer	120	721	339	34	412	450
Total cresol isomers	1056	607	263	288	388	257
Total phenolics	3352	1988	943	984	1312	1064

[a]All measurements are in ppb. T—trace; NC—not calculated; ND—not detected.

low-Btu gasification processes is shown in Table 2. The aqueous samples (24-h integrated) from the *in situ* methods were taken on the seventh day of gasification. The low-Btu sample was an integrated 30-min sample at approximately midway through a 72-h gasification of Montana Rosebud coal. The samples show a striking similarity in oxygenated organics. The principal difference, however, is in their concentrations, which is expected since the processes themselves are different. A few of the oxygenates such as furans and ethers were not detected in the *in situ* processes. Nitrogen-containing organics were detected principally in the *in situ* samples and were absent in the low-Btu sample (Table 2). A number of sulfur compounds such as the thiophenes were detected in all three samples; the principal variation probably was due to the inability to detect trace quantities of some of the higher homologues.

The concentration of aromatics was much higher in the Hanna II sample than in the other two (Table 2). However, the interesting feature is the qualitative similarities for each individual species appearing in all samples. Many alkanes were also identified and quantified.[1]

Table 3 presents the semivolatile (extractable) organics which were found in the same samples listed in Table 2. Many organics were present at trace levels

except for phenols, which constituted the major percentage of all organics. A greater diversity of phenolic isomers was detected in the Hoe Creek I sample than in the others. In contrast, a series of carboxylic acids was detected (and quantified) in the Hanna II sample which was not present in the Hoe Creek or low-Btu samples. A number of pyridines were also identified in the Hanna II sample which were not detected in either of the other coal gasification samples (Table 3).

The principal differences observed among the three coal gasification experiments were expected since the gasification processes were operated using different parameters. For example, in the *in situ* processes, different rates and techniques of air injection and pressures were used, and these possibly were differences in the coal itself.

Organics in Water from Oil-Shale Retorting

Table 4 lists the organic composition of waters from two *in situ* oil-shale retorting processes. The OMEGA-9 retort water was analyzed immediately upon its receipt, and a second identical sample was stored at 24°C for two months and then analyzed. This experiment was performed to determine the effects of storage at room temperature on organic

Table 2. Levels of purgeable organics in aqueous samples from coal gasification[a]

Class	Compound	In-situ gasification[b]		Low-Btu gasification[c]
		Hoe Creek I	Hanna II	MERC
Oxygen	acetaldehyde	59.5 ± 8.6	34 ± 5	92 ± 53
	acetone	2226 ± 334	9330 ± 100	259 ± 50
	C_4H_8O isomer	9 ± 5	ND	42 ± 33
	methyl ethyl ketone	44 ± 5	5000 ± 0	22 ± 8.3
	methyl isopropyl ketone	89 ± 33	T	
	2-pentanone	846 ± 3.6	4700 ± 800	12 ± 5
	3-pentanone	297 ± 3.6	2330 ± 500	T
	4-methyl-2-pentanone	48 ± 8.3	67 ± 7	78 ± 33
	2-methyl-3-pentanone	68 ± 8.3	83 ± 21	93 ± 42
	3-methyl-2-pentanone	79 ± 25	ND	446 ± 222
	3-hexanone	61 ± 6	330 ± 28	86 ± 37
	2-hexanone	7 ± 6	670 ± 55	202 ± 67
	cyclopentanone	T	167 ± 35	T
	$C_7H_{14}O$ isomer	13 ± 6	ND	T
	2-methylcyclopentanone	56 ± 36	670 ± 225	T
	3-methylcyclopentanone	50 ± 21	ND	T
	3-heptanone	62 ± 11	ND	67 ± 33
	2-heptanone	40 ± 7	100 ± 50	195 ± 50
	$C_7H_{12}O$ isomer	14 ± 3.3	ND	ND
	anisole	6 ± 1.7	ND	5 ± 2
	benzaldehyde	20 ± 5	ND	T
	2-octanone	30 ± 5	ND	188 ± 100
	benzofuran	6 ± 1.7	13 ± 5	267 ± 67
	phenol	NQ	NQ	NQ
	acetophenone	21 ± 5	ND	
	methylbenzofuran isomer	29 ± 5.3	3000 ± 117	391 ± 20
	2-decanone	7 ± 2	ND	T
	dimethylbenzofuran isomer	21 ± 3.3	1000 ± 700	ND
	dimethylbenzofuran isomer	31 ± 5	T	ND
	dibenzofuran	T	ND	ND
	furan	ND	ND	7 ± 4.3
	3-methylpentanal	ND	ND	185 ± 42
	diisopropyl ketone	ND	ND	T
	4-methyl-3-heptanone	ND	ND	T
	2-methyl-3-heptanone	ND	ND	13 ± 3.3
	octanone isomer	ND	ND	T
	3-nonanone	ND	ND	T
	2-nonanone	ND	ND	156 ± 83
	3-methyl-2-heptanone	ND	ND	4.3 ± 2
Nitrogen	acetonitrile	1246 ± 150	17 ± 15	ND
	propionitrile	175 ± 22	2000 ± 55	ND
	isobutyronitrile	34 ± 11	100 ± 2	ND
	methacrylonitrile	T		ND
	α-methylbutyronitrile	T	17 ± 8	ND
	N-methylpyrrole	12 ± 4	33 ± 16	ND
	pyrrole	111 ± 11	5000 ± 500	ND
	n-pentylnitrile	38 ± 13	133 ± 33	ND
	3-methylpyrrole	12 ± 6	1667 ± 170	ND
	2-methylpyrrole	27 ± 7.3	1660 ± 1100	ND
	dimethylpyrrole isomer	17 ± 5.7	1000 ± 250	ND
	n-hexylnitrile	41 ± 13.7		ND
	2-methylpyridine	33 ± 3		ND
	methylpyridine isomer	16 ± 2.3	670 ± 225	ND
	dimethylpyrrole isomer	9 ± 4.7	1670 ± 945	ND
	dimethylpyridine isomer	T	T	ND
	methylpyridine isomer	T	33 ± 17	ND
	cyanobenzene	24 ± 5	2670 ± 352	ND
	n-propylamine	ND	17 ± 2	ND
	n-butyronitrile	ND	117 ± 0	ND
	pyridine	ND	T	ND

Table 2. (cont'd)

| Class | Compound | In-situ gasification[b] | | Low-Btu gasification[c] |
		Hoe Creek I	Hanna II	MERC
Nitrogen cont'd	aniline	ND	17 ± 8	ND
	trimethylpyridine isomer	ND	T	ND
	methylethylpyridine isomer	ND	T	ND
	dimethylaniline isomer	ND	58 ± 40	ND
	2-methylbenzimidazole	ND	ND	58 ± 16
	5,6-dimethylbenzimidazole	ND	ND	T
Sulfur	thiophene	241 ± 29	67 ± 18	59 ± 5.7
	2-methylthiophene	56 ± 7	T	134 ± 17
	3-methylthiophene	50 ± 11	300 ± 28	107 ± 23
	ethylthiophene isomer	20 ± 14	ND	83 ± 42
	dimethylthiophene isomer	14 ± 4	67 ± 15	14 ± 3.3
	dimethylthiophene isomer	13 ± 3.3	ND	28 ± 3.3
	dimethylthiophene isomer	5 ± 2.0	ND	14 ± 2.6
	C_3-alkyl thiophene isomer	5 ± 2	1330 ± 279	T
	benzothiophene	39 ± 8.6	ND	52 ± 20
	C_3-alkyl thiophene isomer	ND	ND	T
	C_3-alkyl thiophene isomer	ND	ND	20 ± 5
	C_4-alkyl thiophene isomer	ND	ND	T
	C_4-alkyl thiophene isomer	ND	ND	T
Aromatic	benzene	4318 ± 1154	6670 ± 70	1214 ± 325
	toluene	1081 ± 298	5000 ± 1000	121 ± 44
	ethylbenzene	41 ± 14	1660 ± 10	104 ± 42
	xylenes (p + m)	121 ± 14	2670 ± 25	61 ± 25
	styrene	55 ± 27	1300 ± 210	83 ± 48
	o-xylene	82 ± 22	1300 ± 330	181 ± 17
	isopropylbenzene	T	30 ± 8	26 ± 5
	n-propylbenzene	6.7 ± 1.7	167 ± 25	T
	ethyltoluene (p + m)	55 ± 6	3303 ± 334	278 ± 5.6
	1,3,5-trimethylbenzene	27 ± 7	ND	139 ± 17
	o-ethyltoluene	34 ± 6.7	2000 ± 299	148 ± 17
	1,2,4-trimethylbenzene	35 ± 5	6670 ± 491	198 ± 100
	1,2,3-trimethylbenzene	42 ± 14	2670 ± 811	361 ± 200
	indan	72 ± 29	1000 ± 210	247 ± 47
	indene	140 ± 17	1000 ± 675	363 ± 75
	C_4-alkyl benzene isomer	25 ± 2.3	167 ± 38	T
	C_4-alkyl benzene isomer	3.7 ± 3	20 ± 4	T
	C_4-alkyl benzene isomer	T	500 ± 25	T
	methylindene isomer	38 ± 8.7	T	
	C_4-alkyl benzene isomer	T	670 ± 392	T
	naphthalene	1707 ± 79	13340 ± 134	5121 ± 1336
	C_3-alkyl indan isomer	13 ± 6.7	ND	20 ± 3
	n-hexylbenzene	T	T	18 ± 5
	methyldihydronaphthalene isomer	T	670 ± 70	118 ± 50
	C_2-alkyl indan isomer	7 ± 3.3	1340 ± 888	T
	C_3-alkyl indan isomer	5 ± 3	ND	T
	β-methylnaphthalene	66 ± 12.4	2830 ± 284	967 ± 225
	C_3-alkyl indan isomer	13 ± 5.7	ND	ND
	α-methylnaphthalene	78 ± 15.4	2000 ± 100	278 ± 26
	C_7-alkyl benzene isomer	T	67 ± 15	T
	biphenyl	31 ± 10.3	5330 ± 460	ND
	ethylnaphthalene isomer	20 ± 7	1670 ± 710	ND
	dimethylnaphthalene isomer	55 ± 9	2000 ± 1120	12 ± 3.3
	dimethylnaphthalene isomer	13 ± 1.7	670 ± 300	24 ± 5
	dimethylnaphthalene isomer	49 ± 63	1330 ± 710	9 ± 3.3
	dimethylnaphthalene isomer	74 ± 17	160 ± 130	ND
	biphenylene	21 ± 1.7	1700 ± 491	ND

Table 2. (cont'd)

Class	Compound	In-situ gasification[b]		Low-Btu gasification[c]
		Hoe Creek I	Hanna II	MERC
Aromatic cont'd	acenaphthene	31 ± 5.3	1690 ± 170	ND
	C_3-alkyl naphthalene isomer	T	T	ND
	C_3-alkyl naphthalene isomer	T	ND	ND
	C_3-alkyl naphthalene isomer	6 ± 3	ND	2.6 ± 0.1
	C_3-alkyl naphthalene isomer	12 ± 3.7	ND	11 ± 3.7
	trimethylnaphthalene	48 ± 5.7	ND	5 ± 3
	propylnaphthalene isomer	ND	ND	ND
	methyl indan isomers	ND	T	T
	dimethylindan isomer	ND	T	T
	C_5-alkyl benzene isomer	ND	0.06 ± 0	8 ± 1.7
	C_6-alkyl benzene isomer	ND	T	T
	C_6-alkyl benzene isomer	ND	T	T
	C_6-alkyl benzene isomer	ND	T	T
	C_5-alkyl benzene isomer	ND	T	ND
	dimethylindene isomer	ND	1660 ± 170	ND
	dimethylindan isomers	ND	1000 ± 170	T
	propyltoluene	ND	ND	99 ± 50
Alkanes	many	NC	NC	NC

[a]All measurements are in ppb. ND—not detected; T—trace; NC—not calculated.
[b]Sample of product water taken approximately on 7th day of gasification.
[c]Integrated 30-min sample.

Table 3. Levels of some extractable/semivolatile organics in aqueous samples from coal gasification[a]

Class	Compound	In situ gasification[b]		Low Btu gasification[c]
		Hoe Creek I[d]	Hanna II[e]	MERC
Oxygen	cyclopentanone	110	680	56
	2-methylcyclopentanone	20	110	72
	3-methylcyclopentanone	14	T	ND
	p-methylanisole	22	36	ND
	n-nonanoic acid	5	ND	ND
	$C_8H_{12}O$ isomer	5	ND	ND
	C_6H_8O isomer	7	ND	ND
	$C_7H_{10}O$ isomer	150	36	ND
	$C_8H_{12}O$ isomer	20	ND	ND
	$C_8H_{12}O$ isomer	T	ND	ND
	n-decanoic acid	3	ND	ND
	phenyl acetate	4	ND	ND
	n-undecanoic acid	5	ND	ND
	dimethylphenol isomer	T	ND	2,440
	C_3-alkyl phenol isomer	T	2,560	42
	C_3-alkyl phenol isomer	T	ND	T
	cresol isomers + phenol	5,000	26,800	280
	C_3-alkyl phenol isomer	4,000	ND	84
	$C_{10}H_{14}O$ isomer	10	ND	
	ethylphenol isomer	6,000	42	1,360
	dimethylphenol + cresol isomers	48,000	15,760	9,522
	C_3-alkyl phenol isomer	400	ND	36
	dimethylphenol isomer	425	ND	30
	ethylphenol isomer	455	ND	50
	C_3-alkyl phenol isomer	600	ND	121

Table 3. (cont'd)

| Class | Compound | In situ gasification[b] | | Low Btu gasification[c] |
		Hoe Creek I[d]	Hanna II[e]	MERC
Oxygen cont'd	dimethylphenol isomer	3,800	ND	ND
	ethylphenol isomer	400	ND	ND
	C$_3$-alkyl phenol isomer	350	ND	T
	C$_2$-alkyl phenol isomer	1,600	ND	ND
	C$_3$-alkyl phenol isomer	20	ND	ND
	C$_3$-alkyl phenol isomer	25	ND	ND
	C$_3$-alkyl phenol isomer	T	ND	ND
	C$_2$-alkyl benzaldehyde isomer	850	ND	ND
	C$_2$-alkyl benzaldehyde isomer	450	ND	ND
	n-pentanoic acid	ND	140	ND
	benzoic acid	ND	T	ND
	toluic acid isomer	ND	490	ND
	toluic acid isomer	ND	390	ND
	n-heptaldehyde	ND	48	ND
	n-hexanoic acid	ND	2,680	ND
	n-octanal	ND	36	ND
	n-heptanoic acid	ND	8,870	ND
	cyclohexenone	ND	T	ND
	n-octanoic acid	ND	1,230	ND
Nitrogen	methylpyridine	4		ND
	benzonitrile	40	54	ND
	dimethylpyridine	ND	580	ND
	3-methylpyridine	ND	250	ND
	4-methylpyridine	ND	100	ND
	2-n-propylpyridine	ND	T	ND
	aniline	ND	9	ND

[a]ND—not detected; T—trace.
[b]Samples taken at 7th day of gasification.
[c]Sample integrated 30-min composite.
[d]Values in ppm.
[e]Values in ppb.

composition, as well as any effects from bacterial action.

The boiler-blown water (OCC process) was qualitatively quite different from the OMEGA-9 sample (Table 4). Many organic acids were detected. Furthermore, the concentration of organics decreased considerably in the incubated OMEGA-9 retort water. Many nitrogen- and sulfur-containing organics were identified in the OMEGA-9 water which were absent in the boiler blowdown water.

The results presented on organic constituents of waters associated with coal and shale conversion are useful for preparing identity and concentration profiles for before, during, and after the operation of energy-related processes in order to assess the impact of these processes on the environment. Data of this nature is important if the proper control technology is to be instituted.

Acknowledgment

The cooperation and assistance of the personnel at the Laramie and Morgantown Energy Technology Centers in Laramie, Wyo., and Morgantown, W. Va., and Lawrence Livermore Laboratories, Livermore, Calif., are gratefully acknowledged for providing samples throughout this study. The authors thank Ms. Ann Alford of EPA, Athens, Ga., for her constant encouragement and constructive criticisms throughout the program.

Table 4. Levels of purgeable/volatile and extractable/semivolatile hetero-atom-
containing organics associated with in situ oil-shale retorting[a]

Class	Compound	OCC boiler blowdown water[b]	OMEGA-9 retort water[c]	
			initial	final
Oxygen	methyl ethyl ether	4.56 ± 2.1	ND	ND
	acetaldehyde	2.75 ± 1.1	ND	ND
	acetone + diethyl ether	48.0 ± 16	126.3 ± 40	95 ± 44
	methyl ethyl ketone	T	75.8 ± 3.4	152 ± 67
	n-hexanal	T	ND	ND
	acetophenone	T	ND	ND
	isopropanol	ND	22.8 ± 6	9 ± 4
	t-butanal	ND	17.5 ± 7	8.5 ± 2.5
	methyl isopropyl ketone	ND	20.2 ± 12	9.6 ± 4.8
	2-pentanone	ND	197 ± 72	123 ± 42
	3-pentanone	ND	26 ± 14	11 ± 6
	4-methyl-2-pentanone	ND	105 ± 70	51 ± 20
	2-methyl-3-pentanone	ND	62 ± 30	11.3 ± 1.9
	3-methyl-2-pentanone	ND	12 ± 2.5	0.4 ± 0.2
	3-hexanone	ND	5.3 ± 17	12.8 ± 6.2
	2-hexanone	ND	56.6 ± 22	38 ± 14
	cyclopentanone	ND	5.0 ± 2.9	2.9 ± 1.4
	5-methyl-3-hexanone	ND	T	T
	2-methyl-3-hexanone	ND	T	T
	2-methylcyclopentanone	ND	7.7 ± 2 7	T
	5-methyl-2-hexanone	ND	T	T
	2-methyl-3-hexanone	ND	T	T
	2-methylcyclopentanone	ND	7.7 ± 2.7	T
	5-methyl-2-hexanone	ND	T	ND
	4-heptanone	ND	T	8.2 ± 4
	3-heptanone	ND	28 ± 18	30 ± 24
	2-heptanone	ND	244 ± 100	52 ± 18
	3-methylcyclohexanone	ND	0.96 ± 0.2	ND
	2-methylcyclohexanone	ND	T	T
	3-octanone	ND	T	ND
	2-octanone	ND	126 ± 42	19 ± 12
	trimethylcyclohexanone isomer	ND	0.7 ± 0.3	0.6 ± 0
	C₃-alkyl cyclohexanone isomer	ND	T	ND
	trimethylcyclohexanone isomer	ND	T	ND
	α-methylbutyric acid	1.8	24	ND
	isopentanoic acid	4.2	8	ND
	n-pentanoic acid	3.6	5	1.3
	2-methylpentanoic acid	3.0	4	ND
	β-methylpentanoic acid	3.3	6	ND
	n-hexanoic acid	5.4	38	20
	n-heptanoic acid	54	79	53
	2,4-dimethylheptanoic acid	52	56	T
	n-octanoic acid	115	14	8.4
	n-nonanal	9.6	4	ND
	3,5-dimethylheptanoic acid	4.8	41	T
	3,5-dimethyloctanoic acid	34	47	T
	n-nonanoic acid	132	493	ND
	n-decanoic acid	123	ND	57
	3,5,5-trimethyl-2-cyclohexanone	10.2	ND	ND
	benzoic acid	115	227	111
	cyclohexenedione (tent.)	1.2	ND	ND
	n-undecanoic acid	1.8	ND	ND
	o-toluic acid	3.0	21	4.2
	m- or p-toluic acid	16	T	30
	β-phenylpropionic acid	1.2	4	ND
	2,5-dimethyltoluic acid	10.8	17	ND

Table 4. (cont'd)

Class	Compound	OCC boiler blowdown water[b]	OMEGA-9 retort water[c]	
			initial	final
Oxygen cont'd	dimethylphenol isomer	3.0	51	34
	phenol	459	461	293
	cresol isomer	T	ND	ND
	dimethylphenol isomer	552	15	ND
	cresol isomer	561	779	94
	C_2-alkyl phenol isomer	19.2	69	ND
	C_3-alkyl phenol isomer (tent.)	T	ND	T
	dimethylphenol isomer	9.0	T	ND
	ethylphenol isomer	17.4	T	68
	C_2-alkyl phenol isomer	1.8	T	T
	dimethylphenol isomer	125	ND	T
	C_3-alkyl phenol isomer	9	ND	9
	C_2-alkyl phenol isomer	231	T	14
	C_3-alkyl phenol isomer	107	ND	T
	C_3-alkyl phenol isomer	T	ND	17
	C_3-alkyl phenol isomer	1.2	ND	46
	C_4-alkyl phenol isomer	8.4	ND	ND
	δ-valerolactone	ND	29	16
	dibenzofuran	ND	29	ND
Nitrogen	acetonitrile	ND	42 ± 11	23 ± 11
	propionitrile	ND	27 ± 7	10 ± 4
	isobutyronitrile	ND	3 ± 2	1± 0.4
	n-butyronitrile	ND	13 ± 3.4	T
	α-methylbutyronitrile	ND	30 ± 14	5 ± 3
	pyrrole	ND	23 ± 15	T
	n-pentylnitrile	ND	11 ± 4	5 ± 2
	4-methylpentylnitrile	ND	T	T
	pyridine	ND	152 ± 62	185 ± 121
	2-methylpyridine	ND	41 ± 7	18 ± 9
	3-methylpyridine +	ND	17 ± 6	6 ± 2
	n-hexylnitrile	ND		
	dimethylpyridine isomer	ND	16 ± 6	80 ± 9
	2-ethylpyridine	ND	T	T
	aniline + C_2-alkyl pyridine	ND	45 ± 7	17 ± 8
	C_3-alkyl pyridine isomer	ND	15 ± 4	T
	C_2-alkyl pyridine isomer	ND	35 ± 12	2 ± 1
	n-heptylnitrile	ND		
	C_3-alkyl pyridine isomer	ND	63 ± 14	14 ± 5
	cyanobenzene	ND	235 ± 155	49 ± 20
	C_4-alkyl pyridine isomer	ND	T	T
	2,4,6-trimethylpyridine	ND	T	ND
	C_4-alkyl pyridine isomer	ND	T	T
	C_4-alkyl pyridine isomer	ND	T	T
	C_4-alkyl pyridine isomer	ND	T	6 ± 3
	C_5-alkyl pyridine isomer	ND	11 ± 6	5 ± 3
	C_5-alkyl pyridine isomer	ND	19 ±8	ND
	1,3,5-trimethylpyrazole	ND	14 ± 6	ND
	C_5-alkyl pyridine isomer	ND	4 ± 2	ND
	quinoline	ND	19 ± 8	ND
	C_8-alkyl pyridine isomer	ND	T	ND
	aniline	ND	497	432
	benzylamine	ND	377	228
	methylquinoline	ND	41	70
	C_2-alkyl quinoline	ND	T	T
	C_2-alkyl quinoline	ND	T	T

Table 4. (cont'd)

Class	Compound	OCC boiler blowdown water[b]	OMEGA-9 retort water[c]	
			initial	final
Sulfur	thiophene	ND	20 ± 6	5 ± 3
	2-methylthiophene	ND	T	ND
	3-methylthiophene	ND	T	ND
	thiacyclopentane	ND	6 ± 3	2 ± 0.9
	3-methylisothiazole	ND	7 ± 1	ND
	2-methylthiacyclopentane	ND	3 ± 0.2	T
	dimethylthiacyclopentane	ND	0.34 ± 0.2	T
	thiacyclohexane	ND	5 ± 2	ND
	2,5-dimethylthiacyclopentane	ND	4 ± 1	ND
	methylthiacyclohexane	ND	T	ND

[a]All measurements are in ppb. ND—not detected; T—trace.
[b]OCC = Occidental Oil Co., sample from DeBeque, Colo.
[c]OMEGA-9 retort water prior to and after incubation for 2 months at 75°F.

REFERENCES

1. E. D. Pellizzari, *Identification of Components of Energy-Related Wastes and Effluents,* Research Triangle Institute Report, EPA-600/7-78-004, Research Triangle Park, N.C., January 1978.

2. E. D. Pellizzari, N. P. Castillo, S. Willis, D. Smith and J. T. Bursey, "Identification of Organic Constituents in Aqueous Effluents from Energy-Related Processes," *Fuel Chemistry,* in press.

3. E. E. Pellizzari, N. P. Castillo, S. Willis, D. Smith and J. T. Bursey, "Identification of Organic Components in Samples from Energy-Related Processes," *Am. Soc. Testing Mat.,* in press.

4. J. H. Campbell, E. D. Pellizzari, and S. Santor, *Results of a Groundwater Quality Study near an Underground Coal Gasification Experiment (Hoe Creek I),* UCRL-52405, Lawrence Livermore Laboratory, Livermore, Calif., February 6, 1978.

DISCUSSION

Frank Schweighardt, Department of Energy, Pittsburgh Energy Technology Center: What methods were used to identify the compounds you listed? Did you use mass spectrometry, retention data, or other methods to assign the peaks?

Edo Pellizzari: The methodology will be referenced in our progress reports and in the proceedings of this symposium. Basically we used mass cracking patterns of the unknowns and compared them to library spectra, and we used retention indices. Capillary column gas chromatography and Fourier transform infra-red Spectrometry were used to discern some isomers that mass spectral information could not discriminate between.

Mike Guerin, Oak Ridge National Laboratory: You have listed quite a large number of individual kinds of constituents. We see that kind of thing sometimes too. Do you have any idea what percentage of total organic carbon you are accounting for in any of those samples?

Edo Pellizzari: The range of organic carbon in the series of samples from the gasification processes varies from 20 to 40%. We examined approximately 40 samples, and of the total organic carbon, roughly 80% is phenols. The techniques that we used allow detection and identification of these compounds, as well as quantification down to a one part per billion level. It is for that reason that this list is so extensive. Most of these constituents are present at less than a part per million.

Phil Singer, University of North Carolina: Can you repeat that again—you said 80% of the total organic carbon identified were phenols.

Edo Pellizzari: Up to 80.

Phil Singer: Mike, was your question directed at how much of all the organic carbon is accounted for by phenols plus others?

Mike Guerin: That's correct. Can you comment on that? How much of the other 20% are you catching?

Edo Pellizzari: We don't know. The way the samples were treated did not allow us to do a mass balance. There could be, at most, 50 to 60% that remains unaccounted for in extreme cases. As I observed before, the techniques that we used detect the volatile, purgeable, and extractable compounds that would pass through a gas chromatograph. The semivolatile and the nonvolatile constituents were not determined at all, and they could constitute a major group of compounds like polynuclear aromatics or others. That's a topic for another symposium.

ENVIRONMENTAL PROTECTION AGENCY LEVEL-1 CHEMISTRY: STATUS AND EXPERIENCES

L. D. Johnson*   J. A. Dorsey*

ABSTRACT

Increased awareness of the need for reliable and comprehensive knowledge about the emissions from industrial and energy related processes has resulted in a series of major environmental assessment projects within the Environmental Protection Agency. To carry out these projects effectively, it has been necessary to develop new sampling and analysis strategies. These strategies are reviewed, and data from environmental assessment projects are discussed.

INTRODUCTION

Increased awareness of the need for reliable and comprehensive knowledge about the emissions from industrial and energy related processes has resulted in a series of major environmental assessment projects within EPA. These projects are designed to determine the short-term needs for currently regulated pollutants, the long-term needs for future regulatory programs, and the associated control-technology development information requirements. To assess these goals adequately and also to provide a useful data base for other research and development projects is obviously a very ambitious undertaking. A great deal of careful planning has been necessary to maximize utilization of funds available. The four major components of an environmental assessment are:

1. a systematic evaluation of the physical, chemical, and biological characteristics of all streams associated with a process;

2. predictions of the probable effects of those streams on the environment;

3. ranking of those streams relative to their individual hazard potential; and

4. identification of any necessary pollution control technology programs.

To carry out effectively all of these elements of an environmental assessment, it has been necessary to develop new sampling and analysis strategies. Initial studies demonstrated the superior information and cost effectiveness of a phased approach over more direct strategies.[1,2] The cost advantages of the phased approach were found to be approximately proportional to the complexity of the process being sampled. In the case of a coal gasifier, costs were nearly doubled by a direct attack upon the problem. It is also unlikely that the single-phase method can produce data of comparable quality in any event, since unpredictable source-specific problems interfere more strongly with such a strategy.

The phased approach[3-7] uses three levels of sampling and analysis. Level 1 is a survey (Fig. 1), Level 2 is a directed detailed analysis of Level-1 information, and Level 3 is a monitoring of selected pollutants over long periods.

Level-1 methods have been developed and applied to a variety of industrial and energy sources. Further research into the limitations of the techniques has been done in conjunction with the early field use, and revisions to the protocols have been made where necessary. Level-2 sampling and analytical methods are nearly complete in conception, but require more work on details and on field and laboratory verification. Projects to develop guidelines for use of Level-3 methods are in the planning stage.

---

*Industrial Environmental Research Laboratory, U.S. Environmental Protection Agency, Research Triangle Park, North Carolina, 27711.

70

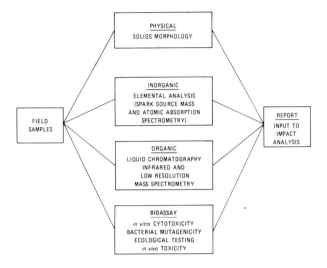

Fig. 1. Flow chart of Level-1 scheme.

LEVEL-1 SAMPLING PROGRAM

It is not sound practice to try to define a detailed sampling program until (1) the general characteristics of the streams in question have been evaluated and (2) any unfavorable sampling system/sample interactions have been found (e.g., chemical reaction, volatility loss). Hence, an effective sampling program involves a series of reiterative tests in which each iteration enhances the source assessment by focusing resources and efforts on the pollutants and streams of concern and improves the accuracy of the sampling program.

Level-1 sampling stresses the concept of completeness by presuming that all streams leaving the process will be sampled. Level-1 sampling is not based on prior judgements about stream composition, since whatever prior knowledge is available is, at best, incomplete. Predictive and extrapolative techniques used during source assessments check the empirical data but do not replace them. Level-1 sampling systems sample all substances in the stream sufficiently for both chemical and biological tests. These tests do not necessarily specify substances or their chemical form. Furthermore, Level-1 sampling programs use existing stream access sites. Although care must be used to ensure that the samples are not biased, the commonly applied concepts of multiple point, isokinetic, or flow-proportional sampling are not rigidly followed. Normally, sampling is over one full cycle of each desired set of process operating conditions. Values for individual samples in a set are combined to produce a single average value for the total process cycle.

Level-1 Analytical Method

For an environmental source assessment, the analytical methods will vary from relatively simple, wet chemistry to highly complex instrumental techniques. Analyses proceed from general, broadly applicable, survey methods to more specialized techniques to measure specific components. This broad range of analysis is based on the same 3 level concept as the sampling program. At each phase of the analytical program, the exactness and sophistication of the techniques are appropriate to the quality of the samples taken and the information required. Hence, use of analytical resources on screening samples from streams of unknown pollution potential is minimized.

Level-1 sampling provides a set of samples which represents the average composition of each stream. This sample set is separated, either in the field or in the laboratory, into solid-, liquid-, and gas-phase components. Each component is evaluated with survey techniques which define its basic physical, chemical, and biological characteristics. The survey methods selected cover a wide range of materials with sufficient sensitivity to ensure a high probability of detecting environmental problems.

For Level 1, the analytical techniques and instrumentation have been kept as simple as possible to provide an effective level of information at minimum cost. Each individual piece of data developed adds a relevant point to the overall evaluation. Conversely, since the information from a given analysis is limited, all the tests must be done to provide a valid assessment of the sample.

Physical analysis of solid samples is included in Level-1 analysis because the size and shape of the particles have a major effect on their behavior in process streams, control equipment, atmospheric dispersion, and the respiratory system. In addition, some materials have characteristic physical forms which can aid in their identification. Chemical analyses to determine the types of substances present give information for predicting control approaches, potential toxicity, and atmospheric dispersion and transformation. Finally, because prediction of hazard based on physical and chemical analyses alone is subject to many uncertainties, biological assay techniques measure potential toxicity (Table 1).

Table 1. Basic Level-1 analyses

Measurement type	Analytical technique
Physical	Cyclone particle size
	Optical microscopy
Chemical	Spark source mass spectrometry
	Wet chemical (selected anions)
	Gas chromatography
	Elution chromatography
	Infrared spectrometry
	Low resolution mass spectrometry
Biological	Rodent acute toxicity
	Microbial mutagenesis
	Cytotoxicity
	Fish acute toxicity
	Algal bioassay
	Soil microcosm
	Plant stress ethylene

DISCUSSION

The major emphasis in environmental measurements over the past two years has been on development of procedures for Level 1. As presently conceived, Level-1 procedures will produce a cost-effective information base upon which to rank streams for planning any subsequent programs. The specific procedures applied are detailed in two documents: the first covers sampling and chemical analysis,[8] and the second, biological tests.[9] Use of the complete battery of procedures helps answer six questions:

1. Do streams leaving the processing unit have a finite probability of exceeding existing or future air, water, or solid waste standards?

2. Do any of the streams leaving the processing unit contain any classes of substances that are known or suspected to have adverse environmental effects?

3. Into what general categories (classes) do these adverse substances fall?

4. What are the most probable sources of these substances?

5. On the basis of their adverse effects and mass output rates, how should streams be ranked?

6. For streams with potentially harmful environmental effects, what basic direction is likely for control strategies?

The overall accuracy goal of Level-1 analysis is to report the true value within a factor of 3. This means that a reported value between 30 and 300 is acceptable when the true value is 100. Many of the Level-1 sampling or analytical techniques are more accurate than this goal, but the relaxation of traditional limits allows significant cost savings. In general, grab sampling is used when feasible to minimize sampling costs.

Particulate-laden stack gases require sophisticated and relatively expensive equipment for collecting valid samples even at Level 1. To provide unbiased sampling of this type requires the use of a sampling train designed specifically for environmental assessment sampling. The source assessment sampling system (SASS) has been designed to operate at 150 liters/min and collect both solids and vapors.[10,11] The entrained particles are sorted by diameter into four groups: >10 μm, 3–10 μm, 1–3 μm, and <1 μm. Vapor-phase organic materials are adsorbed on a solid sorbent (XAD-2), and the inorganic vapors are retained in the chemically active impinger solutions. Because it is impossible to predict the concentration of a constituent in the gas stream at the start, sampling is based on the minimum volume of gas necessary to provide detection of materials in the analytical scheme. A minimum of 30 m^3 (1100 ft^3) is sufficient to ensure detection of materials at approximately 1 mg/m.^{3}

Sampling fugitive air emissions presents problems similar to those of sampling stack gases; specialized equipment is necessary for this operation. A fugitive assessment sampling train (FAST) has been designed which operates at 5.2 m^3/min. Entrained particles are fractionated into >15 μm, 3–15 μm, and <3 μm. Organic vapors from a side stream are trapped in an XAD-2 sorbent trap. The primary techniques applied for inorganic analysis are gas chromatography for the gaseous components and spark source mass spectrometry (SSMS) for elemental analysis of solids and liquids (Fig. 1). Certain anions (including sulfate, nitrate, and phosphate) are presently determined by wet chemical techniques. However, future analysis will be based on ion chromatography to increase the number of anions detectable. Atomic absorption (AA) has been applied only for mercury, arsenic, and antimony because they were thought to be lost by vaporization during SSMS analysis. Recent data indicate that SSMS values for arsenic are within Level-1 accuracy requirements and that antimony values may also be acceptable. It is likely that AA will be retained in future schemes only for mercury determination. Inorganic gas analysis needs to be more comprehensive.

Level-1 organic analysis uses five operations, one done in the field and four in the laboratory. At the sampling site gas chromatography analyzes for organic substances with boiling points below 100°C. The separation is essentially by boiling point with detection by flame ionization. The column does not separate individual compounds but does group all materials into seven boiling point ranges.

Laboratory organic analysis uses a seven step liquid elution chromatography separation on silica gel. It is an analytical step (in that behavior of a given class of compounds is predictable) as well as a separation step (since the fractions may be further analyzed much more readily than the original mixture).

The second laboratory operation is determination of total organic content. This operation measures the organic content of each of the chromatographic fractions, which allows proper aliquot size selection for optimum column operation. The original Level-1 scheme depended entirely on reduction to dryness and weighing for determination of total organic. Recent data show that many materials boiling in the range below 275°C may be partially lost by that approach.

Accordingly, a gas chromatography procedure for volatile organic compounds has been included in the Level-1 procedure. Total organic content in the laboratory samples is the sum of the results of the gravimetric and the chromatographic analyses. It should be recalled that organics in the boiling range below 100°C are not collected but are analyzed in the field by gas chromatography.

The third laboratory analysis operation is infrared absorption spectrophotometry. Infrared spectra of the seven chromatographic fractions may be used to confirm the absence or presence of certain compound classes or functional groups as indicated by the chromatographic data. It is occasionally possible to obtain specific compound identification from the infrared spectra; but, as previously mentioned, the complexity of most environmental samples makes this the exception rather than the rule.

Finally, low resolution mass spectrometry (LRMS) is applied to each separation fraction which exceeds a specified concentration when referred to the sample source. The present levels are 0.5 $\mu g/m^3$ for gas streams, 1 mg/kg for solids, and 0.1 mg/liter for liquids.

The original Level-1 procedure did not contain LRMS,[12] but it was included in the modified procedure[8] to prevent potential triggering of Level-2 analyses by large amounts of suspicious, but innocuous, organics. LRMS can be a very powerful tool, especially when combined with the other Level-1 components. In many cases, specific compounds may be identified and measured when the entire scheme is applied.

CURRENT STATUS AND RESULTS

All the elements in the Level-1 scheme have been used in studies of discharges from industrial and energy processes. The data developed is far too voluminous to be presented here; in fact, it has not yet been evaluated enough to demonstrate the full usefulness of the approach. The results presented are, therefore, more a demonstration of the usefulness of selected elements within the scheme.

In a study of the precision of the SASS train[13] for the collection of particles and organic matter from gas streams, total particulate loadings determined with two independent SASS trains agreed within 20% of the value determined with the standard EPA Method 5. The size distributions determined by the two trains also agreed closely (Fig. 2). The quantity and distribution of extractable organic material between the two trains was also well within Level-1 tolerances (Table 2).

Additional analysis of these samples using the liquid chromatographic separation total chromatographable organics, infrared, and low resolution mass spectrometry showed that the sampling and analytical procedures gave reproducible data.

The samples acquired from these tests were also used to evaluate inter- and intralaboratory precision of the various steps in the analytical scheme. In general, the analysis for total extractable organic content and the distribution between chromatographic and gravimetric fractions were well within the precision (factor of 3) required by Level 1. Triplicate analysis by four laboratories gave an average total organic content of 480 mg with a coefficient of variation of 4%. However, the liquid chromatographic separation gave a wider range of values in the distributions by fractions, indicating a need for more careful column preparation and sampling.

74   L. D. Johnson and J. A. Dorsey

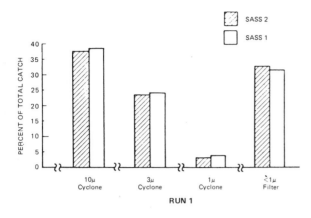

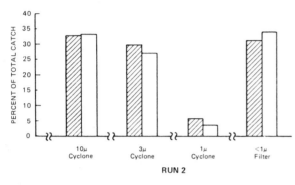

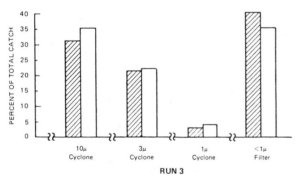

Fig. 2. Size distributions for source assessment sampling system (SASS).

An illustrative sample of data comes from a ferroalloy plant (Table 3). The concentrations shown in Table 3 represent total SASS-train organic catch before and after a scrubber.[14] The two runs were not simultaneous, and process feed mixes were changed between runs. It is therefore not entirely correct to calculate scrubber efficiency from this data; however, the striking reduction in certain compound categories is almost certainly beyond that to be expected from the feed changes. It is immediately obvious that 450 mg/m^3 of high-molecular-weight fused aromatics shown in Table 3 represents a very serious potential problem. It appears that the scrubber reduced these and other compound classes significantly. The data from samples taken after the scrubber will require further evaluation before a judgement is reached.

A brief example of Level-1 biological test data (Table 4) was derived from a pilot scale fluidized bed combustor[15] and put into the present form as a result of a second study.[16] The responses ranged from not detectable to high, with medium well represented also. These data, along with the associated chemical data, show that several of the sample types are of interest for further work.

ONGOING WORK AND FUTURE TRENDS

Although considerable effort has gone into development of the phased approach to environmental assessment, there are still many areas which require further investigation and improvement. This is especially true with respect to Level 2 and Level 3, but work is still needed on Level-1 techniques. Just a few of the more promising projects are:

1. use of porous polymer resins for sampling organic material in water;

Table 2. Organic extractables from SASS evaluation (mg/m^3)

Method	Cyclone		XAD-2 (Extract)		XAD-2 (Module)	
	SASS-1	SASS-2	SASS-1	SASS-2	SASS-1	SASS-2
Gas chromatography	0.03	0.01	3.41	3.58	(Rinse)	
Gravimetric analysis	1.65	1.58	10.2	8.99	69	81
Total	1.7	1.6	13.6	12.6	69	81

Table 3. Organics from ferroalloy plant

Compound category	Concentration[a] (mg/m^3)	
	Before scrubber	After scrubber
Aliphatic hydrocarbons	6.9	2.5
Aromatic hydrocarbons	3.7	23
Fused aromatics <216	380	26
Fused aromatics >216	450	0.02
Ethers		0.1
Ketones	67	0.8
Alcohols	<0.1	<0.3
Esters	0.9	1.8
Amines		<0.1
Heterocyclic N	80	0.1
Heterocyclic S	37	1.7
Carboxylic acids	4.3	0.7
Sulfides		<0.1
Amides		<0.1
Sulfur		0.2
Nitrites		<0.1
Silicones		0.2

[a]Total SASS train organic catch.

Source: Rudolph, J. L., and Levins, P. L., "Ferroalloy Sample Analysis," A. D. Little, Inc., Draft EPA Report, Contract 68-02-2150, March 1978.

Table 4. Bioassay data

Sample	Salmonella mutagenicity	RAM	Rodent toxicity	Algal (FW)	Daphnia	Fish (FW)
Dolomite	ND	ND	ND	ND	ND	ND
Fly ash	ND	ND	ND			
Spent bed	ND	ND	ND			
Coal	ND	ND	ND	M	L	L
Fine (1μ) particles	L	L				
Coarse (3μ) particles	L	ND				
Slurry		L	ND	H	M	L
Spent bed leachate		ND	ND	M	M	M
Fly ash leachate		ND	ND	ND	ND	ND

ND = Not detectable.
Blank = Test not performed.
M = Medium response.
L = Low response.
H = High response.

2. sampling and analysis of highly water-soluble organics;

3. interpretation and synthesis of Level-1 chemical, biological, and engineering data;

4. increased accuracy of SSMS analysis and extension to As and Sb;

5. use of ion chromatography for anion analysis; and

6. revision of the Level-1 Procedures Manual.[8]

CONCLUSIONS

The phased approach developed for environmental assessments has been successfully used in a number of programs. The Level-1 results have been useful for defining potential problems for further study and directing additional resources into the areas of greatest need. Even though not all parts of the scheme have been found effective on each study, every part has produced needed data in at least one individual study. In considering the usefulness of the approach, it is therefore necessary to consider the overall results rather than the value of any single component.

Although the procedures selected initially have been constantly improved in detail, no significant changes in the basic techniques have been necessary. The phased approach is, therefore, a practical and cost-effective method for identifying potential environmental problems associated with discharges from industrial and energy processes.

REFERENCES

1. J. W. Hamersma and S. L. Reynolds, *Field Test Sampling/Analytical Strategies and Implementation Cost Estimates: Coal Gasification and Flue Gas Desulfurization,* EPA-600/2-76-093b (NTIS No. PB 254166), U.S. Environmental Protection Agency, Washington, D.C., April 1976.

2. J. Vlahakis and H. Abelson, *Environmental Assessment Sampling and Analytical Strategy Program,* EPA-600/2-76-093a (NTIS No. PB 261259), U.S. Environmental Protection Agency, Washington, D.C., May 1976.

3. L. D. Johnson, "Procedures for Environmental Assessment of Industrial Wastewater," in *Symposium Proceedings: Third Annual Conference on Treatment and Disposal of Industrial Wastewaters and Residues, Houston, Texas, April 1978.*

4. J. A. Dorsey et al., *Environmental Assessment Sampling and Analysis: Phased Approach and Techniques for Level 1,* EPA-600/2-77-115 (NTIS No. PB 268563), U.S. Environmental Protection Agency, Washington, D.C., June 1977.

5. L. D. Johnson and R. G. Merrill, "Organic Analysis for Environmental Assessment," in *Symposium Proceedings: Environmental Aspects of Fuel Conversion Technology, III (Hollywood, Florida, September 1977),* EPA-600/7-78-063 (NTIS No. PB 282429), U.S. Environmental Protection Agency, Washington, D.C., April 1978.

6. C. H. Lochmuller, "Analytical Techniques for Sample Characterization in Environmental Assessment Programs," in *Symposium Proceedings: Environmental Aspects of Fuel Conversion Technology, II (December 1975, Hollywood, Florida),* EPA-600/2-76-149 (NTIS No. PB 257182), U.S. Environmental Protection Agency, Washington, D.C., June 1976.

7. R. M. Statnick and L. D. Johnson, "Measurements Programs for Environmental Assessment," in *Symposium Proceedings: Environmental Aspects of Fuel Conversion Technology, II (December 1975, Hollywood, Florida),* EPA-600/2-76-149 (NTIS No. PB 257182, U.S. Environmental Protection Agency, Washington, D.C., June 1976.

8. J. W. Hamersma, S. L. Reynolds, and R. F. Maddalone, *IERL-RTP Procedures Manual: Level 1 Environmental Assessment,* EPA-600/2-76-160a (NTIS No. PB 257850), U.S. Environmental Protection Agency, Washington, D.C., June 1976.

9. K. M. Duke, M. E. Davis, and A. J. Dennis, *IERL-RTP Procedures Manual: Level 1 Environmental Assessment Biological Tests for Pilot Studies,* EPA-600/7-77-043 (NTIS No. PB 268484), U.S. Environmental Protection Agency, Washington, D.C., April 1977.

10. D. B. Harris, W. B. Kuykendal, and L. D. Johnson, "Development of a Source Assessment Sampling System," presented at Fourth National Conference on Energy and the Environment, Cincinnati, Ohio, October 1976.

11. D. E. Blake, *Source Assessment Sampling System: Design and Development,* EPA-600/7-78-018 (NTIS No. PB 279757), U.S. Environmental Protection Agency, Washington, D.C., February 1978.

12. P. W. Jones et al., *Technical Manual for Analysis of Organic Materials in Process Streams,* EPA-600/2-76-072 (NTIS No. PB 259299), U.S. Environmental Protection Agency, Washington, D.C., March 1976.

13. F. Smith, E. Estes, and D. Wagoner, "Field Evaluation of the SASS Train and Level 1 Pro-

cedures," in *Symposium Proceedings: Process Measurements for Environmental Assessment (February 1978, Atlanta, Georgia),* EPA-600/7-78-168, August 1978.

14. J. L. Rudolph and P. L. Levins, "Ferroalloy Sample Analysis," A. D. Little, Inc., Draft EPA Report, Contract 68-02-2150, March 1978.

15. J. Allen et al., "Comprehensive Analysis of Emissions from Exxon Fluidized Bed Combustion Miniplant Unit," Battelle Memorial Institute, Draft EPA Report, Contract 68-02-2138, U.S. Environmental Protection Agency, Washington, D.C., September 1977.

16. D. Brusick, "Level 1 Bioassay Data Reporting Format," Litton Bionetics, Draft EPA Report, Contract 68-02-2681, U.S. Environmental Protection Agency, Washington, D.C., August 1978.

DISCUSSION

Ron Neufeld, University of Pittsburgh: This is just a comment, basically. The level-one approach is supposed to be a screening operation, and from a pragmatic point of view, I am struck by the diversity of tests that you have outlined in the level-one technology. For a level-one survey to be useful, in my opinion, not only to coal conversion but to industrial wastes and perhaps hazardous wastes as well, I think it would be appropriate to develop a simpler set of analyses than can be carried on by other individuals or by other industries. Perhaps you would like to call that a level zero. That's something that is a more easily done survey to give you some better indication of potential hazards.

Larry Johnson: Yes, we initiated this work with the intent that the methods be very easy to execute and most of the methods are. The liquid chromatography is very easy: it involves small columns and traditional technology. Low-resolution and spark-source mass spectrometry are not so straight-forward. We were lured into adopting spark source because it provides so much information for the money: you can determine many elements that you cannot with other survey techniques. Actually, a good portion of even this work can be done by basic laboratories. We have established one laboratory that is staffed by a small number of engineering technicians. They are able to do all of the workup—the sampling and processing of the samples to the point where spectral analyses must be carried out—and then provide the samples to a contractor for final measurement.

It is possible to use level one and find much that just about any laboratory can do. You do have to be careful, however, because even simple methods have to be executed properly. The simplicity of the methods, the crude information being sought, and the factor of 3 tolerance in quantification can lull people into complacency, and suddenly you can find uncertainties of a factor of 20.

There are a number of methods in level one that have evolved from starting with even simpler methods. Organic characterization, for example, initially involved no fractionation at all—total organics were all that was to be measured. We felt that this was too crude because you might have a relatively large quantity of an innocuous material (or even a contaminant of the workup such as grease), which would be detected as total carbon and initiate a level-two assessment. We also hesitated to incorporate low-resolution mass spectrometry, but it has proven very useful several times in avoiding a more detailed assessment. The liquid chromatographic fractionation step provides useful data-weight distribution of the fractions—even in the absence of low-resolution mass spectrometry.

You raise a refreshing point of view. We are generally pressured for a more extensive initial assessment, a "level 1.5", rather than going back to a "level 0.5".

SAMPLING OF PROCESS STREAMS FOR PHYSICAL AND CHEMICAL CHARACTERIZATION OF RESPIRABLE AEROSOLS*

G. J. Newton[†]   S. H. Weissman[†]
R. L. Carpenter[†]   R. L. Hanson[†]
H. C. Yeh[†]   C. H. Hobbs[†]

ABSTRACT

Similar aerosol sampling systems have been designed, constructed, and used to obtain respirable samples from process streams of two advanced coal utilization technologies. An experimental low-Btu stirred-bed coal gasifier (LBTUG) and an experimental atmospheric-pressure fluidized-bed coal combustor (FBC) were studied. In both cases the sampling system withdrew a sample from the process stream, cooled and diluted it, and transported the aerosol produced to a sampling chamber. In all cases the sampling system collected only respirable-sized particles (<10 μm). A number of aerosol sampling devices were used. In addition to filter samples for bulk aerosol properties, cascade impactors were used to obtain samples that yielded data on size distribution and chemical composition as a function of particle size. The Lovelace Aerosol Particle Separator, a type of aerosol centrifuge, was used to obtain size-differentiated samples for electron microscopy and comparison with aerodynamic-sized particles for determination of effective particle densities. Point-to-plane electrostatic precipitators were also used to obtain samples for transmission and scanning electron microscopy as well as x-ray energy spectroscopy. Tenax-GC samples were used to determine vapor-phase hydrocarbons in the aerosol. In a collaborative program with the Morgantown Energy Technology Center, aerosols at two locations in the LBTUG process streams and four positions along the FBC exhaust stream have been sampled. Both coal-use technologies produce respirable-sized particles that are significantly different from those produced by conventional pulverized coal combustors.

INTRODUCTION

Low-Btu gasification is scheduled for a significant role in this nation's future energy options. There are a great number of uncertainties surrounding coal gasification ranging from questions concerning economic viability of processes to questions of environmental impact. In large part, these uncertainties arise from the fact that a number of different gasification processes are competing with one another for marketplace dominance. More importantly, process streams in each gasification scheme are only partially characterized, making environmental concerns difficult to elucidate and plant waste treatment difficult to define.

These uncertainties have generated considerable interest in process stream samplings.[1] Because respirable aerosols generated by coal conversion technologies constitute a potentially significant health hazard, we have intentionally designed a process stream sampling system with an upper size cutoff of approximately 10 μm. Characterization of fine particle fractions of process streams is also of great interest to those designing coal gasifiers, because these particles have been implicated in gas turbine blade erosion observed when low-Btu gas is used to fire turbines. Fine particles may also cause methanation catalyst poisoning and increase gas distribution system maintenance costs if released from high-Btu gasification plants.

---

*Research performed under U.S. Department of Energy Contract Number EY-76-C-04-1013.

†Inhalation Toxicology Research Institute, Lovelace Biomedical and Environmental Research Institute, Albuquerque, N.M. 87115.

Although this symposium is concerned with synthetic fuel phases of coal use, we have included our experience in sampling effluents from an experimental fluidized-bed coal combustor (FBC) for three reasons:

1. Sampling systems used in both FBC and gasifier studies are similar, differing only slightly to meet space and operating environment limitations.

2. The FBC unit studied was an experimental system, and the limitations we observed in sampling are probably representative of those which others will encounter in sampling with similar scale equipment designed for other coal conversion processes.

3. Finally, the FBC is often included in conceptual designs of large synfuel plants to produce raw product gases, to burn char and other wastes, or to be a utility boiler for plant process stream production.

In any case, effluents from these combustors in a coal conversion plant must be included in the total plant effluent. For coal conversion plants, auxiliary combustor effluents often represent a major portion of the total effluent outfall.

The data presented in this work are the result of a cooperative program between the Inhalation Toxicology Research Institute and the Morgantown Energy Technology Center (METC). Samples obtained in this study are being evaluated for both toxicological and process characterization purposes.

The METC 18-in. Fluidized-Bed Combustor

The 18-in.-ID atmospheric-pressure fluidized-bed combustor (Fig. 1) has been under development at METC since 1967. Various fuels, including coals, lignite, and culm, have been burned in the combustor. Bed materials have also been varied in efforts to enhance SO_2 capture. The fluidized bed is supported by a conical plate perforated by $1/8$-in. holes fitted with stainless steel elbows. These elbows direct fluidizing air towards the center of the vessel parallel to the conical surface. Water passing through the heat exchanger bank, which consists of U-shaped tubes ($1/4$ in. ID), extracts heat from the bed. An additional heat exchanger at the top of the combustor vessel controls temperature of exiting gases.

A screw conveyor meters presized coal into a pneumatic feed tube which conveys fuel into the combustor near the bottom of the bed. Limestone or

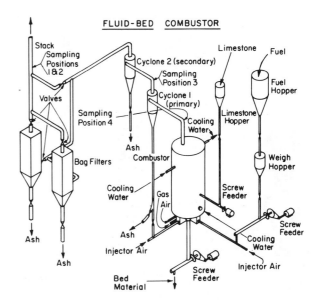

Fig. 1. Schematic diagram of 18-inch fluid bed combustor, fuel and limestone feed systems, and cleanup devices.

other sorbent may be fed into the combustor by a screw conveyor above the fluidized bed. Bed level is controlled by periodic removal of material from the bottom with a 3-in. screw conveyor or via an overflow and lock hopper. Combustion products pass through two cyclone separators. Partially cleaned combustion gases pass through a baghouse filter system for final particle removal. Solids from the primary cyclone contain enough unburned fuel to justify reinjection into the combustor. A system of valves (Fig. 1) allows the baghouse to be placed in the cleanup system or to be bypassed.

The METC Low-Btu Coal Gasifier

The METC coal gasifier is a pressurized version of the McDowell-Wellman atmospheric stirred-bed gasifier (Fig. 2). A grate in the lower end of the pressure vessel supports the coal. The vessel bottom is sealed with lock hoppers through which spent ash may be removed. The gasifier top is closed with a hemispherical dome through which gas exits and coal is fed via a lock hopper. The gasifier uses the Lurgi process for low-Btu gasification which requires air, steam, and coal. Once inside the gasifier, coal undergoes a complex series of reactions beginning with devolatilization, which produces tars and oils, and ending with combustion of coal char to CO. This mixture of vapors and gases along with ash particles exits the gasifier and passes through a large cyclone to remove ash particles (Fig. 2).

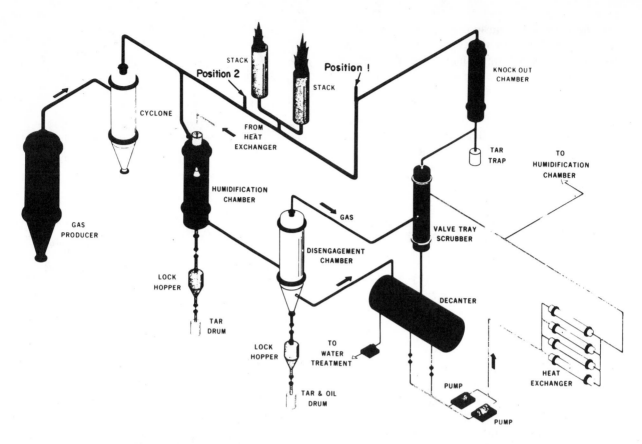

Fig. 2. Schematic diagram of low-Btu gasifier and partial flow gas cleanup system.

After exiting the main cyclone, the low-Btu gas is divided. One-fourth of the total flow enters a gas cleanup system (Fig. 2), and the remaining three-quarters of the gas passes through a pressure letdown nozzle and a set of mufflers. This uncleaned gas is burned in twin flares. Raw gas is cleaned by saturating it with water vapor and subjecting the mixture to a series of discrete temperature and pressure drops. This procedure transfers material from vapor to liquid state. Resulting tar and oil droplets are moved by disengagement chambers and scrubbers. After cleanup, gas is tested and the unused portion returned to the main gas stream just before the flare.

Sampling Locations

The FBC was sampled to determine what changes took place in the exhaust stream as it moved through the cleanup system. These data on fine particles were considered useful to those concerned with exhaust cleanup as well as those concerned with source characterization and fugitive emissions. Sampling probes were placed before and after each stage of the effluent cleanup system (Fig. 1).

Low-Btu gasifier sampling was approached from a somewhat different viewpoint. Unlike the FBC, there is no representative exhaust aerosol from the gasifier itself since no use is presently made of the produced gas. However, if low-Btu gas is depressurized, diluted, and cooled, such as would happen if leaks occurred, an aerosol is formed. To study characteristics of this aerosol, samples of gas were withdrawn from both clean and raw gas process streams (Fig. 2). These samples were cooled (raw gas) or depressurized (clean gas) and diluted with N_2. Aerosol so formed was then sampled to determine its physical and chemical properties.

MATERIALS AND METHODS

Basic Considerations

A major limitation in studying coal combustion effluents is an inability to sample the aerosols adequately. Since no single aerosol instrument in existence is capable of meeting the stated goals, a series of instruments and techniques was used to characterize coal combustion aerosols. Flow conditions within ducts were turbulent, with flow

Reynolds numbers on the order of 10^4–10^5. These turbulent flow conditions have no defined streamlines and compromised the concept of isokinetic sampling. As a compromise, sample velocities were selected so that a range of one to two times duct-to-sample velocities was achieved. In addition to turbulent flow, temperatures in the FBC ranged from 800 to 150°C at Positions 4 to 1 in Fig. 1. Two approaches were considered: (1) sample at stream temperature within ducts, or (2) extract a portion and dilute, age, and sample it under more optimum conditions. Sampling at temperature within ducts was rejected due to small duct size, material corrosion and erosion, potential "cloud effects" on size-selective samplers from high mass loadings, and high-temperature effects on aerosol parameters which are not completely understood. Basic strategy involved extracting a small portion (15–25 liters/min) with duct-to-sample velocity ratios of about 1.2:1, diluting with 100 liters/min of dry, cool air, and piping the diluted sample to a chamber from which samples could be obtained under milder flow conditions.

A similar sampling method was used for the low-Btu stirred-bed coal gasifier (LBTUG). Once the coal gas aerosol was formed, problems associated with transporting and sampling it were found to be similar to those presented by the FBC aerosol. Thus the basic system design is similar. Since thermophoretic effects were expected to be significant during the dilution and aerosol formation phase, a radially injected diluter was used. Because low-Btu gas will form flammable and explosive mixtures with air, N_2 was used as a diluent rather than air to avoid the hazard presented by corona discharges present in the sampling electrostatic precipitators. Since low-Btu gas is corrosive, the entire system was constructed of stainless steel, but the FBC system used brass, copper, and stainless steel. Space and sample volume considerations limited the sampling chamber to a cylinder approximately 1 m long and 34 cm in diameter.

A probe (similar to the FBC probes) was used to sample raw coal gas at Position 2 (Fig. 2). At this position exhaust gas from sampling could be released to the atmosphere (63 ft above ground level). At Position 1 (Fig. 2) clean gas was at 42 psig and a critical orifice was used. The orifice was placed immediately in front of the diluter. This sampling location was inside the METC pilot plant house, requiring that the sampling system exhaust be ducted outdoors

Sampling System Design

Figure 3 is a schematic diagram of major components of the sampling train. For the FBC, sampling probes were 3-ft lengths of $\frac{1}{2}$-in.-OD (0.062-in. wall) stainless steel tubing with inserts at the probe tip to reduce the inside bores to $\frac{3}{8}$-in. (0.95 cm) at Positions 1 and 2 and $\frac{1}{4}$-in. (0.64 cm) at Positions 3 and 4. Edges were chamfered to reduce edge effects. At Positions 1 and 2 a 90° bend aligned the probe in the 8-in.-ID stack, while at Positions 3 and 4, probes were installed on the outer radius of an elbow so that about 24 in. of the sample probe was axially located inside a straight

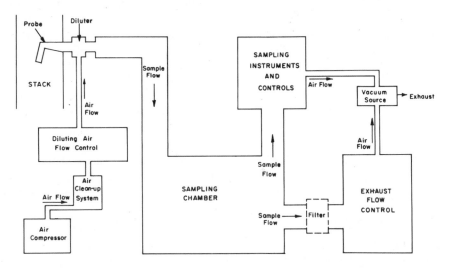

Fig. 3. Schematic diagram of major sampling system components.

section of the 4-in.-ID duct. A ball valve was placed immediately after the sample port, permitting the sampling system to be detached while maintaining FBC exhaust integrity.

Probe Design

The primary goal in designing an extractive sampling system is to obtain representative samples. Under laboratory conditions where gas flow is laminar, the recommended technique is to obtain samples under isokinetic sampling conditions (i.e., the velocity of flow entering the probe is equal to gas flow velocity in the duct). In the FBC exhaust streams at flow volumes of 200–500 scfm, the flow Reynolds numbers are on the order of 10^4–10^5 at gas temperature ranges of 150 to 800°C. This means that defined streamlines do not exist and flow is turbulent. Under turbulent flow conditions matching mean velocity will no longer be isokinetic and the process probably should be called iso-mean-velocity (IMV) sampling.

Conventional sampling probes are usually L-shaped so that a particle-laden gas stream entering a probe makes a 90° change in direction within a short distance and is then transported to a collecting or detecting instrument. When gas stream velocity is very high and/or flow is turbulent, IMV sampling will reduce sample bias as compared to aniso-mean-velocity (AIMV) conditions, but bias due to particle losses at the bend may be so high that the overall bias is worse than that encountered in a well-designed AIMV sampling system (Fig. 4).

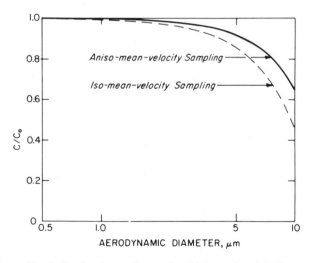

Fig. 4. Predicted sampling probe efficiency (inertial effects only) for aniso-mean-velocity and iso-mean-velocity sampling conditions.

Aerosol penetration was predicted from a method for calculating sample entry efficiency[2] and an equation for calculating aerosol losses at a 90° bend.[3]

In addition to sampling efficiency, three other factors were also considered in the design of this sampling system: (1) adaptability for sampling under a variety of operating conditions of the experimental FBC, (2) probe installation within existing exhaust ducts, and (3) off-the-shelf availability of probe components. Table 1 lists probe sizes used, sampling conditions, and an estimated sampling efficiency using Davies' equation for AIMV and Yeh's equation for aerosol losses at a 90° bend. The probe design was a compromise to reduce total bias due to entrance bias effects and inertial losses at a 90° bend.

A radially injected diluter (Fig. 5) was designed to reduce thermophoretic losses during cooling and dilution. Aerosol enters the diluter through a ½-in. pipe. A high-temperature gasket separates the dilution region from the support flange. Clean diluting air enters through a porous stainless steel cylinder perpendicular to sample flow direction. This arrangement provides a layer of sheath air and tends to compress the sample, allowing cooling and mixing to occur away from cooler walls, thereby reducing thermophoretic losses. Examination of the diluter after several days of operation demonstrated that losses were minimal. Four compressed air lines, each containing its own flow controls (Fig. 6), were connected to the diluter assembly. Final filters were used to ensure that no particulate matter was introduced into the sample by diluting air. After exiting the diluter, the sampling train was expanded to 1.5-in.-ID copper pipe for transfer to the sampling chamber, which was a specially adapted 55-gal steel drum with appropriate fittings.

Sample and exhaust flows and pressure were controlled by rotameters and pressure gauges. A portable air compressor system with associated air cleanup and drying units provided diluting air. Two rotary vacuum pumps were used to exhaust the sampling chamber. By proper adjustment of dilution and exhaust flow controls, the sampling chamber was maintained at slightly negative pressure and the amount of effluent drawn from the stack was controlled.

Immediately downstream from the diluter, a Tenax-GC (Altech Associates, Arlington, Illinois) sampling system was installed in a tee of the sampling line (Fig. 6). Tenax, a gas chromatography column

Table 1. Probe sizes, sampling conditions, and estimated probe
sampling efficiency (C/C_0)

	Duct size (cm)			
	20.32	20.32	10.16	10.16
Total flow rate, scfm	200	500	200	500
Stream temperature, °C	150	150	810	810
Reynolds no. (duct)	2.82×10^4	7.04×10^4	5.63×10^4	1.41×10^5
Probe size, cm	0.95^a	0.95^a	0.635^b	0.635^b
Sampling flow rate, std liters/min	25	25	25	25
Reynolds no. (probe)	2.65×10^3	2.65×10^3	2.13×10^3	2.13×10^3
U (probe), cm/sec	793	793	4551	4551
U_0 (duct)/U (probe)	0.50	1.25	0.89	2.21
	C/C_0			
D_{ae}, (μm)				
0.1	1.0	1.0	1.0	1.0
0.5	1.0	1.0	1.0	1.01
1.0	0.99	1.0	1.0	1.04
3.0	0.94	0.97	0.98	1.20
5.0	0.86	0.92	0.96	1.41
7.0	0.74	0.84	0.94	1.60
10.0	0.52	0.65	0.92	1.80

[a] L-shaped probe.
[b] Straight probe.

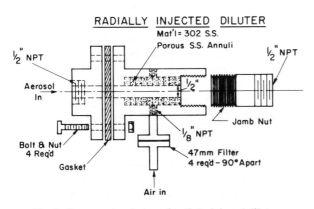

Fig. 5. Cross-sectional view of radially injected diluter.

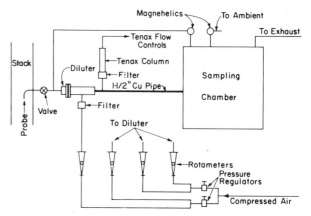

Fig. 6. Schematic diagram of diluting air flow controls, diluter, and Tenax sampler.

support material, adsorbs gaseous hydrocarbons larger than the C^4 isomers. Since its first use,[4] this material has been increasingly used to sample for gaseous hydrocarbons in exhaust streams. The Tenax side of the sampling line contained a cutoff valve to maintain integrity of the sampling system, a filter to separate particles, and a brass cylinder containing the Tenax adsorbent. Exhaust from the Tenax adsorbent line was passed through the

sample cleanup filter before exhausting into the main vacuum line.

The sampling chamber (Fig. 7) served several functions. First, the sample entry line was constructed so that particles larger than 10 μm would be removed by impact on the sampling chamber walls. The sampling chamber was fitted with ports to permit

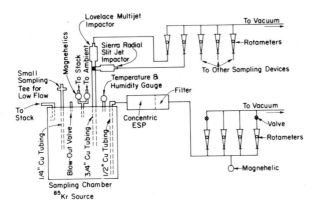

Fig. 7. Schematic diagram of sampling chamber and sample flow controller.

samples to be connected in such a way that each instrument would receive a representative sample. Excess sample was pulled through the chamber, then passed through a concentric electrostatic precipitator and an absolute filter and vented outside. The sampling line from the probe entered vertically through the top of the chamber and then turned perpendicularly so that the aerosol stream was aimed at one wall which served as an impact surface for particles larger than 10 μm. Provisions were made for both low-flow (0.1–1.0 liters/min) and high-flow (25 liters/min) sampling instruments. Sample lines extended through the top of the barrel and part way down into the chamber. These lines were positioned so that the incoming sample was pulled up past them by the chamber exhaust. The sampling chamber contained a [85]Kr deionizer source to reduce electrostatic charge on particles to near Boltzmans equilibrium. There were a blowout port, for safety purposes, and two sampling chamber pressure gauges. One gauge measured sampling chamber pressure with respect to the sampling probe, while the other measured sampling chamber pressure with respect to atmospheric pressure. These two gauges were necessary to ensure transport of sample from the effluent line to the sampling chamber. Pressure gauges also ensured that the entire system operated at desired pressure relative to atmospheric pressure.

The airflow control system had three distinct parts: (1) the diluter flow section (Fig. 6) to ensure that an accurately metered flow of air was supplied to the diluter, (2) the main exhaust control (Fig. 7), which consisted of a group of pressure gauges and flow controllers in parallel as well as associated cutoff valves to measure flow leaving the sampling chamber accurately, and (3) the sampling instrument control-

lers (Fig. 7), a series of parallel flow controllers and pressure gauges.

Sampling instruments themselves withdrew varying amounts of aerosol from the sampling chamber. To maintain a constant flow out of the sampling chamber, it was necessary to set up exhaust flow controllers such that several of them could be preset to control the same flow that each of the sampling instruments used. Each sampling instrument had its own flow controller which was preset to the appropriate flow. The valve system permitted one to begin sampling with an instrument and cut off the appropriate flow from the main exhaust control section simultaneously, thereby maintaining constant chamber flow conditions.

Diluter airflow controls and chamber controls were set so that 100 liters/min of air was injected into the diluter, while 125 liters/min was withdrawn from the sample barrel. The difference, 25 liters/min, represented the probe sampling rate. Flow controls were adjusted so that the sampling chamber was negative with respect to the probe and to the atmosphere. To sample the effluent aerosol, the appropriate instrument was installed in a tee, connected to the proper flow controller, and the appropriate fraction of flow was switched from the chamber main exhaust to the sampling system, starting a flow through the sampler.

Aerosol Sampling Instrumentation

A battery of sampling instruments (Table 2) was selected to (1) determine aerodynamic size distribution measurements among particles less than 10 μm in diameter, (2) compare count and aerodynamic size distributions to obtain information on the density of collected particles, (3) obtain transmission and scanning electron microscope samples for subsequent analysis, (4) obtain sufficiently large samples for both spark source mass spectrometry and x-ray fluorescence spectroscopy, and (5) obtain samples for organic constituent determinations.

The Lovelace Multi-Jet Impactor (LMJ)[5] and the Sierra Radial Slit Jet Impactor (SRSJ) (Sierra Instruments, Inc., Carmel, California) were chosen as the main size-selective samplers. Criteria used for selection were low wall losses, operation within the Marple regime (S/W and T/W $\cong$ 1–5 and flow Reynolds numbers of 500–3000),[6] leak-proof operation, and suitable sample collection capacity.

Choice of these impactors does not imply that other impactors were unacceptable. Although many impactors can separate particles and have been

Table 2. Samplers used on effluent streams from METC-FBC

Sampling device	Flow rate (liters/min)	Total sample weight	Comments
Lovelace Multi-Jet Cascade Impactor	25	400 mg	Aerodynamic size fractions 0.6 to $\cong$ 10 μm
Sierra Radial Slit Jet Cascade Impactor	21	500 mg	Aerodynamic size fractions 0.6 to $\cong$ 10 μm
Mercer Cascade Impactor	0.5	1.0 mg	Aerodynamic size fractions 0.3 to 5.0 μm
Lovelace Aerosol Particle Separator	0.3	~10 mg	Aerodynamic size fraction ($\cong$ 26) 0.5 to 3.0 μm. Grids for transmission electron microscopy as a function of aerodynamic size
Filter (47 mm)	25	~1 g	On Selas Ag membrane to help determine mass loading
Point-to-plane electrostatic precipitator	0.5	~1 ng	Samples for transmission electron microscopy for geometric size analysis and scanning electron microscopy for morphological and elemental analyses
Tenax sampler	2.0	~10 mg	Samples for determination of vapor-phase polycyclic aromatic hydrocarbons above C_5
Concentric electrostatic precipitator	100–125	~10 g	Device is used as a sample chamber exhaust cleanup. Samples are used for major elemental analysis and particulate hydrocarbon analysis.

widely used by other investigators, the fact that they do not satisfy Marple's criteria suggests that extensive calibration is required. Cascade impactors operated within the Marple regime can be calibrated easily using computational methods. This approach permits accurate stage calibrations under operating conditions within the impactor design range.

Reliance in the field on calculated calibrations requires laboratory validation of that calibration. Equally important is a laboratory demonstration that all impactors used follow the same calibration curves. To ensure that computed calibration factors would be accurate, a comparison was made among the laboratory standard Mercer,[7] LMJ, and SRSJ impactors. When the impactors simultaneously sampled a uranine (sodium fluorescein) CsCl aerosol (10 mg/ml uranine, 10 mg/ml CsCl, pH 11), they agreed almost perfectly (Fig. 8). Agreement among these three cascade impactors indicates that well-designed, correctly operated impactors can be

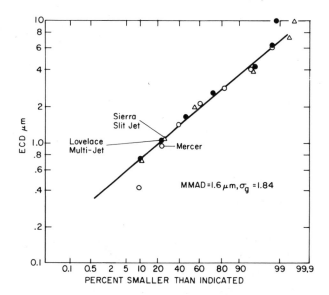

Fig. 8. Comparison of three cascade impactors simultaneously sampling a test aerosol.

adapted to field sampling situations with confidence using computed stage characteristics.

Aerosol collection with cascade impactors involves compromises in factors leading to a choice of collection substrates. Trade-offs made for the METC field studies were unusual. Rationale for the choices made is as follows: Often ideal collection substrates for cascade impactors are hard surfaces coated with a silicone grease which acts as a cushion rather than a glue.[8] Anticipated chemical analyses (spark source and x-ray fluorescence spectrometry) and large (mg) sample requirements led to the conclusion that silicone greases were not acceptable for this sampling effort. Many investigators have used fiber filter material for collection substrates. Consequently, tests of available filter media were made to determine the best compromise collection surface (Table 3). Wall losses, an unexpected adverse effect, were seen using all filter media while sampling the uranine-CsCl test aerosol. Use of filter media also compromised determination of correct particle size distributions. Filter media caused wall losses ranging from 7 to 18%. The same aerosol collected on brass shim stock (0.005 in.) coated with Dow-Corning Anti-Foam A indicated wall losses of 3 to 5% and 5 to 10% for the LMJ and SRSJ, respectively. Although many investigators have described particle bounce-off and its correction using greasy substrates, many of these studies were conducted using polystyrene latex spheres which may not represent characteristics of

coal combustion effluent aerosols. A series of tests evaluated uncoated 5-mil brass shim stock as a collection surface. For both the LMJ and SRSJ, wall losses ranged from ≅5-10% and were not significantly different from those obtained while using Anti-Foam A as a coating. From these tests, it was decided that uncoated brass or stainless steel shim stock was a reasonable compromise and that fly ash samples from the FBC and LBTUG could be collected without contaminating samples with silicone greases.

Sample purity considerations also apply to total filter samples taken from the FBC and LBTUG. A filter medium was required which could withstand sampling conditions present in the FBC but which would not contaminate the sample with significant quantities of the elements or organics of interest. A comparison of trace element content of various filter media showed that silver membrane filters were extremely pure elementally and contained few organic contaminants.[9] Silver membrane filters are also stronger than many other filter media and more able to withstand conditions encountered in sampling the FBC. Silver membranes were used for all filter samples from the FBC, but fiberglass filters were employed on LBTUG samples due to H_2S-silver incompatibility.

Three versions of point-to-plane electrostatic precipitators were used to obtain samples for electron microscopy. The standard unit[10] was used to collect

Table 3. Summary of performance comparisons of collection substrates in cascade impactors[a]

Collection substrate	Number of runs	MMAD (µm)	σ_g	Wall losses (% of sample)	Deviation (%) from Mercer Impactor results	
					MMAD	σ_g
Gelman type A fiberglass	2	2.25	2.06	10.25	41	12
5-mil brass shim stock and Anti-Foam A	4	1.51	1.89	4.78	−6	3
5-mil brass shim stock	4	1.58	1.89	7.98	1	3
Millipore fluoropore, white, 0.5-µm pore size	3	1.72	1.96	6.87	8	7
Nucleopore clear, plain, regular, 0.4-µm pore size	2	1.4	2.13	11.8	13	16
Millipore membrane mixed cellulose centers, white, 0.8-µm pore size	1	1.5	2.31	17.4	−6	26
Selas flotronics silver membrane filter, 0.8-µm pore size	1	0.95	2.21	18.2	−41	20

[a]Measured with Mercer Impactor. MMAD = mass median aerodynamic diameter; σ_g = geometric standard deviation.

material for examination by transmission electron microscopy, providing data on count statistics of the collected aerosol. Two modified versions of the standard device were built. One modified electrostatic precipitator was adapted for holding JEOL scanning electron microscope (SEM) stubs. This permitted samples of aerosol to be precipitated directly onto SEM stubs, minimizing sample contamination. The METC scanning electron microscope uses Cambridge SEM stubs. To provide samples for their instrument, a second modification of the point-to-plane electrostatic precipitator was made.

The Lovelace Aerosol Particle Separator (LAPS)[11] is a spiral duct aerosol centrifuge which collects aerosols separated according to aerodynamic diameter onto a stainless steel foil. The foil has transmission electron microscope grids placed along it to provide data on the count distribution of particles collected having the same aerodynamic diameter, permitting an estimate of the range of densities found in effluent aerosol particles.[12]

Sampling System Performance

The sampling system was designed to collect particles under 10 μm in diameter. For both the FBC and gasifier, sampling system losses were found to be low. The 10-μm upper limit exceeded the median aerosol diameter by a sufficient margin so that little aerosol deposition within the sampling system was found except at Position 4 (Fig. 1) on the FBC. This was not unanticipated since there is significant carry-over of coarse material from the top of the fluid bed. This carry-over material is collected in a cyclone and returned to the combustor (Fig. 1).

Significant losses were also observed in the probe at Position 2 in the raw gas stream of the gasifier. Examination of the probe after 30 h of operation revealed that it was coated with a char which probably was deposited by thermophoresis since an unheated section of the probe penetrated insulation, pipe, flange, and lagging. Sampling lines after the diluter as well as the diluter itself were found to be relatively clean.

Additional efforts to evaluate sampling train performance involved construction of a similar system and laboratory testing. Laboratory testing duplicated most system components including all 90° bends and horizontal runs. Test conditions did not simulate elevated temperature. A test aerosol was generated by nebulizing a suspension of montomorillonite clay and sodium fluorescein using standard techniques. Generator concentrations were 20 mg of clay and 10 mg of uranine per cubic centimeter, which when dry produced an aerosol that was approximately lognormal with a mass median aerodynamic diameter of 2.5 μm and with a geometric standard deviation of 1.8. The test aerosol was sampled with the mock FBC sampling train, and system losses were determined for individual components in the train. Analyses were by standard fluorometric techniques of material rinsed out of the sampling train components and compared to the aerosol sampled. These results showed that maximum system losses were less than 17%, with 5% being lost in the probe at the 90° bend. Diluter losses were less than 3%, with the remainder of the losses diffusely distributed.

RESULTS

FBC Sampling

Aerosol samples obtained from the FBC exhaust have been obtained from the burning of Montana Rosebud coal(subbituminous), a Western Kentucky high-sulfur bituminous coal, and Texas lignite. In addition to particle samples, the aerosol vapor phase was sampled for high-molecular-weight organic materials using a Tenax sampler. These samples have been used to deduce aerosol size distribution data and examine particle morphology. These samples have been examined by spark source mass spectroscopy (SSMS) for inorganic trace element composition and for organic content by extraction with methylene chloride or n-pentane followed by gas chromatography–mass spectrometry and/or high-performance liquid chromatography. Samples of the effluents from burning different coals are in various stages of analysis. A detailed description of sampling and analytical procedures has been completed for the Montana Rosebud combustion samples.[1]

A wide range of aerosol size distributions was observed at various positions in the combustor exhaust for both Texas lignite and Montana Rosebud coals (Table 4). Figure 9 shows typical cumulative size distribution plots of samples obtained from Position 2. Size distribution parameters were fitted to the data using a nonlinear least squares method.

Examinations of LAPS and point-to-plane electrostatic precipitator samples were by transmission and scanning electron microscopy. As shown in Fig. 10, FBC fly ash is not composed of spherical

Table 4. Particle size and mass loading in FBC effluent streams

Position	Montana Rosebud coal				Texas lignite			
	Particle size[a] (μm)	Std. dev.	Mass loading (g/m³)	Std. dev.	Particle size[a] (μm)	Std. dev.	Mass loading[b] (g/m³)	Std. dev.
1	2.3	0.3	7.0	0.038	2.5	0.2	1.3	0.4
2	2.2	0.2	3.8	0.15	2.8	0.3	2.2	0.6
3	2.2	0.1	1.2	0.3	3.0	0.2	2.8	0.6
4	3.9	0.8	4.7	1.8	5.6	2.6	18.9	5.2

[a]Mass median aerodynamic diameter. The geometric standard deviations were fairly similar in all instances (1.6 to 1.8).
[b]Verified leak in baghouse filters.

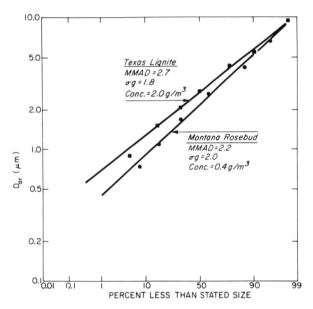

Fig. 9. Typical cumulative size distribution plots of cascade impactor data obtained for Montana Rosebud and Texas Lignite (Position 2).

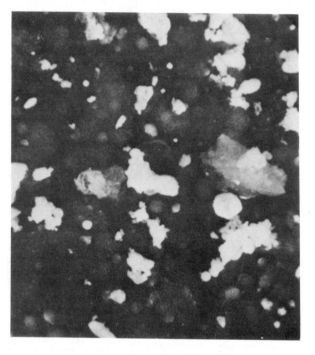

Fig. 10. Scanning electron micrograph of FBC fly ash from Montana Rosebud coal (Position 1).

particles as are normally found in ash from pulverized coal power plants. Comparison of the physical size range of collected particles having the same aerodynamic diameters indicated that ash particles vary in density. Reduction of this observation to a range of numerical values is complicated by the irregular particle shapes.

The FBC ash from the effluent cleanup devices and from aerosol samples were analyzed by SSMS. Samples from impactors operated at Position 1, the stack breech, produced information on the size dependence of elemental concentrations as a function

of particle mass median aerodynamic diameter. Figure 11 shows typical results obtained at this location.

Determination of the mass of organic material present before and after each cleanup device was done using Tenax-GC adsorbent traps to sample the effluent vapor phase and the concentric electrostatic precipitator to sample the particle phase. Established extraction and analysis procedures were used.[13] Table 5 shows results obtained for vapor-phase

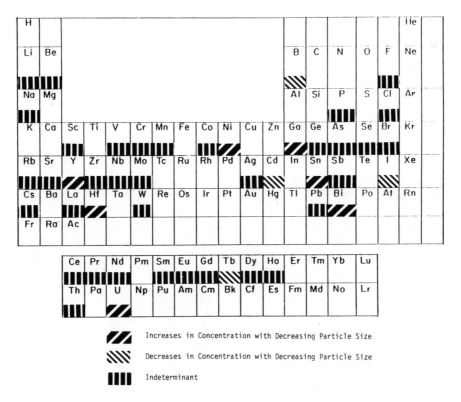

Increases in Concentration with Decreasing Particle Size

Decreases in Concentration with Decreasing Particle Size

Indeterminant

Fig. 11. Modified periodic chart showing distribution of the elements in FBC fly ash as a function of particle size.

Table 5. Estimated vapor-phase hydrocarbon concentration in FBC effluent[a]

Sampling effort[b]	Position	Vapor-phase hydrocarbon effluent concentration (mg/m^3)
1	1–After bag filter	33
1	2–Before bag filter	33
1	3–Between cyclones	16
1	4–Above bed	13
2	1–After bag filter	5.1
2	2–Before bag filter	3.0
2	3–Between cyclones	4.9
2	3–Between cyclones (undiluted)	3.0
3	4–Above bed	9.2
3A	1–After bag filter	1.3
3A	2–Before bag filter	12
3A	3–Between cyclones	1.1
3A	4–Above bed	1.6

[a]Vapor-phase hydrocarbons were collected on Tenax.
[b]Fuel for efforts 1 and 2 was Montana Rosebud, for 3 and 3A, Texas lignite and Texas lignite refuse respectively.

hydrocarbons while the FBC was burning Montana Rosebud coal.

Gasifier Sampling

Similar determinations were made for gasifier process streams. Samples were taken for bulk- and size-fractioned analysis of aerosol particles after diluting and cooling or depressurizing, cooling, and diluting. These samples were collected for both raw and clean gas. Aerosol size distribution and concentration data are shown in Table 6. Aerosol concentrations are *not* referred to process stream conditions since, for raw gas at least, most of the aerosol was produced by the cooling and diluting step in the sampling system.

Point-to-plane electrostatic precipitator samples and LAPS samples were examined by scanning electron microscopy for both the clean and raw gas. In general, there were few ash particles. The majority of aerosol particles were tars and oils. There was no evidence of sample changes from the high vacuum in the electron microscope. Tar and oil particles spread on impact on the LAPS foil, so it was not possible to

Table 6. Aerosol characteristics of cooled and diluted low-Btu gas[a]

Gas	Location	Number of observations	MMAD ± S.D. (μm)	σ_g ± S.D.	Coal gas aerosol concentration (g/m^3)
Clean	1	11	0.69 ± 0.16	1.81 ± 0.33	2.08 ± 0.41 × 10^{-2}
Raw	2	8	1.35 ± 0.15	1.25 ± 0.05	0.32 ± 0.10

[a]MMAD = mass median aerodynamic diameter; σ_g = geometric standard deviation.

 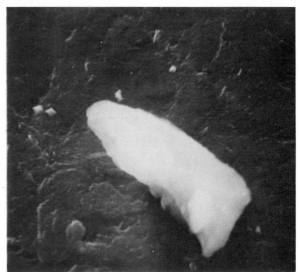

Fig. 12. Scanning electron micrograph of fly ash particles from the low-Btu gasifier raw gas stream (left) and clean gas stream (right).

determine particle density. Figure 12 shows ash particles typical of those found in both raw and clean gas.

The SSMS analysis of gasifier aerosol samples showed many elements are present in milligram quantities (Table 7). Coal gas aerosols also show that amount of elements present depends partly on particle size. The SSMS analysis of small-particle samples shows the presence of a variety of trace elements (Table 8).

Some preliminary characterization of coal gas aerosol particulate and vapor-phase hydrocarbon content has been done using gas chromatography-mass spectroscopy (GC-MS) (Tables 9–12). Samples were analyzed after extracting and concentrating the extract. Methylene chloride was used for extraction of particle samples, and n-pentane was used for extracting Tenax samples. Following extraction and concentration, 1 μl was injected into the GC-MS. A 30-m glass capillary column coated with Dexsil 300

was used. Helium was used as a carrier gas with a constant temperature increase of 2°C/min from 60 to 270°C.

DISCUSSION AND CONCLUSIONS

The sampling system used on the FBC and LBTUG has operated successfully on a number of occasions. Few problems were observed with system operation. The sampling point-to-plane electrostatic precipitator exhibited anomalous behavior when used to sample coal gas. The device arced before establishing a corona discharge. Increasing the needle to collection-plane spacing reduced the tendency to arc.

Coal gas aerosol produced from LBTUG process streams appears to be composed mainly of tar and oil droplets. These materials are chemically labile, changing from a liquid to a "varnish" overnight. Implications of this change and sample preservation

Table 7. Elements present[a] in clean
coal gas at concentrations
greater than 1 mg/g

Element	Number of samples[b]
Aluminum	All
Barium	Some
Boron	Some
Calcium	All
Chlorine	Some
Copper	Most
Fluorine	All
Iron	All
Magnesium	Most
Manganese	Some
Phosphorus	All
Potassium	Most
Sodium	Most
Silicon	All
Strontium	Some
Sulfur	All
Titanium	All

[a]Three small filter samples analyzed using SSMS. Samples obtained under different operating conditions.

[b]"All" signifies presence at >1 mg/g in three out of three samples. "Most" signifies presence at >1 mg/g in two out of three samples. "Some" signifies presence at >1 mg/g in one out of three samples.

Table 8. Trace elements enriched in
small particles in raw and clean coal gas

Raw	Clean
Aluminum	Barium
Arsenic	Bromine
Barium	Cadmium
Cobalt	Chlorine
Gallium	Iodine
Germanium	Lanthanum
Magnesium	Molybdenum
Rubidium	Phosphorus
	Potassium
	Silver
	Strontium

Table 9. Tentative identification of particulate
phase hydrocarbons from raw gas

Scan	M.W.	Hydrocarbon
1241	156	Dimethylnaphthalene isomer
1251	156	Dimethylnaphthalene isomer
1304	152	Biphenylene or acenaphthylene
1370	154	Acenaphthene or 1,4-dihydro-1,4-ethenonaphthalene
1427	168	Dibenzofuran
1467	170	Trimethylnaphthalene isomer
1548	170	Trimethylnaphthalene isomer
1572	166	Fluorene or phenalene
1593	170	Trimethylnaphthalene isomer
1630	182	Xanthene or a methyldibenzofuran
1673	182	Xanthene or a methyldibenzofuran
1778	180	Methylfluorene isomer
1902	184	Dibenzothiophene
1961	178	Phenanthrene
1975	178	Anthracene
2168	192	Methylphenanthrene isomer
2175	192	Methylphenanthrene isomer
2189	192	Methylphenanthrene isomer
2216	192	Methylphenanthrene isomer
2406	206	Dimethylphenanthrene isomer
2460	202	Fluoranthene
2539	202	Pyrene
2675	216	Benzofluorene or methylpyrene isomer
2734	216	Benzofluorene or methylpyrene isomer
3036	228	Benzo(c)phenanthrene or a benzofluoranthene
3051	228	A benzofluoranthene or benzo(a)anthracene
3059	228	Benzo(a)anthracene or chrysene
3096	228	Chrysene or triphenylene
3225	242	Methylbenz(a)anthracene isomer
3497	252	Benzofluoranthene isomer
3546	252	Benzofluoranthene isomer
3624	252	Benzo(e)pyrene or benzofluoranthene isomer
3642	252	Benzo(a)pyrene or benzo(e)pyrene
3696	252	Perylene or benzo(a)pyrene

methods to eliminate it have not yet been adequately studied. The initial liquid aerosol material presents a problem in collecting filter samples of the aerosol. No filter material has been found suitable for collecting samples of these aerosols.

Data obtained thus far from FBC sampling permit several preliminary observations. The FBC effluent particle size distribution appears to be independent of coal type or combustor operating conditions. The effluent aerosol is respirable and could be deposited in deep lung tissue. Effluent aerosol particle mass loading is a function of both coal type and combustor operating conditions. Effluent vapor-phase hydrocarbon levels range between 0.5 and 50 mg/m,^{3} but particle-borne hydrocarbon levels are observed to be <0.5 mg/m^3. Hydrocarbon levels are apparently a function of coal source and operating conditions. Effluent particle trace element profiles differ from those observed for pulverized coal combustion (PCC) and are a function of both operating conditions and coal source. Compared with PCC, fewer elements in the FBC exhaust show size-dependent concentrations. Finally, particle morphology of FBC effluent aerosols is different from that observed in PCC effluent aerosols. The FBC-generated particles show an irregular structure, but whereas PCC-generated particles are fused or spherical, presumably because of higher combustion temperatures in pulverized coal combustors.

Table 10. Tentative identification of vapor-phase
hydrocarbons in raw gas stream

Scan	M.W.	Hydrocarbon
366	116	Indene or 1-ethynyl-4-methylbenzene
537	122	Dimethylphenol isomer
591	130	Methylindene isomer
600	122	Dimethylphenol or ethylphenol isomer
673	128	Azulene or naphthalene
682	128	Naphthalene or azulene
902	117	Indole or benzeneacetonitrile
940	142	Methylnaphthalene isomer
948	148	Methylbenzothiophene isomer
968	142	Methylnaphthalene isomer
1082	154	Acenaphthene or biphenyl
1140	156	Ethylnaphthalene isomer
1180	156	Dimethylnaphthalene isomer
1200	156	Dimethylnaphthalene isomer
1216	156	Dimethylnaphthalene isomer
1221	154	1,4-Dihydro-1,4-ethenonaphthalene or acenaphthene
1250	156	Dimethylnaphthalene isomer
1260	152	Acenaphthylene or biphenylene
1270	152	Acenaphthylene or biphenylene
1277	156	Dimethylnaphthalene isomer
1314	158	Methylbiphenyl isomer
1326	154	Acenaphthene or 1,4-dihydro-1,4-ethenonaphthalene
1378	170	Trimethylnaphthalene isomer
1388	168	Dibenzofuran
1393	170	Trimethylnaphthalene or methylethylnaphthalene
1407	170	Methylethylnaphthalene isomer
1420	170	Trimethylnaphthalene isomer
1438	170	Trimethylnaphthalene isomer
1462	170	Trimethylnaphthalene isomer
1475	170	Trimethylnaphthalene isomer
1503	170	Trimethylnaphthalene isomer
1511	166	Fluorene or phenalene
1527	166	Fluorene or phenalene
1547	170	Trimethylnaphthalene or methylethylnaphthalene isomer
1564	168	Propenylnaphthalene or methylbiphenyl isomer
1584	182	Xanthene or methyldibenzofuran isomer
1626	182	Biphenyl-4-carboxaldehyde, anthene, or methyldibenzofuran isomer
1905	178	Phenanthrene
1918	178	Anthracene

Table 11. Tentative identification of particulate-phase
hydrocarbons in cleaned gas stream

Scan	M.W.	Hydrocarbon
1901	178	Phenanthrene
1916	178	Anthracene
2111	192	Methylphenanthrene isomer
2117	192	Methylphenanthrene isomer
2349	206	Dimethylphenanthrene isomer
2398	202	Fluoranthene
2479	202	Pyrene
2640	216	Benzofluorene or methylpyrene isomer

Data from gasifier sampling permit several observations about coal gas aerosols. When low-Btu gas is cooled or depressurized and diluted, an aerosol is formed. The aerosol is composed of respirable-sized oil and tar droplets. These aerosols are chemically complex and sufficiently labile to change overnight at room temperature (at least when collected on a stainless steel surface). Clean gas aerosol contains a ratio of high-molecular-weight hydrocarbons of about 1 part particle-borne material to 20 parts vapor-phase material under the conditions

Table 12. Tentative identification of vapor-phase
hydrocarbons in clean gas

Scan	M.W.	Hydrocarbon
343	118	2,3-Dihydroindene
390	116	Indene
528	134	4-Carbon alkylbenzene isomer
594	132	Methylbenzofuran isomer
696	132	Methyl-2,3-dihydroindene isomer
737	130	Methylindene isomer
755	130	Methylindene isomer
841	128	Naphthalene
1139	142	Methylnaphthalene isomer
1175	142	Methylnaphthalene isomer
1308	154	Biphenyl or 1,2-dihydroacenaphthylene
1369	156	Ethylnaphthalene isomer
1400	156	Dimethylnaphthalene isomer
1404	156	Dimethylnaphthalene isomer
1429	156	Dimethylnaphthalene isomer
1438	156	Ethylnaphthalene isomer
1441	156	Dimethylnaphthalene isomer
1494	152	Acenaphthylene or biphenylene
1562	154	Acenaphthene or 1,4-dihydro-1,4-ethenonaphthalene
1623	168	Dibenzofuran
1766	166	Fluorene

in the sampling system. Rapid cooling and diluting process stream gas appears to preferentially transfer high molecular weight polynuclear aromatic compounds (PNA) from vapor to particulate phases (Tables 9–12). Raw and clean gas aerosols differ distinctly in elemental composition. To some unknown extent this is probably due to removal of ash by the gas cleanup system. Size of the aerosol particles also differs between raw and clean gas aerosols. Even after cleanup there are still potentially toxic elements present in the aerosol whose concentration is enriched relative to the feed coal. The toxicological significance of these findings is as yet unknown.

Sampling of FBC and LBTUG processes indicate that both processes produce respirable-sized aerosols. Analyses for organic constituents have revealed the presence of high molecular weight organic compounds in the FBC. Passage of the effluent through a baghouse filter considerably reduces hydrocarbon levels. Coal gas aerosols are a complex mixture of tars and oils which have been poorly characterized. The biological hazard, if any, associated with these aerosols is unknown.

The SSMS analysis of effluent and process streams gives particle-size-dependent trace element profiles for aerosol particles from both FBC and gasifier. Neither of these aerosols is similar to the aerosols released by conventional coal combustion sources. The FBC ash shows fewer elemental concentration size dependencies, but the gasifier appears to form particles enriched in toxic metals.

Sampling of these coal utilization processes via extractive techniques followed by dilution has proven to be a useful method. Studies will be continued on both FBC and LBTUG to determine aerosol characteristics as a function of coal source and operating conditions.

ACKNOWLEDGMENTS

The authors thank METC personnel who supplied FBC and LBTUG design and operating conditions, especially J. J. Kovach for coordination of efforts and W. M. Wallace, G. D. Case, L. C. Headly, J. Wilson, R. Rice, and U. Grimm for constructive criticism and review. We also acknowledge the efforts of the ITRI sampling team, R. Tamura, D. Horinek, R. Peele, T. Stephens, and E. Barr. Finally, we thank our scientific colleagues at ITRI for review and support, especially Dr. R. O. McClellan for encouragement, review, and support.

REFERENCES

1. R. L. Carpenter et al., *Characterization of Aerosols Produced by an Experimental Fluidized Bed Coal Combustor Operated with Sub-Bituminous Coal,* Report LF-57, Lovelace Inhalation Toxicology Research Institute, 1978.

2. C. N. Davies, "The Entry of Aerosols into Sampling Tubes and Heads," *Br. J. Appl. Phys.* Ser. 2, 1, 921 (1968).

3. H. C. Yeh, "Use of a Heat Transfer Analogy for a Mathematical Model of Respiratory Tract Deposition," *Bull. Math. Biol.* 36, 105 (1975).

4. A. Zlatkis et al., "Profile of Volatile Metabolites in Urine by Gas Chromatography-Mass Spectrometry," *Anal. Chem.* 45(4), 763 (1973).

5. G. J. Newton, O. G. Raabe, and B. V. Mokler, "Cascade Impactor Design and Performance," *J. Aerosol Sci.* 8, 339 (1977).

6. V. A. Marple, "A Fundamental Study of Inertial Impactors," Ph.D. thesis, University of Minnesota, Minneapolis, 1970.

7. T. T. Mercer, M. L. Tillery, and G. J. Newton, "A Multi-Stage, Low Flow Rate Cascade Impactor," *J. Aerosol Sci.* 1, 9 (1970).

8. T. T. Mercer and R. G. Stafford, "Impaction from Round Jets," *Ann. Occup. Hyg.* 12, 41 (1969).

9. J. J. Dulka and T. H. Risby, "Ultratrace Metals in Some Environmental and Biological Systems," *Anal. Chem.* 48, 640a (1976).

10. P. E. Morrow and T. T. Mercer, "A Point-to-Plane Electronic Precipitator for Particle Size Sampling," *Am. Ind. Hyg. Assoc. J.* 25, 8 (1964).

11. P. Kotrappa and M. E. Light, "Design and Performance of the Lovelace Aerosol Particle Separator," *Rev. Sci. Instrum.* 43, 1106 (1972).

12. P. Kotrappa and C. J. Wilkinson, "Densities in Relation to Size of Spherical Aerosols Produced by Nebulization and Heat Degradation," *Am. Ind. Hyg. Assoc. J.* 33, 449 (1972).

13. R. L. Hanson et al., "Characterization of Potential Organic Emissions from a Low-Btu Gasifier for Coal Conversion," in *Proceedings of Third International Symposium on Polynuclear Aromatic Hydrocarbons, (October 25-27, 1978),* Battelle Columbus Laboratories, in press.

DISCUSSION

Bob Thomas, Los Alamos Scientific Laboratory: George, I was just curious, it's almost an academic question, but when you find a σ_g of approximately 4, doesn't that mean that you don't even have a distribution to work with? What does that mean to you, isn't that just a smear of particle sizes?

George Newton: That's a good question. I think that's due either to the operating conditions of that gasifier being temporarily erratic, or the operators may have just charged coal in at that time. When they dump coal in the gasifier, the whole thing vibrates, and big sparks fly up. A few of those may have been sampled by the cascade impactor, and that would affect over-size distribution.

Jim Payne, Science Applications, Inc.: When you were doing your Tenax sampling, what was the temperature of the gas that was being sampled? Did you observe any thermal desorption that would effect your trapping efficiency? Secondly, how long did you pump samples through your Tenax columns and what size were they?

George Newton: The Tenax samplers were operated at essentially ambient temperature. Considering that these samples have been diluted, I would say they were always less than 100° F except at positions that involved particularly hot operations. We did not make any attempt to reduce the temperature of the Tenax columns. Further improvements in the sampler will involve steps to reduce the temperature of the Tenax. The samplers were operated at about 2 liters/min, and we sampled for up to 90 min.

Incidentally, the sample sizes were on the order of 10 g in a column approximately 6 in. long. We have since gone to some effort to determine how long it takes to load these columns. We need serial samples, at least three of them in a row so that we can determine how much is getting through.

COMPARATIVE COMPOSITION OF PETROLEUM AND SYNTHETIC CRUDE OILS

J. E. Dooley* S. E. Scheppele*
G. P. Sturm, Jr.* P. W. Woodward*
J. W. Vogh*

ABSTRACT

Five coal liquids prepared from four coals of varying rank were characterized by procedures previously developed for petroleum. The coal liquids were distilled into four fractions and separated by chromatography for subsequent spectral study. The data reported confirm the more aromatic, cyclic, and polar character of coal liquids and the higher nitrogen content compared with petroleum. To our knowledge, this study constitutes the most extensive systematic comparison of coal liquids with petroleum.

INTRODUCTION

Coal-derived liquids, shale oil, tar sand liquids, and heavy petroleum constitute alternatives to conventional petroleum crude oils for refinery feedstocks. Conventional crude-oil-assay procedures and simple heavy-oil characterization schemes provide analytical data of value for these alternative feedstocks. However, such limited data are not adequate to predict with any degree of accuracy the refining characteristics of such materials. Consequently, detailed compositional data must be acquired for these materials and compared with corresponding data for the conventional feedstocks. This paper specifically relates the compositions of the products obtained from the liquefaction of four coals via three different liquefaction processes to the compositions obtained for six petroleum crude oils.

In general, the composition of coal-derived liquids will depend on a number of factors such as the rank of coal and the process conditions.[1] Available data demonstrate that oils derived from various coal liquefaction processes and from coal pyrolysis contain appreciable quantities of polynuclear hydrocarbons and heteroatom-containing polynuclear compounds.[1] Thus if coal-derived liquids are to serve as sources of liquid fuels and as refinery feedstocks, they will generally require an initial upgrading to reduce significantly the concentrations of both heteroatom-containing compounds and polynuclear aromatic hydrocarbons.

Because of the complex chemical compositions of fossil energy fuels, their detailed molecular characterization is facilitated by separating the material into less complicated fractions prior to instrumental analysis. Consequently, both the petroleum crude oils[2-8] and the coal-derived liquids[9-12] were separated, using the methodologies developed at the Bartlesville and Laramie Energy Technology Centers for the American Petroleum Institute Research Project 60.[13,14] Low-resolution low-voltage (11-eV)[14-17] and low-resolution high-voltage (70-eV)[18] electron-impact mass spectrometry (EI/MS) were the principal techniques used in group-type analyses for the fractions obtained from the separations. In the latter stages of these investigations, both high-resolution[19] 70-eV EI/MS and low- and high-resolution field-ionization mass spectrometry[20] (FI/MS) were also employed to assist in the identification and quantification of compound types.

CHARACTERIZATION METHODS

The first step in the separation involves distillation of the total sample. The approach is illustrated in

---

*Department of Energy, Bartlesville Energy Research Center, P.O. Box 1398, Bartlesville, Oklahoma 74003.

Fig. 1. A rotating-drum isothermal still was used to produce the fraction of the crude oil boiling up to about 240° C. The residue from this distillation was then processed through a wiped-wall Rota-Film still at progressively higher temperatures and lower pressures to obtain the desired distillates. In the heavy ends work on petroleum, the distillates of primary interest were the 370 to 535° C and the 535 to 675° C boiling ranges. Distillation of the coal liquids used a Perkin-Elmer* spinning-band column in later work in order to acquire distillate fractions boiling over a more narrow temperature range. Use of these limited-boiling-range distillates enhances the correlation of composition results with results obtained from refining studies but still permits reasonable comparison with earlier petroleum work that employed a Rota-Film still.

The distillates are separated into saturate, monoaromatic, diaromatic, and polyaromatic-polar concentrates by dual silica/alumina column chromatography.[21] Gradient elution is accomplished using blends of pentane, benzene, and methanol ether. The technique has been shown to be applicable to petroleum, coal liquids, and other fossil-energy materials. The fact that a separation requires 50 continuous hours constitutes a disadvantage of the method. It is thus important to note that recent research at BETC shows that the same separation can be accomplished in less than 2 h using high-pressure liquid chromatography (HPLC). The HPLC column is regenerated after each run so that no repacking is necessary as is the case with the older method. Although the amount of material processed per pass through an HPLC column is less than with an open column, multiple columns and passes provide as much material in an 8-h period as in a 50-h run. Results obtained from the newer technique are compared with those obtained with the older method in Table 1.

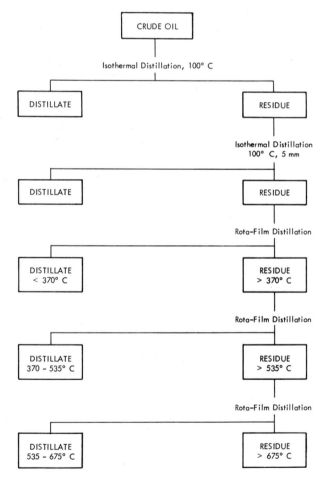

Fig. 1. Preparation of distillates for Swan Hills (south field), Alberta, Canada, crude oil.

Table 1. Saturate and aromatic concentrates from South Swan Hills distillate at 370 to 535°C (Acid and base free)

	API-60 method	HPLC method
Saturate	65.94	64.75
Monoaromatic	12.55	13.65
Diaromatic	6.27	6.67
Diaromatic-2		3.02
Polyaromatic-polar (backflush)	10.82	9.99
Backflush (THF)		0.26
Total	95.58	98.34

The saturate concentrates are analyzed using 70-eV EI-MS.[18] Aromatic concentrates are further separated by gel permeation chromatography (GPC).[16,22] A single preparative scale column is used for this purpose and is packed with two different sizes of polystyrene gel. About 2 g of sample may be charged to the column, and about 40 fractions are collected using an automatic fraction collector.

Additional separations included removal of the acids and bases from the whole distillates of the petroleum samples by ion exchange resins. In the coal

*Reference to specific brand names is made for identification only and does not imply endorsement by the Department of Energy.

liquid characterization research, the time required for removal of the acids and bases was reduced significantly by extraction using solutions of NaOH and HCl in methanol.

Low-resolution low-voltage EI/MS data were acquired for the various aromatic GPC fractions using a CEC 21-103C mass spectrometer. The compound types present in a given GPC fraction were identified by using GPC-elution volume/molecular-structure correlations to interpret the nominal m/e values of the molecular ions observed.[15,16] The molecular-ion intensities were used to quantify the amounts of the various compound types present in a given GPC fraction. The quantitative distributions for the various GPC fractions were also converted to quantitative distributions for both the distillates and the total sample. In the later stages of the coal-liquid characterization, both the identification and quantification of the compound types in a given GPC fraction were complemented by the acquisition of high-resolution 70-eV EI/MS data and low- and high-resolution FI/MS data, using a CEC 21-11-B double focusing mass spectrometer. Proton nuclear magnetic resonance (NMR) spectra were also acquired for the same GPC fractions to provide additional interpretation. For later eluting GPC fractions, uv fluorescence was occasionally used to identify individual compounds such as benzopyrenes, pyrenes, chrysenes, and benzoperylenes. Mass spectral data for these same fractions were consistent with the presence of significant amounts of these compounds (40% or more in some fractions).

This briefly summarizes the methods used in characterization studies at BETC to determine composition of fossil energy materials. For more detailed discussions of techniques, see the references given below.

RESULTS AND COMPARISONS

Techniques were developed at BETC for analyzing the fraction of petroleum boiling in the range 350 to 550°C. Subsequently, however, these methods were used successfully to characterize the 200 to 550°C distillates from the coal liquids. Consequently, comparisons of the compositions of coal liquids and petroleum, based on our current data, must necessarily be limited to physical properties of total liquids and of the detailed molecular compositions for the 370 to 535°C distillates of each material.

The physical property data in Table 2 were prepared from routine crude oil assays conducted at BETC. These data provide the analyst preliminary information on the materials that will assist in planning further separations and analyses. Of the six petroleum crude oils, the Bartlett, Kansas, one is an example of a very heavy crude located in southeast Kansas, southwest Missouri, and northeast Oklahoma; the oil was extracted from the ground using a solvent injection process. The other five petroleum crudes were those selected for study during the API-60 project conducted at the Bartlesville and Laramie Energy Technology Centers during the 1966-1974 period. Of the coal liquids, the Utah and Western Kentucky syncrudes were prepared from Utah and

Table 2. Physical properties of selected petroleum crude oils and liquids produced from coal

Crude oil	Specific gravity	SSU viscosity at 77°F, (sec.)	SSU viscosity at 100°F, (sec.)	Pour point (°F)	Sulfur (wt %)	Nitrogen (wt %)	Carbon residue (%)
Prudhoe Bay	0.893	111	84	15	0.82	0.230	4.7
Gach Saran	0.880	91	72	<5	1.57	0.226	4.6
Swan Hills	0.826	39	37	<5	0.11	0.056	1.2
Recluse	0.815	42	36	60	0.08	0.031	0.6
Wilmington	0.938	470	229	<5	1.59	0.631	4.9
Bartlett	0.963		>6,000	40	0.89	0.321	9.0
Utah syncrude	0.947	112	63	64	0.05	0.478	2.0
Western Kentucky syncrude	0.923	48	43	<5	0.08	0.226	0.0
Synthoil	1.081		2,026	40	0.42	0.786	11.2
H-coal syncrude	0.953	59	46	<5	0.27	0.633	2.3
H-coal fuel oil	0.935	38	35	<5	0.21	0.440	0.8

Western Kentucky coals by the FMC Corporation using the COED pyrolysis process. The synthoil product was prepared from a West Virginia coal by the Pittsburgh Energy Technology Center. Liquids obtained from Illinois no. 6 coal by operation of the H-Coal process, in both the syncrude and fuel oil modes of operation, were supplied by Hydrocarbon Research, Inc.

A cursory examination of the Table 2 data could lead one to believe that some of these coal liquids might not be any more difficult to process than some of the petroleum crudes shown. However, the higher content of both aromatic hydrocarbons and aromatic compounds containing heteroatoms in the coal liquids, as shown in the following detailed compositional studies, would present problems in processing. Specific gravities are generally higher for the coal liquids (probably indicative of higher aromatic or heteroatom content). The nitrogen content is generally higher for the coal liquids than for the petroleum crudes. The percent nitrogen ranges from 0.226 for the Western Kentucky syncrude to 0.786% for synthoil compared with a range of 0.031 for Recluse to 0.631% for Wilmington. The amount of sulfur in coal liquids is usually low (0.5% or less, see Table 2) resulting from the liquefaction process or some upgrading of the liquid. On the other hand, petroleum crudes may contain sulfur up to about 14%, but the amount is generally 3% or less.[23] The high percentage of nitrogen compounds complicates the processing of coal liquids because nitrogen is more difficult to remove by hydrogenation than sulfur.

Table 3 provides information on the distribution of distillates prepared at BETC for some petroleum crudes and coal liquids. As indicated in the table, both the liquefaction process and related hydrotreating have a pronounced effect on the coal liquids produced. In some instances these distillate distributions are similar to those for petroleum crudes. Since distillations were aimed at providing the 370 to 535°C distillates, other distillates and residuals prepared for each oil have been summed together to show comparative information. Exact conditions used to produce the coal liquids are not known, and comparisons must recognize these unknown factors. Note that, with the exception of synthoil, most of the coal liquids had boiling ranges ending at about 535°C.

Chromatographic separations of the 370 to 535°C distillates from each of the materials (with the exception of H-coal products) provided the data shown in Table 4. Amounts given for each concentrate are calculated on the basis of the whole crude. As seen in column 6 of Table 4, aromatic compounds account for a significantly greater percentage of the coal liquids than of the petroleum crudes. It is important to note that the polyaromatic-polar compounds are more abundant in the coal liquids than in the petroleum crudes.

Table 5 is a tabulation of hydrocarbon compound types in the polyaromatic concentrates from the 370 to 535°C distillates. Types are shown as mass Z series, Z in the formula C_nH_{2n+z}. The $-16Z$ series, for example, represents such types as acenaphthalenes; phenanthrenes/anthracenes contribute to the $-18Z$

Table 3. Distillate distributions of selected petroleum crude oils and liquids produced from coal

Crude oil	<370°C (wt %)	370 to 535°C (wt %)	535°C+ (wt %)	<200°C (wt %)	>200°C (wt %)	<240°C (wt %)	>240°C (wt %)
Prudhoe Bay	50.9	21.6	27.5			17.5	82.5
Gach Saran	52.3	19.7	28.0			22.2	77.8
Swan Hills	72.6	18.4	9.0			37.0	63.0
Recluse	72.5	18.6	8.9			39.9	60.1
Wilmington	38.7	24.7	36.6			13.5	86.5
Bartlett	17.6	28.9	52.1	0	100	a	100
Utah syncrude	59.3	40.3	b	13.9	86.1		
Western Kentucky syncrude	75.7	24.2	b	21.5	78.5		
Synthoil	47.0	27.3	25.7	4.4	95.6		
H-coal syncrude				34.6	65.3		
H-coal fuel oil				35.6	63.7		

aBartlett heavy oil contained some material boiling below 240°C as indicated by simulated distillation data. Boiling range of the first distillate prepared ranged from 225°C to about 360°C.

bUtah and Western Kentucky syncrudes had some material boiling above 535°C, but amounts were small because simulated distillation data indicated boiling range ended in this region.

Table 4. Concentrates from distillates at 370 to 535°C

Crude oil	Saturates (wt %)	Monoaromatics (wt %)	Diaromatics (wt %)	Polyaromatics polar[a] (wt %)	Aromatics + polars (total amount)
Prudhoe Bay	10.4	3.7	2.6	4.7	11.0
Gach Saran	9.6	3.3	2.3	4.2	9.8
Swan Hills	12.1	2.3	1.2	2.7	6.2
Recluse	13.8	2.1	0.9	1.6	4.6
Wilmington	9.1	4.1	3.1	8.4	15.6
Bartlett	15.5	5.3	2.7	4.8	12.8
Utah syncrude	10.4	6.9	6.4	15.8	29.1
Western Kentucky syncrude	5.7	6.1	5.6	6.4	18.1
Synthoil	2.8	2.2	5.1	17.0	24.3
H-coal syncrude[b]	4.8	10.6	9.1	33.1	52.8
H-coal fuel oil[b]	7.7	20.3	10.6	18.8	49.7

[a]Includes acids and bases removed in separations.

[b]H-coal materials include all material boiling above 200°C; other materials limited to 370 to 535°C. End of boiling range for H-coal syncrude was about 490°C, for H-coal fuel oil about 410°C.

Table 5. Mass spectral Z series distributions of polyaromatic-polar concentrates from distillates at 370 to 535°C of selected petroleum crude oils and coal liquids

Crude oil	Z series								
	−16 (wt %)	−18 (wt %)	−20 (wt %)	−22 (wt %)	−24 (wt %)	−26 (wt %)	−28 (wt %)	−30 (wt %)	−32 (wt %)
Prudhoe Bay	0.11	0.26	0.43	0.40	0.31	0.20	0.14	0.05	0.03
Gach Saran	0.03	0.15	0.25	0.17	0.14				
Swan Hills	0.05	0.14	0.23	0.31	0.27	0.20	0.13	0.08	0.07
Recluse	0.02	0.07	0.14	0.18	0.17	0.14	0.11	0.07	0.06
Wilmington	0.17	0.36	0.37	0.36	0.28	0.11	0.06	0.01	0.02
Bartlett	0.11	0.28	0.36	0.44	0.33	0.19	0.10	0.03	0.02
Utah syncrude	0.91	1.88	2.30	1.96	1.22	0.60	0.09	0.02	
Western Kentucky syncrude	0.37	0.76	0.93	1.27	0.97	0.41	0.22	0.05	0.01
Synthoil	0.18	0.75	1.46	2.34	2.24	1.48	1.29	0.86	0.54
H-coal syncrude[a]	0.54	0.54	1.53	2.88	2.52	2.09	1.80	1.05	0.70
H-coal fuel oil[a]	0.97	1.28	1.56	0.85	0.14	0.15	0.08	0.02	0.01

[a]Not comparable to other materials because boiling range is significantly different.

series; pyrenes are found in the −22Z series; and chrysenes are found in the −24Z series, etc. Previous publications have in some instances specified individual compound types within given Z series. However, for purposes of this discussion hydrocarbon types in a Z series are not differentiated. For example, benzopyrenes and perylenes occur in the same Z series. As seen in Table 5, the −16 through −26 series hydrocarbons account for a greater percentage of the Utah and Western Kentucky syncrudes than of the petroleum crudes. It should also be noted that both the synthoil and H-coal syncrude contain appreciable amounts of polynuclear aromatic hydrocarbons in the −28 through −32Z series.

Clearly the fractions prepared at BETC are ideally suited for characterization. Actually, closer examination of some of our GPC fractions, especially later eluting ones, has been used to identify individual components such as benzopyrene isomers and chrysene. Other observations, in comparing petroleum and syncrude derived data, include information concerning alkyl substitutions, ring distributions, etc. In the early eluting GPC separations from the petroleum 370 to 535°C distillates, molecular-weight distributions exceeding mass 650 were observed. In contrast, the highest molecular-weight material from the coal-liquid distillates of the same boiling range occurred at larger elution volumes than for the petroleum distillates and possessed molecular-weight distributions extending up to about mass 500. Since GPC separates essentially by molecular size, this result implies a higher degree of aromatic ring condensation and/or a greater number of, and shorter, alkyl side chains substituted on aromatic rings in the coal liquids compared to the petroleum crudes.

SUMMARY AND CONCLUSIONS

Several coal liquids and petroleum crude oils have been compared in regard to both their physical properties and the molecular compositions of the 370 to 535°C fractions obtained from their distillation. The distillates were separated using distillation, dual silica/alumina column chromatography, and GPC. The separated fractions were characterized using mass-spectrometric techniques, NMR, and uv fluorescence.

The data obtained in these studies demonstrate that coal liquids and petroleum crude oils differ in composition. The high aromatic and polar content of the coal liquids will present environmental and refining problems not common to petroleum. The relatively high nitrogen content will be difficult or costly to remove by hydrogenation, but sulfur concentrations should be reduced to acceptable levels during nitrogen removal. Comparison of the physical property data and the detailed molecular compositions for coal liquids and crude oils shows that the former data by themselves are insufficient for determining processing conditions for upgrading coal liquids or for evaluating their environmental impact. Consequently, the contribution of detailed separations and instrumental analyses to the development of economically acceptable alternate fossil energy technology cannot be overemphasized.

REFERENCES

1. For a recent review see: S. E. Scheppele, *Mass Spectrometry and Fossil-Energy Conversion Technology,* FE 2537-7, U.S. Department of Energy, Washington, D.C., 1978.

2. H. J. Coleman, J. E. Dooley, D. E. Hirsch, and C. J. Thompson, *Anal. Chem.* **45**(9), 1724 (1973).

3. C. J. Thompson, J. E. Dooley, D. E. Hirsch, and C. C. Ward, *Hydrocarbon Process.* **52**(9), 123 (1973).

4. J. E. Dooley, C. J. Thompson, D. E. Hirsch, and C. C. Ward, *Hydrocarbon Process.* **53**(4), 93 (1974).

5. J. E. Dooley, D. E. Hirsch, and C. J. Thompson, *Hydrocarbon Process.* **53**(7), 141 (1974).

6. C. J. Thompson, J. E. Dooley, J. W. Vogh, and D. E. Hirsch, *Hydrocarbon Process.* **53**(8), 93 (1974).

7. J. E. Dooley, D. E. Hirsch, C. J. Thompson, and C. C. Ward, *Hydrocarbon Process.* **53**(11), 187 (1974).

8. G. P. Sturm, Jr., P. W. Woodward, J. W. Vogh, S. A. Holmes, and J. E. Dooley, BERC RI-77/7, U.S. Department of Energy, Washington, D.C., 1977.

9. J. E. Dooley, G. P. Sturm, Jr., P. W. Woodward, J. W. Vogh, and C. J. Thompson, BERC RI-75/7, U.S. Department of Energy, Washington, D.C., 1975.

10. G. P. Sturm, Jr., P. W. Woodward, J. W. Vogh, S. A. Holmes, and J. E. Dooley, BERC RI-75/12, U.S. Department of Energy, Washington, D.C., 1975.

11. P. W. Woodward, G. P. Sturm, Jr., J. W. Vogh, S. A. Holmes, and J. E. Dooley, BERC RI-76/2, U.S. Department of Energy, Washington, D.C., 1976.

12. S. A. Holmes, P. W. Woodward, G. P. Sturm, Jr., J. W. Vogh, and J. E. Dooley, BERC RI-76/10, U.S. Department of Energy, Washington, D.C., 1976.

13. W. E. Haines and C. J. Thompson, LERC RI-75/5–BERC RI-75/2, U.S. Department of Energy, Washington, D.C., 1975.

14. For a review see: J. E. Dooley, S. E. Scheppele, and C. J. Thompson, "Characterizing Syncrudes from Coal," chap. 32 in *Analytical Methods for Coal and Coal Products,* ed. by C. Karr, Jr., Academic Press, in press.

15. D. E. Hirsch, J. E. Dooley, and H. J. Coleman, *Correlations of Basic Gel Permeation Chromatography Data and Their Applications to High-Boiling Petroleum Fractions,* RI 7875, U.S. Department of the Interior, Bureau of Mines, Washington, D.C., 1974.

16. D. E. Hirsch, J. E. Dooley, H. J. Coleman, and C. J. Thompson, *Qualitative Characterization of 370° to 535°C Aromatic Concentrates of Crude Oils from GPC Analyses,* RI 7974, U.S. Department of

the Interior, Bureau of Mines, Washington, D.C., 1974 .

17. H. E. Lumpkin and T. Aczel, *Anal. Chem.* **36,** 181 (1964).

18. "Petroleum Products and Lubricants, II," *Book ASTM Stand.* **24,** 700 (1977).

19. T. Aczel and H. E. Lumpkin, *Am. Chem. Soc., Div. Pet. Chem., Prepr.* **22**(3), 911 (1970).

20. S. E. Scheppele, P. L. Grizzle, G. J. Greenwood, T. D. Marriott, and N. B. Perreira, *Anal. Chem.* **48,** 2105 (1976).

21. D. E. Hirsch, R. L. Hopkins, H. J. Coleman, F. O. Cotton, and C. J. Thompson, *Anal. Chem.* **44,** 915 (1972).

22. H. J. Coleman, D. E. Hirsch, and J. E. Dooley, *Anal. Chem.* **41,** 800 (1969).

23. H. T. Rall, C. J. Thompson, H. J. Coleman, and R. L. Hopkins, "Sulfur Compounds in Crude Oil," Bull. 659, U.S. Department of the Interior, Bureau of Mines, Washington, D.C., 1972.

TRACE ELEMENTS IN THE SOLVENT REFINED COAL PROCESSES, SRC I AND SRC II*

Royston H. Filby†   Samir R. Khalil†

INTRODUCTION

Trace Elements in Coal Conversion

Coal conversion processes are under concerted development because of the prospect of alleviating the U.S. dependence on imported oil by exploiting the abundant U.S. coal reserves. Coal liquefaction, which generally involves hydrogenation of coal at elevated temperatures and pressures, shows promise of providing utility boiler fuels, chemical feedstocks, and synthetic crude oils (syncrudes) which have lower sulfur, trace element, and mineral matter contents than the coals from which they are derived. Although the primary objective of coal liquefaction is to produce low sulfur and low ash fuels, there are significant reductions in trace element contents concomitant with the reduction in ash because most trace elements are predominantly associated with the mineral matter of coal and only to a lesser degree with the organic matrix.

Coals of a given rank exhibit a relatively narrow range of major element (C, H, N, O) composition but often vary widely in trace element contents. Such variations are due less to variations in the trace element composition of the original plant material than to the geochemistry of the depositional environment and epigenetic processes occurring after formation of the coal beds. Data on the trace element composition of three bituminous coals used in liquefaction (Table 1) may be compared with the average trace element contents of coals from the Illinois Basin.[1] It can be seen from Table 1 that there are large variations in contents of most elements in the coals, even in coals from the Illinois Basin, indicated by the standard deviation values.[1]

The fate of trace elements present in the coal during coal conversion processes is of concern since large amounts of some toxic elements or species will be generated in such processes should a large-scale coal conversion industry develop. To illustrate this with an example, the daily and annual production rates of some toxic elements (As, Se, Sb, Hg, Co, Ni, Cr) and the most abundant metal normally found in coals (Fe) have been calculated for a 6000 ton per day coal liquefaction plant using Illinois bituminous coal (Table 2). Demonstration plants of this size are being designed for the Solvent Refined Coal processes. In addition to the problem of the disposal of 2.19 million metric tons of ash, there is the problem of the fate upon disposal of elements such as Hg, As, Se, and Sb, which can form volatile species and can be discharged into the environment to cause significant, although localized, health hazards.

The problems involved in determining trace element behavior in coal gasification and in coal liquefaction are significantly different, and this paper concerns only the latter technology. The discussion will be concerned with data collected for the Solvent Refined Coal Processes SRC I and SRC II, developed by the Pittsburg & Midway Coal Mining Company under contract with the U.S. Department of Energy. The present status of major coal liquefaction processes (including SRC I in which the product is a solid) is shown in Table 3.

Knowledge of the behavior and fate of trace elements in the Solvent Refined Coal processes (SRC I and SRC II) is important for two broad reasons: (a) the effects of trace elements on the liquefaction process itself and on the products; and (b) the environmental aspects of trace element transformations and releases during liquefaction and disposal of wastes.

*Work performed under Subcontract No. 8 with the Pittsburg & Midway Coal Mining Co. under contract with the Department of Energy, Contract EX-76-C-01-496.

†Nuclear Radiation Center and Chemistry Department, Washington State University, Pullman, Washington 99164.

Table 1. Concentrations (μg/g) of trace elements in three coals used in liquefaction
(SRC II) compared with average of Illinois Basin coals

Element	Kentucky[a] coal	Pittsburg[a] seam coal	Illinois[a] coal	Illinois Basin average	Illinois Basin coals[b] standard deviation
Ti	430	822	655	600	200
V	108	21.3	34.3	32	13
Ca	2660	5760	4230	6700	4800
Mg	1070	1930	1580	500	200
Al	10,000	13,800	12,600	12,000	3900
Fe	13,800	11,800	14,600	20,000	6300
Na	208	775	753	500	400
K	1870	1270	2260	1700	700
As	6.7	6.7	2.46	14	20
Sb	2.24	0.32	0.48	1.3	1.4
Se	3.00	1.38	2.66	2.2	1.0
Hg	0.237	0.146	0.200	0.2	0.19
Co	2.95	2.23	3.07	7.3	5.3
Ni	8.2		14.0	21	10
Cr	14.6	11.2	26.6	18	9.7
Cu	6.7	0.38		14	6.6
Sc	2.19	2.07	2.85	2.7	1.1
La	5.30	5.37	6.15	6.8	2.8
Rb	34.5	11.3	13.4	19	9.9
Cs	0.85	0.76	0.98	1.4	0.73

[a]R. H. Filby, M. L. Hunt, and S. R. Khalil, *Trace Elements in the Solvent Refined Coal (SRC) Process,* Technical Progress Report for the Period Sept. 1977–April 1978, Washington State University, Pullman, Wash., May 1978; and R. H. Filby, S. R. Khalil, and M. L. Hunt, *Trace Elements in the Solvent Refined Coal (SRC) Process,* Final Interim Report for the Period 1976-77, in preparation.

[b]H. J. Gluskoter et al., Illinois Geological Circular 499, 1977.

Table 2. Production of some important trace
elements in the 6000 ton/day plant

Element	Concentration in coal (ppm)	Production	
		Daily (kg)	Annual (metric tons)
As	11.6	63.1	230
Sb	0.98	5.33	19.5
Se	2.18	11.9	43.3
Hg	0.113	0.615	2.25
Ni	18.0	98.0	358
Co	5.25	28.6	104
Cr	10.4	56.6	207
Fe	24000	1.31×10^5	4.77×10^5
Ash	11%	5.99×10^5	2.19×10^6

Source: R. H. Filby et al., *Solvent Refined Coal (SRC) Processes: Trace Elements,* Research and Development Report No. 53, vol. III, part 6, U.S. Department of Energy Report FE/496-T17, Washington, D.C., 1978.

Table 3. Status of coal liquefaction processes

Process	Description	Status
Solvent Refined Coal (SRC)	Hydrogenation with solvent; no catalyst; solid (SRC I) or liquid (SRC II) product	50 ton/day pilot plant operating; 6000 ton/day to be constructed
H-Coal	Hydrogenation with solvent and catalyst; liquid product	3 ton/day pilot plant; 600 ton/day under construction
Exxon Donor Solvent	Hydrogenation with solvent; no catalyst; liquid product	250 ton/day pilot plant under construction
Gulf Catalytic Coal Liquids	Hydrogenation with solvent and catalyst product; liquid product.	120 lb/day pilot plant

Process-related aspects of trace element behavior can be categorized as

1. the distribution of trace elements in process streams and fractions;

2. the effects of trace elements on product yields;

3. the effects of process conditions (e.g., coal type, temperature, hydrogen pressure, etc.) on trace element distributions;

4. the effects of trace element contents on the composition and burning characteristics of products;

5. the effects of the nature of trace elements in feed coals on the formation of organometallic species; and

6. the effects of trace element species, particularly organometallic compounds, on catalysts used in upgrading of products (e.g., further hydrogenation of coal-derived products or desulfurization).

Some of the important environmental aspects of trace element behavior in the SRC process are

1. the release of volatile toxic trace element species from gas streams (e.g., H_2Se, Hg, AsH_3);

2. the formation of volatile carbonyls [e.g., $Ni(CO)_4$] and their release to the plant environment;

3. the formation of toxic organometallic species and incorporation in products (e.g., metallocenes);

4. the release of toxic trace elements in wastewaters or other waste products; and

5. the effects of trace metal forms in fuels on the incorporation into different fly ash particle sizes during combustion.

Trace Elements in Coal

To understand the behavior of trace elements in the Solvent Refined Coal processes fully, it is necessary to know in what form such elements exist in coal and how these species behave during liquefaction. Unfortunately, the nature of metal species in coal and how they react under liquefaction conditions are only partly known.

Although there are aspects of the origin of coal that are incompletely understood, it is generally recognized that coal is derived from plant remains laid down in a sedimentary-swamp type of environment under reducing conditions. The plant material was converted to coal via formation of humic materials and lignite under the influence of compaction, temperature, and pressure.

Plant remains (cellulose, lignin) $\longrightarrow$ Humic materials $\longrightarrow$ Lignite $\longrightarrow$ Coal

There are several modes of origin (Table 4) by which trace elements could have been incorporated into coals, and these can be conveniently classified depending on whether the elements are organically bound in the coal (in macerals) or inorganically bound (in minerals).

Although it is impossible to identify by visual methods the trace element species which are organically bound, many mineral species have been observed in coals, and these can be separated by physical or chemical methods from the organic matrix. Of the most important minerals that occur in coal (Table 5), the most important quantitatively are silica, pyrite (or other iron-sulfur compounds), and the clay minerals. For elements that are present in

Table 4. Origins of metal species in coal

Organically bound (in macerals)	Inorganically bound (in minerals)
Originally present in the source plant material	Brought in as detrital minerals during coal formation (e.g., silica, zircon, accessory minerals)
Complexed by amino acids and humates during deposition of the organic source materials	Formed in situ during deposition of plant material (syngenetic processes), e.g., formation of clay minerals
Later fixed by adsorption or ion exchange on functional groups in lignite ($-COOH$ or phenolic $-OH$) or coal (phenolic $-OH$)	Formed during movement of solutions after formation of coal (epigenetic processes), e.g., pyrite veins

Table 5. Major minerals in coal[a]

Mineral type	Syngenetic		Epigenetic
	Detrital	Deposited	
Clays	Kaolinite Illite Montmorillonite Sericite Chlorite	Kaolinite Illite Montmorillonite Sericite Chlorite	Illite
Carbonates		Calcite Siderite Ankerite Dolomite	Calcite Ankerite Dolomite
Sulfides		Pyrite Marcasite Chalcopyrite Sphalerite	Pyrite Marcasite Chalcopyrite Sphalerite Galena
Oxides	Rutile Ilmenite Quartz	Haematite Quartz	Hydrated Fe Oxides Quartz
Phosphates	Apatite	Phosphorite	
Accessory minerals	Zircon Feldspars		Chlorides NaCl Sulfates Nitrates

Source: M-Th. Mackowsky, "Mineral Matter in Coal," in *Coal and Coal-Burning Strata,* ed. D. G. Murchison and T. S. Westall, Oliver and Boyd, Edinburgh, 1968, pp. 309–21.

coals in minor to major amounts (0.1 to 10%), it is possible to identify specific mineral species with which the elements are predominantly associated. For example, in most coals Fe is present either as pyrite (FeS_2 plus oxidized forms, $FeSO_4$ or $Fe_2O_3 \cdot nH_2O$) or other sulfide species such as marcasite (FeS_2) or chalcopyrite ($CuFeS_2$). For most of the trace metals in coal present at concentrations less than 0.1%, the actual chemical moieties with which the elements are associated are unknown, and often the species must be inferred from indirect evidence. Several authors have attempted to distinguish between inorganic and organic affinities of trace elements in coals on the basis of analyses of specific gravity fractions of whole coals (float-sink methods). Elements associated predominantly with the organic matrix are found in the lowest specific gravity fractions, and elements associated inorganically are found in the higher gravity fractions. The data obtained provide information on those elements that have a predominantly organic affinity, a predominantly inorganic affinity, and those elements that appear to be both inorganically and organically combined (Table 6).

Another approach to the problem is to determine trace elements in a large number of coals from a particular coal basin and to attempt to determine organic or inorganic associations by statistical analysis of the data. Multiple correlation matrices show that for Illinois Basin coals some elements are organically associated and some are inorganically bound and associated with specific minerals[1] (e.g., Cd in sphalerite, ZnS). Multivariate factor analysis of the trace element data reported[1] on the Illinois Basin coals has shown that twelve factors can be extracted which account for nearly all of the

observed variance in the sample set.[2] The geochemical interpretation of the extracted factors in an oblique solution has been suggested,[2] and of the twelve factors, ten may be regarded as inorganic (Table 7). The predominantly inorganic affinity of Zn, Cd, As, Mn, Mo, and Fe reported[1,3] is confirmed by the factor analysis data (Tables 6 and 7). No significant conclusions concerning either B or Ge were obtained from the factor analysis although neither element appeared to be strongly associated with any of the inorganic factors; hence these elements may be organically bound.[1,3,4] From float-sink data[1,4] and factor analysis data[2] probable associations of some environmentally important elements in coal can be tentatively assigned (Table 8).

It should be emphasized that the evidence so far collected concerning trace elements in coals may not apply to the elemental associations of particular coals used in coal liquefaction. Therefore, it is necessary to determine trace element species in the actual coals used in coal liquefaction processes. Progress has recently been made in this direction using low temperature ashing techniques to separate mineral matter from coal.[5]

Fate of Trace Elements in the Solvent Refined Coal Processes

The SRC I and SRC II processes are at an advanced stage of development, and a 50 ton per day pilot plant has been operated by The Pittsburg & Midway Coal Mining Company since 1976 at Fort Lewis, Washington. In the SRC I process coal is pulverized and mixed with a recycle solvent (with donor hydrogen properties) to form a coal slurry which goes

Table 6. Affinities of trace elements in coal

Elemental affinity	Zubovic[a]	Gluskoter et al.[b]	Swaine[c]
Predominantly organic	Ge, Be, Ga, Ti, B, V	Ge, Be, B, Sb	B, Ge
Organic + inorganic	Ni, Cr, Co, Mo, Cu	Co, Ni, Cu, Cr, Se, V, Ti	Hg, Se
Predominantly inorganic	La, Zn	Zn, Cd, As, Mn, Mo, Hg, Pb, Fe	As, Cd, Zn, F, Zr, Fe, Mn, Mo

[a]P. Zubovic, "Physiocochemical Properties of Certain Minor Elements as Controlling Factors in Their Distribution in Coal," in *Coal Science,* ed. P. H. Given, Advances in Chemistry Series No. 55, Amer. Chem. Soc., Washington, D.C., 1966, pp. 221–31.

[b]H. J. Gluskoter et al., Illinois Geological Survey Circular 499, 1977.

[c]D. J. Swaine, "Trace Elements in Coal," in *Recent Contributions to Geochemistry and Analytical Chemistry,* ed. A. E. Tugarinov, Wiley, New York, 1976, pp. 539–50.

Table 7. Geochemical interpretation of factors
from factor analysis of Illinois Basin goals

Factor	Associated elements	Mineral or process[a]
1	Ti, Si, Al, K, Mg, V	Silicates; detrital minerals (Syngenetic process)
2	As, Pb, Sb, (Ge)	As, Sb in host, galena
3	Fe, PYS,[b] Mo	Pyrite (syngenetic or epigenetic?)
4	SUS,[c] Be, Ga, Cu, Cr, ORS[d]	Organic factor?
5	Ca, Mn	Calcite
6	Zn, Cd	Sphalerite
9	Co, Ni	Millerite or cobalt mineral
10	Se, Cr, V	Organic factor?
11	F, P, Mo	Fluorapatite
12	Na, Cl	NaCl – saline factor

[a]Probable associated mineral host or geochemical process.
[b]PYS = pyritic sulfur.
[c]SUS = sulfate sulfur.
[d]ORS = organic sulfur.

Table 8. Probable association of metals in coal

Element	Probable association
Ti	TiO_2, $FeTiO_3$ most probable but organotitanium complexes possible
As	PbS host mineral (FeAsS possible but less likely)
Sb	PbS host mineral – inorganic
Cd	ZnS host mineral – inorganic
Ni, Co	NiS or cobalt sulfide
Pb	PbS
Hg	Probably inorganic sulfide
Se	Possibly organic and partly sulfide
Cr	Doubtful

to a preheater together with H_2 at 1500 psig. The slurry mixture is then allowed to react in a reactor at 850°F and 1500 psig in which the coal has an approximate residence time of 0.5 h. Although the liquefaction is complex and incompletely understood, the coal matrix is depolymerized; the free radicals formed are hydrogenated via the donor solvent molecules (e.g., tetralin and decalin) to give a mixture of low and high molecular weight hydrocarbons, CO, CO_2, H_2O, and H_2S, plus unreacted mineral matter, and the donor solvent is regenerated. After the gases have been removed through pressure let-down, the slurry is filtered to remove mineral matter and unreacted coal, the solvent is distilled off the coal filtrate to recycle in the process, and the resulting product is a low sulfur, low ash, solid product, SRC I (Fig. 1).

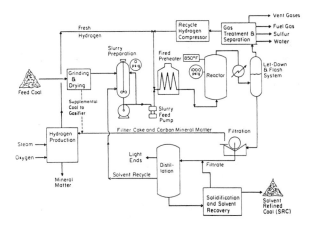

Fig. 1. The Solvent Refined Coal Process-I (SRC I).

In the SRC II mode, the process is similar except that a higher hydrogen partial pressure is used and part of the reacted coal solution is recycled to the reactor to give a longer effective residence (or reaction) time (48–60 min) for the coal than in SRC I. After reaction, light gases are removed by pressure reduction, the liquid (+ mineral matter) stream is fractionated into light distillate (naphtha), middle distillate, and heavy distillate fractions, and a residual material (vacuum bottoms) that contains most of the mineral matter of the coal plus high molecular weight organic material. In a commercial-scale SRC II process (Fig. 2), the vacuum bottoms will be gasified to produce H_2 for the hydrogenation reaction. For SRC I and SRC II a comparison may be made of the process conditions (Table 9) and the product yields (Table 10).

In the SRC I and SRC II processes the sulfur content of the original coal is significantly reduced. In

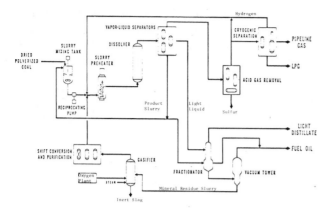

Fig. 2. The Solvent Refined Coal Process-II (SRC II).

Table 9. Process run conditions for SRC I
and SRC II[a]

Conditions	SRC I[b]	SRC II[c]
Mode	No recycle	Slurry recycle
Average dissolver temperature (°C)	499	456
Dissolver pressure (psig)	1500	1883
Nominal residence time (h)	0.5	0.87

[a]Using Illinois coal.
[b]R. H. Filby et al., *Solvent Refined Coal (SRC) Processes: Trace Elements,* Research and Development Report No. 53, vol. III, Part 6, U.S. Department of Energy Report FE/496-T17, Washington, D.C., 1978.
[c]Pittsburg & Midway Coal Mining Co., *Solvent Refined Coal (SRC) Processes: Monthly Report for March 1978,* FE/496-148, Pittsburg & Midway Coal Mining Company, Merriam, Kans., March 1978.

Table 10. SRC I and SRC II product yields

(% moisture free coal)

Component	SRC I[a]	SRC II[b]
H_2	−2.8	−3.1
C_1–C_4 gases	5.9	11.0
CO	0.02	0.2
CO_2	1.4	1.90
H_2O	5.00	9.8
H_2S	1.7	2.8
Light oil	2.4	6.9 Naphtha
Wash solvent	7.8	17.4 Middle distillate
Process solvent	−8.9	9.6 Heavy distillate
Solvent refined coal	69.5	27.6
Ash	11.9	11.5
Insoluble organic material and coal	6.1	4.4
Total	100.0	100.0

[a]R. H. Filby et al., *Solvent Refined Coal (SRC) Processes: Trace Elements,* Research and Development Report No. 53, vol. III, Part 6, U.S. Department of Energy Report FE/496-T17, Washington, D.C., 1978.
[b]Pittsburg & Midway Coal Mining Co., *Solvent Refined Coal (SRC) Processes: Monthly Report for March 1978,* FE/496-148, Pittsburg & Midway Coal Mining Company, Merriam, Kans., March 1978.

the process most of the pyrite (FeS_2) is converted to pyrrhotite (FeS):

$$FeS_2 + H_2 \rightarrow FeS + H_2S$$

and some of the organically bound sulfur reacts to form H_2S:

$$R\text{-}S\text{-}R' + 2H_2 \rightarrow R\text{-}H + R'\text{-}H + H_2S$$

where R and R' represent organic moieties. Although many minerals (e.g., SiO_2) should be inert under the process conditions, others may undergo chemical reactions (as does pyrite, and possibly other sulfides), and volatile species of toxic elements may be formed. In addition, organically bound metals in the coal may react to form simpler organometallic species during the depolymerization and hydrogenation of the coal. Many of these species may be toxic, volatile, or potentially harmful to catalysts and furnace linings, and hence undesirable. Many possible volatile forms and organometallic species can readily form from either organic or inorganic compounds under the conditions used in the process (Table 11). For example, Hg, $Ni(CO)_4$, $(C_6H_6)_2Cr$, and many other species (Table 11) are quite stable under the process conditions or may form at lower temperatures on cooling. Reactions may occur at lower temperatures to form substituted species, for example, substituted carbonyls.

Table 11. Possible environmentally important forms of
some trace elements during liquefaction

Element	Volatile species	Organic species
As	AsH_3, $AsCl_3$, $AsBr_3$	$RAsH_2$, $RR'AsH$, R_3As, $R_4As^+X^-$
Sb	SbH_3, $SbCl_3$, $SbBr_3$	$RSbH_2$, RR^1SbH, R_3Sb
Hg	Hg, HgS	R_2Hg, RHg^+X^-
Se	H_2Se, Se	$R\text{-}Se\text{-}R^1$; $R\text{-}SeO_3H$
Fe	$Fe(CO)_5$, $Fe_2(CO)_9$	$Fe(C_6H_5)_2(CO)_x$, $Fe(C_5H_5)_2$
Ni	$Ni(CO)_4$	Ni-asphaltene
Cr	$Cr(CO)_6$	$Cr(C_6H_6)(CO)_3$, $(C_6H_6)_2Cr$
Ti	$TiCl_4$	$Ti(C_5H_5)_2Cl_2$

R and R' are aliphatic or aromatic groups.

The investigation reported here is a study of trace element behavior in the SRC I and SRC II processes with emphasis on the determination of trace element materials balances in the processes as the preliminary step to understanding some of the trace element chemistry in liquefaction. Preliminary results on SRC I have been reported.[6-8] In this study, neutron activation analysis was used to measure trace element contents in the important process streams of the SRC I and SRC II processes and trace element materials balances were computed from product and process stream yield data.

ANALYTICAL METHODS

Samples were collected (Table 12) from the 50 ton/day Solvent Refined Coal pilot plant at Fort Lewis while it was operating both in SRC I and SRC II modes. Care was taken to ensure that the plant had operated for several days without interruption before sampling and that the samples taken were reasonably representative of steady state conditions. Further details concerning the collection of samples and sample treatment prior to analysis are discussed elsewhere for SRC I (Ref. 6) and SRC II (Ref. 9).

Neutron activation analysis[10] was used to determine the concentration of V, Ti, Ca, Mg, Al, Na, K, Rb, Cs, Sr, Ba, Sc, Tb, Eu, Ce, La, Zr, Hf, Ta, Th, Sm, Mn, Fe, Co, Ni, Cu, Cr, As, Se, Hg, and Sb.

Atomic absorption spectroscopy was used to measure Pb, Cd, and Be in process waters and aqueous effluents from the pilot plant operating in the SRC I mode.[7]

RESULTS AND DISCUSSION

Trace Element Behavior in the SRC I Process

Analytical data for one set of samples collected to establish a trace element materials balance in the

SRC I process were obtained for ground coal, SRC I, pyridine insolubles, wet filter cake, process recycle solvent, wash solvent, sulfur, process water, and effluent water (Table 13). The pyridine insolubles (Table 13) were derived from the solid material (wet filter cake) filtered from the reacted coal solution by removal of admixed soluble organic matter with pyridine. The pyridine insolubles thus represent the insoluble material from the process and contain mineral matter plus a small amount of unreacted coal and high molecular weight organic material. All trace elements are in the pyridine insoluble fraction, although the ratio of the concentration in the pyridine insoluble to the concentration in SRC I depends upon the element (Table 13). Elemental concentrations in the distillate fractions (light oil, wash solvent, and process recycle solvent, are low compared with SRC I and the pyridine insolubles. Comparison of elemental concentrations in SRC I with concentrations in the ground coal show that except for Br, all elements are depleted in the SRC I compared to coal (Table 14). The reason for the increase in Br concentration relative to coal is not known. Of the other elements that are depleted in SRC I, Ti is unique in that the reduction in concentration is small (12%). The reasons for the high retention of Ti compared with other metals present in the coal in the SRC I are discussed below. For elements other than Br, Ti, and Cl, there is a significant reduction in trace element content in SRC I relative to coal.

The concentration data (Table 13) together with the yields of the individual materials as a fraction of the moisture free coal (Table 10) were used to compute an overall elemental materials balance for each element in the SRC I process (Table 15). Given the difficulty of obtaining representative samples from an operating pilot plant, the data (except perhaps for Mg, Br, and Rb) indicate that satisfactory closure has been obtained. The low value for Mg (53%) is probably due to the poor precision of the INAA method for determining Mg, and the high Br value probably indicates an unidentified source of Br. The high value for Rb (259%) is difficult to explain, given the good closure for the other alkali metals (Na, K, and Cs) and probably represents an anomalous analysis since other equilibrium set data give good closure for Rb (Ref. 6). Data obtained on other materials balance for runs in the SRC I mode confirm these findings [6,7] (Tables 13-15).

Trace Elements in the SRC II Process

Preliminary data for some environmentally important elements obtained for SRC II process fractions show trace element concentrations in the feed

Table 12. Sampling points for SRC I and SRC II processes

SRC I		SRC II	
Type	Sampling point	Type	Sampling point
Ground coal	Preparation area	Ground coal	Preparation area
Pyridine insolubles	Filter residue	Vacuum bottoms	Vacuum flash
Wet filter cake	Filter residue		
Light oil	Distillate	Separator 2 oil	Overhead condensate
Recycle solvent	Vacuum flash overhead		
Wash solvent	Light distillate	Total solvent accumulator oil	Total distillate product
Process water	Water hold tank	Process water	Plant water
Effluent water	Plant outlet	Effluent water	Plant outlet
Solvent refined coal	Product area	Vacuum flash condensate oil	Overhead vacuum flash

Table 13. Trace element concentrations in SRC I process streams
(ppm except as noted)

Element	Ground coal	SRC	Pyridine insolubles	Wet filter cake	Light oil	Process recycle solvent	Wash solvent	Sulfur	Process water	Effluent water
As	12.5	2.00	85.7	62.1	0.011	0.24	0.011	<2.0	0.006	<0.001
Sb	0.76	0.06	7.21	5.35	<0.4[a]	8.2[a]	<0.4[a]	<0.1	0.66[a]	2.0[a]
Se	2.0	0.12	16.5	11.3	51.6[a]	24.0[a]	14.4[a]	<1.5	0.16	0.0012
Hg (ppb)	113	39.6	508	346	18.5	1.45	10.5	<100	106	3.2
Br	4.56	7.74	12.0	20.7	0.015	1.0	0.02	<3.0	15.6	31.8
Ni	14.9	<3.0	142	82.4	<0.03	0.4	<0.03	<28.0	<0.004	0.013
Co	5.88	0.22	40.7	26.5	<3.0[a]	40.7[a]	1.43[a]	110	0.2[a]	0.41[a]
Cr	13.7	1.64	106	69.2	37.3[a]	3590[a]	41.3[a]	<2.0	0.007	0.15
Fe (%)	2.11	0.03	16.8	11.7	2.90[a]	211[b]	11.2[a]	<0.1	0.30[a]	1.25[b]
Na	137	4.23	1020	623	0.60	0.50	0.45	3120	0.70	8.3
Rb	<4.0	<0.5	66.5	37.1	<0.01	0.02	<0.01	<9.0	0.78	0.52
Cs	0.75	0.02	5.08	3.20	1.06[a]	<1.2[a]	0.91[a]	<0.2	0.04[a]	0.02[b]
K	1550	4.72	11100	6660	<0.1	0.25	<0.1	179	0.2	1.26
Sc	2.59	0.57	14.8	9.26	0.15[a]	32.8[a]	0.19[a]	<0.02	0.13[a]	0.01[a]
Tb	0.39	0.045	2.06	1.34	0.13[a]	3.75[a]	<0.13[a]	<0.1	0.01[a]	0.01[a]
Eu	0.26	0.055	1.48	0.96	0.01	<0.01	<0.01	<0.01	0.01[a]	0.01[a]
Sm	2.62	0.29	16.9	8.16	<0.01	0.02	<0.01	0.61	0.08[a]	<0.06[a]
Ce	20.9	0.45	156.0	102	<0.004	<0.004	<0.003	<2.0	<0.2[a]	<0.2[a]
La	7.55	0.13	59.8	35.2	<0.01	0.01	<0.01	1.80	0.27[a]	0.5[a]
Sr	88.6	<6.0	456.0	453	<0.6	<0.2	0.74	<45.0	<0.01	<0.04
Ba	53.0	5.75	347.0	185.0	<0.1	1.14	<0.07	<39.0	<0.02	<0.04
Th	2.00	0.22	12.8	7.70	<0.001	0.012	<0.001	<0.2	0.05[a]	<0.01[a]
Hf	0.51	0.084	3.30	2.20	<0.001	0.003	<0.001	<0.2	0.02[a]	<0.01[a]
Ta	0.14	0.046	0.71	0.42	<0.4[a]	2.53[a]	<0.3[a]	<0.2	0.02[a]	0.01[a]
Ga	3.56	1.79	19.4	11.3	<0.01	0.06	<0.01	<1.5	<1[a]	<4[a]
Zr	62.9	16.0	500.0	246	0.07	0.71	<0.1	<61.0	0.02	0.04
Cu	19.9	2.07	189	138	0.03	0.68	0.03	<1.0	<12[a]	<10[a]

[a]Values in ppb.
[b]Values in ppm.

Source: R. H. Filby et al., *Solvent Refined Coal (SRC) Processes: Trace Elements,* Research and Development Report No. 53, vol. III, Part 6, U.S. Department of Energy Report FE/496-T17, Washington, D.C., 1978.

Table 14. Trace element reduction SRC compared with coal

Element	Equilibrium Set 1		Equilibrium Set 2	
	SRC/coal	% Reduction	SRC/coal	% Reduction
Ti	0.88	12	0.74	16
V	0.15	85	0.47	53
Ca	0.22	78	0.22	78
Mg	0.08	92		
Al (%)	0.02	98	0.03	97
Cl	0.61	39	0.34	66
Mn	0.60	40	0.40	60
As	0.16	84	0.07	93
Sb	0.08	92	0.04	96
Se	0.06	94	0.03	97
Hg	0.35	65	0.41	59
Br	1.70	+70	1.33	+33

Source: R. H. Filby et al., *Solvent Refined Coal (SRC) Processes: Trace Elements,* Research and Development Report No. 53, vol. III, Part 6, U.S. Department of Energy Report FE/496-T17, Washington, D.C., 1978.

Table 15. Trace element material
balance for SRC I process

Element	Material balance (percent of moisture free coal)
Ti	149
V	101
Ca	146
Mg	53
Al	92
Cl	85
Mn	129
As	106
Sb	137
Se	119
Hg	98
Br	172
Ni	133
Co	129
Cr	117
Fe	112
Na	142
Rb	259
Cs	97
K	100
Sc	95
Tb	81
Eu	94
Sm	97
Ce	105
La	112
Ba	99
Th	97
Hf	101
Ta	94
Ga	110
Zr	128
Cu	140

Source: R. H. Filby et al., *Solvent Refined Coal (SRC) Processes: Trace Elements,* Research and Development Report No. 53, vol. III, Part 6, U.S. Department of Energy Report FE/ 496-T17, Washington, D.C., 1978.

Table 16. Material balance data for important elements in SRC II process – solid and liquid phases[a]

Element	Concentration (ppm)			Percent of moisture free coal	
	Total distillate[a]	Vacuum bottoms[b]	Coal	Total	Normalized
As	0.0216	4.30	2.02	93.7	98.4
Se	0.258	7.26	3.56	91.2	95.8
Sb	0.00213	0.909	0.420	94.7	99.4
Hg	0.00758	0.237	0.129	79.9	83.9
Br	0.0846	5.40	3.29	72.3	75.9
Ni	0.0668	36.2	24.2	65.2	68.5
Co	0.00847	8.34	3.72	97.6	102
Cr	0.127	69.6	28.9	105	110
Fe	26.7	36800	15400	104	109
Na	2.41	1560	684	99.3	104
K	4.11	4860	2220	95.2	100

[a]Total distillate: 33.9% MFC.
[b]Vacuum bottoms: 43.5% MFC.

Source: R. H. Filby, M. L. Hunt, and S. R. Khalil, *Trace Elements in the Solvent Refined Coal Process,* Technical Progress Report, July–August 1978, November 1978.

and the total distillate, 33.9% MFC, the remaining 22.6% being accounted for by light hydrocarbons (C_1–C_4), H_2O, CO, CO_2, and H_2S. Elemental materials balances have been established for each element based on *only* the vacuum bottoms and total distillate fractions (Table 16). Thus contributions from sulfur, gases, and process waters have been ignored. The data have also been normalized to K = 100%, K being a representative nonvolatile element. The trace elements are concentrated in the vacuum bottoms, with very low amounts in the total distillate compared with coal (Table 16).

The trace element materials balances for As, Se, Sb, Co, Fe, Cr, and Na are satisfactory, but the low values for Hg, Br, and Ni may indicate either losses in the process or significant concentrations in fractions not analyzed (e.g. gases or process waters). Analyses of other materials balance sample sets indicate that Hg, in particular, fails to close and that there is evidence for concentration of Hg in other process streams through volatilization of Hg from the coal during the liquefaction process. No evidence has been found, however, to show that Hg is lost from the plant in flare gases. Complete materials balance calculations for all elements can only be made when all process fractions (e.g., process waters) have been analyzed.

Comparison of Trace Elements in SRC I and SRC II Products with Other Fossil Fuels

Both SRC I and distillate products from the SRC II process are low sulfur, low ash fuels compared with the original feed coal and thus are desirable boiler fuels. A comparison of the concentrations of some environmentally important elements in

coal, vacuum bottoms (residual material after removal of distillate fractions) and the total distillate fraction (Table 16). Data on individual distillate fractions and condensates that comprise the total distillate sample and data on aqueous phases (process waters) are not yet available. The vacuum bottoms account for 43.5% of the moisture free coal (MFC)

SRC I, a blend of heavy and middle distillates (1:5.75) from the SRC II process that resembles a residual fuel oil with an Illinois No. 6 coal, and a typical residual fuel oil derived from a relatively low trace element crude oil[11] shows that for the trace elements As, Sb, Se, V, Ni, and Co, the concentrations in the SRC II product are significantly lower than the corresponding concentrations in residual fuel oil and coal (Table 17). For Hg and Fe the SRC II product and the residual fuel oil are similar but much lower than for coal while Cr is higher in the SRC II blend than in residual fuel oil. Although SRC I is significantly higher in Ti, As, Sb, Se, Hg, Cr, and Fe than the residual fuel oil, it is lower in V and Ni. The SRC I has lower concentrations of all elements studied compared to the Illinois feed coal. From the standpoint of concentrations of toxic trace elements, the SRC II fuel blend appears to be an excellent substitute for petroleum derived fuel oil.

Table 17. Important trace elements in SRC I and SRC II materials and in coal and fuel oil

(parts per million)

Element	SRC I[a]	SRC II blend[b]	Illinois[a] coal	Residual fuel oil[c]
As	2.10	0.0148	16.3	0.055
Sb	0.136	0.00259	1.10	0.04
Se	0.123	0.0284	2.52	0.09
Hg	0.0432	0.00339	0.114	0.004
V	9.51	0.179	29.7	87
Ti	326	<10	595	<10
Ni	1.82	0.0992	13.7	12.5
Co	0.314	0.00304	5.54	0.32
Cr	4.19	2.57	14.3	0.07
Fe	438	4.14	22500	5.0

[a]Thirty-five production runs of SRC I from R. H. Filby et al., *Solvent Refined Coal (SRC) Processes: Trace Elements,* Research and Development Report No. 53, vol. III, Part 6, U.S. Department of Energy Report FE/496-T17, Washington, D.C., 1978.

[b]Heavy/middle distillate ratio 1:5.75.

[c]R. A. Cahill, *A Study of the Trace Element Distribution in Petroleum,* M.S. Thesis, University of Maryland (1974).

Behavior of Trace Elements During Hydrogenation in the SRC I and SRC II Processes

Trace element materials balance data indicate that, except for Hg in the SRC II process, the elemental balances close very well when the major output streams of the processes are considered. Thus there is no direct evidence for the release of significant amounts of trace elements in fugitive emissions (i.e., flares) in either SRC I or SRC II. However, although the Hg balance in SRC I closes well, Hg does not close in the SRC II process when only the total distillate product and the vacuum bottoms are considered. Preliminary results[12] indicate that Hg is partially volatilized from the coal in the SRC II process into an overhead stream and is condensed in one of the light oils of a gas-liquid separator. This can be seen clearly when the concentrations of Hg in the separator No. 2 oil from the SRC II process is compared with other SRC II oils and a light distillate oil from the SRC I process (Table 18). Although the

Table 18. Trace elements in SRC I and SRC II distillate fractions (SR 11)

(ppb, except as noted)

Element	Separator #2 oil[a]	Separator #3 oil[a]	Solvent acc. oil[a]	Light oil SRC I[b]
As	57.7	32.9	12.7	2.85
Sb	4.45	<0.2	0.550	1.86
Se	214	57.1	115	23
Hg	776	4.6	27.3	3.40
Br	11.1	8.71	6.65	18.0
Ni	189	<20	30.3	40
Co	31.1	1.45	2.34	1.8
Fe (ppm)	120	0.73	3.39	0.30
Cr	14.0	14.5	81.8	<5

[a]R. H. Filby et al., *Solvent Refined Coal (SRC) Processes: Trace Elements,* Research and Development Report No. 53, vol. III, Part 6, U.S. Department of Energy Report FE/496-T17, Washington, D.C., 1978.

[b]R. H. Filby, M. L. Hunt, and S. R. Khalil. *Trace Elements in the Solvent Refined Coal Process,* Technical Progress Report for the period July–August 1978, November 1978.

concentration of Hg in the separator 2 oil (776 ppb) is greater than that of the original coal (129 ppb), it is apparently present in the oil as particles of either Hg or HgS.[13] Of the three proposed mechanisms for the volatilization of Hg in the SRC II process (Table 19), the second mechanism appears more likely than the first from previous work on the mechanism of sublimation of HgS.[14] These proposed mechanisms assume that Hg is present in the coal as a sulfide.[15] The volatilization of organic Hg species (mechanism 3) cannot be ruled out, particularly if Hg is bound organically in the coal.

The behavior of Ti in the SRC I process is particularly interesting because of the anomalously high Ti content of SRC I. Several explanations are

Table 19. Possible mechanism for Hg volatilization in SRC II process

Dissolver reaction (450–500°C)	Volatile species	Cool zone condensate
$HgS_{(s)} \longrightarrow$	$HgS_{(g)} \longrightarrow$	$HgS_{(s)}$
$HgS_{(s)} + H_2 \longrightarrow$	$Hg_{(g)} + H_2S \longrightarrow$	$Hg_{(1)}$
	$\longrightarrow$	$HgS_{(s)}$
Hg (in coal) + solvent $\longrightarrow$	$R\text{-}Hg\text{-}R'_{(g)} \xrightarrow{H_2S}$	$HgS_{(s)}$

possible; e.g., Ti occurs in the coal as a finely divided oxide (TiO_2) that passes through the filters which retain coarser size fractions of other minerals or that organotitanium species are present. The nature of Ti and other elements in SRC I is being investigated with chromatographic separation procedures (GPC and HPLC), and preliminary results indicate that both organic and inorganic forms of Ti occur in SRC I, which would be consistent with independent conclusions[5] that Ti may occur in an organically bound form in coal. Preliminary evidence also indicates that other elements (such as V, As, Sb, and Se) may exist partially as organometallic species in SRC I whereas Na, K, Mg, Ca, and Al appear to be inorganically bound.[16] The inorganic nature of Na, K, Mg, Ca, and Al species in SRC I would be consistent with the fact that these elements appear to be associated with the mineral component of the coal.

Trace Elements in Process and Effluent Waters

In both SRC I and SRC II processes, water is produced as a reaction product in the hydrogenation of the coal, and it is also used as a quenching agent. These process waters contain trace metals from a number of possible sources (original coal as well as corrosion products) in addition to H_2S and organic compounds soluble in water. Because such process waters contain phenols, sulfur compounds, and trace metals, they must be treated to remove such toxic compounds before discharge to the environment. The treatment process used in the SRC I and SRC II processes is flocculation with aluminum hydroxide, biodegradation of organic species, and then filtration through activated charcoal. Although primarily designed to remove organic species in process waters, the treatment process is effective in reducing toxic

trace element levels (Table 20). The concentrations of Hg, Se, As, Sb, Fe, Co, Cr, Ni, Fe, Ce, Pb, and Be shown in Table 20 for a process water and the corresponding effluent water from the SRC I process are not directly comparable since the effluent water represents run-off water, wash water, and cooling waters used in the pilot plant in addition to process waters. What is significant, however, is that the process water contains elevated levels of Hg, Se, and As and that these elements are reduced to essentially background levels (Hamer Marsh water) by the treatment process.

Table 20. Trace element levels in process water, effluent water, and Hamer Marsh water for SRC I process

Element	Concentration (µg/liter)		
	Process water	Effluent water	Hamer Marsh water
As	6.3	<1	<1
Sb	0.66	2.0	2.6
Se	159	1.20	0.90
Hg	106	3.2	1.0
Fe	300	1250	820
Co	0.20	0.41	0.34
Cr	7.4	153	137
Ni	4.0	13.0	12.0
Cd		2.2	<0.3
Pb		<7	<10
Be		<0.4	<0.4

As a better indication of the effectiveness of the treatment process, the concentrations of As, Sb, Hg, Se, Pb, Cd, Be, Co, Cr, Fe, Ni, and Zn were measured in 35 samples taken daily during production runs of SRC I.[16] The data, shown in Table 21, when

Table 21. Important trace elements in SRC I production effluent waters (µg/liter)

Element	N	$\overline{X} \pm S^a$	Hamer[a] Marsh	Median U.S. rivers[b]	EPA 1972 drinking water criteria[c]
As	36	<10	<5	0.4	100
Sb	35	0.63 ± 0.38	0.50		50
Se	34	0.64 ± 0.32	0.45	<20	10
Hg	35	0.54 ± 0.39	0.38	0.08	0.2
Pb	10	1.4 ± 0.6	2	5	30
Cd	36	2.7 ± 2.0	0.15	80	10
Be	36	<0.2	<1		100
Co	35	0.70 ± 0.47	0.26	0.9	50
Cr	35	4.9 ± 5.0	6.2	0.2	50
Fe	34	180 ± 97	360	670	300
Ni	34	40 ± 26	7.0	10	1000
Zn	26	310 ± 250	90	100	200

[a]R. H. Filby, M. L. Hunt, and S. R. Khalil, unpublished data.
[b]J. F. Kopp and R. C. Kroner, *Trace Metals in Waters of the United States*, Federal Water Pollution Control Association Report, Cincinnati, Ohio, 1970.
[c]Environmental Protection Agency, *National Interim Primary Drinking Water Regulations, Fed. Regist.*, 59566–88 (1975).

compared with Hamer Marsh water, median values for U.S. rivers,[17] and EPA 1972 Drinking Water Criteria,[18] show that the effluent water exceeds EPA criteria only marginally for Hg and Zn and that the treatment process is efficient in reducing trace element levels of effluent waters to acceptable levels.

ACKNOWLEDGMENTS

The authors acknowledge Dr. Marjorie Hunt for her careful and time-consuming determinations of Pb, Cd, and Be in SRC I effluent waters. The help of Ms. Cathy Grimm, Naeem Hashmi, and Curtis Palmer is also gratefully acknowledged. The efforts of John Frame in keeping instrumentation on-line and the W.S.U. reactor staff are sincerely appreciated.

REFERENCES

1. H. J. Gluskoter et al., Illinois Geological Survey Circular 499, 1977.

2. R. H. Filby and R. C. Brown, "Use of Factor Analysis in Interpretation of Coal Trace Element Data," in preparation.

3. D. J. Swaine, "Trace Elements in Coal," in *Recent Contributions to Geochemistry and Analytical Chemistry,* ed. A. E. Tugarinov, Wiley, New York, 1976, pp. 539–50.

4. P. Zubovic, "Physiocochemical Properties of Certain Minor Elements as Controlling Factors in Their Distribution in Coal, in *Coal Science,* ed. P. H. Given, Advances in Chemistry Series No. 55, Amer. Chem. Soc., Washington, D.C., 1966, pp. 221–31.

5. R. N. Miller and P. H. Given, "Variations in Inorganic Constituents of Some Low Rank Coals," in *Proc. International Conference on Ash Deposits and Corrosion Due to Impurities in Combustion Gases,* ed. R. W. Bryers, Hemisphere Publishing Corp., Washington, D.C., 1978, pp. 39–50.

6. R. H. Filby et al., *Solvent Refined Coal (SRC) Processes: Trace Elements,* Research and Development Report No. 53, vol. III, part 6, U.S. Department of Energy Report FE/496-T17, Washington, D.C., 1978.

7. R. H. Filby, K. R. Shah, and C. A. Sautter, "Trace Elements in the Solvent Refined Coal Process," in *Proc. Symposium on Environmental Aspects of Fuel Conversion Technology III,* ed. F. A. Ayer and M. F. Massoglia, U.S. Environmental Protection Agency Report EPA-600/7-78-063, Washington, D.C., 1978.

8. R. H. Filby, S. R. Khalil, and C. A. Sautter, "Fate of Trace Elements in the SRC I Process," submitted to *Fuel.*

9. R. H. Filby, S. R. Khalil, and M. L. Hunt, *Trace Elements in the Solvent Refined Coal Process,* Final Interim Report for the period 1976–77, in preparation.

10. R. H. Filby, K. R. Shah, and C. A. Sautter, "A Study of Trace Element Distribution in the Solvent Refined Coal (SRC) Process Using Neutron Activation Analysis," *J. Radioanal. Chem.* 37, 693 (1977).

11. R. A. Cahill, *A Study of the Trace Element Distribution in Petroleum,* M.S. Thesis, University of Maryland, 1974.

12. R. H. Filby, M. L. Hunt, and S. R. Khalil, *Trace Elements in the Solvent Refined Coal Process,* Technical Progress Report for the period July–August 1978, November 1978.

13. R. H. Filby and S. R. Khalil, unpublished data.

14. K. C. Mills, *Thermodynamic Data for Inorganic Sulfides, Selenides, and Tellurides,* Butterworths, London, 1974.

15. D. J. Swaine, "Trace Elements in Coal," presented at the Eleventh Trace Substances Conference, Columbia, Mo., June 1977.

16. R. H. Filby, M. L. Hunt, and S. R. Khalil, unpublished data.

17. J. F. Kopp and R. C. Kroner, *Trace Metals in Waters of the United States,* Federal Water Pollution Control Association Report, Cincinnati, Ohio, 1970.

18. Environmental Protection Agency "National Interim Primary Drinking Water Regulations," *Fed. Regist.* 59566–88 (1975).

DISCUSSION

Don Gardiner, Union Carbide Corporation, Nuclear Division, Mathematics and Statistics: First let me compliment you, Dr. Filby, on a very smooth presentation. As you might expect, my ears perked up when you talked about the factor analysis. Do I assume correctly that the factors that were listed 1 through 12 were in order of the amount of variance that they explained?

Roy Filby: Yes, that's correct.

Don Gardiner: And there were two, I think arsenic and cadmium, that you were particularly interested in that didn't appear until further down the list. I was wondering whether that by the time they did appear, the vectors in which they appear were explained in an appreciable amount of the variance, that is, appreciable enough to be of interest to you.

Roy Filby: I think so. We extracted, actually, 24 factors originally from the data, of which most were quite insignificant. There was not a large difference between the first three factors and the next five. We also tried different methods of massaging the data by going to oblique solutions and several other different methods of looking at it because we felt that the factors would be correlated. No matter which way this was done, the first six or seven factors appeared to be the same in any solution, whereas some of the later factors were switched around in importance and even changed, so we couldn't really give much significance to those. There are a number of other studies that indicate, for example, that cadmium and zinc are correlated.

Jim Evans, Enviro Control: You were talking about titanium. Have you perhaps talked to anybody at Conoco Coal Development Corporation about titanium in the CSF process which is very similar to the SRC? I know that they did considerable work with the coal minerals, particularly titanium, because it kept plating out on the catalyst.

Roy Filby: No, we haven't talked with anybody there. We have done a lot of work on trying to look at titanium in SRC I, and by using a combined chromatography absorption-type resin, we can distinguish two very distinct types of titanium in coal.

There is a form that simply washes right through the column, which appears to be submicron titanium dioxide. There is another form of titanium that is held by the column and appears in the median molecular weight material of the SRC, which appears to be organically associated. There do appear to be two forms of titanium in SRC.

Jim Evans: I had thought that the titanium would be present as a fairly large molecule because it always turned up on the outside of the catalyst; it never penetrated it.

Roy Filby: Yes, we really have not calibrated the molecular weight of the species yet. This is because we are not looking at a simple GPC separation but at GPC plus absorption. It makes it difficult.

Phil Singer, University of North Carolina: On that last table, were those concentrations always soluble-metal concentrations or are they total metal concentrations?

Roy Filby: Those were "soluble." All of those samples were filtered through a 0.4-micron nucleopore filters.

Phil Singer: So the total metal concentration in those streams could have been several times larger.

Roy Filby: We analyzed some of the filters and found very little material. These effluents have gone through a tertiary treatment process with the final step being carbon filtration, and we found very little material on the filters themselves. They were very clean.

THE NATURE OF STACK EMISSIONS FROM THE EXPERIMENTAL-SCALE COMBUSTION OF CRUDE SHALE OIL*

B. R. Clark[†]   C. E. Higgins[†]
A. D. Horton[†]   M. R. Guerin
E. H. McBay[†]

J. Ekmann[‡]   C. R. McCann[‡]
D. Bienstock[‡]   M. P. Mathur[‡]
C. M. White[‡]   F. K. Schweighart[‡]
A. G. Sharkey[‡]

R. E. Poulson[§]

ABSTRACT

A series of four combustion tests was made with two crude shale oils. Various combustion parameters were measured to assess the combustion behavior by comparison with similar petroleum fuels. Stack emissions were measured with permanent on-line equipment as well as some specially designed apparatus for collecting volatile organic compounds on solid adsorbents for subsequent analysis. The shale oils burned like typical distillate fuel oils with boiler efficiencies comparable to a No. 6 oil. The stack emissions were considerably different in content from those resulting from the combustion of petroleum fuels. Overall hydrocarbon emissions were quite low, but NO/NO_x emissions were high, reflecting the large nitrogen content of these oils.

INTRODUCTION

The combustion of fossil-derived fuels continues to supply the major portion of all the energy consumed in homes, industry, and transportation. Less expensive fossil fuels, e.g., raw or beneficiated coal, are used in large utility boilers to generate electrical energy, while the more expensive fuels, e.g., derivatives of petroleum, are largely combusted in propulsion engines. In using any fuel, two essential concerns must be addressed: (1) the suitability of the fuel for its intended use (specifications) and (2) the nature of the combustion products added to the environment.

A relatively long history in the use of petroleum provides substantial information on the combustion behavior of petroleum-derived fuels. Even with this information, the fate of petroleum combustion effluents (e.g., their role in the formation of smogs) is not entirely understood. The difficulty of predicting the nature of combustion effluents and their environmental and health effects will be increased greatly when fuels from coal and oil shale largely supplant petroleum.

Early in the stages of development of boilers and engines especially designed to burn nonpetroleum ("synthetic") fossil fuels, the combustion behaviors of synthetic crudes and their derived products ought to be carefully measured. This activity should focus on

*Research sponsored jointly by the U.S. Environmental Protection Agency under Interagency Agreement D6-0129 and the Division of Biomedical and Environmental Research, U.S. Department of Energy, under Contract W-7405-eng-26 with the Union Carbide Corporation.

[†]Oak Ridge National Laboratory, Oak Ridge, Tennessee 37830.

[‡]Pittsburgh Energy Technology Center, Bruceton, Pennsylvania 15213.

[§]Laramie Energy Technology Center, Laramie, Wyoming 82071.

combustion behaviors from both the engineering and the health effects viewpoints with respect to the combustion products. Acceptable use of synthetic fuels will most likely result from the coordinated research and development of combustion devices (stationary and mobile) with improved efficiency and reduced emission of objectionable products. This is in contrast to the current state of affairs with the reciprocating motion, internal combustion engine which is neither efficient, compared to other designs, nor acceptable with regard to emissions. "Patchwork devices" by the automotive industry have reduced emissions levels, but clearly engine design and emission behavior never enjoyed the benefit of common consideration during designing stages.

Synthetic fuels are still in the very early development stages. This situation provides a unique opportunity to simultaneously consider the design of better suited combustion devices,[1] the nature of combustion products, and even the desired specifications for fuels. Perhaps timely feedback to a growing synthetic fuels production industry can be used as a guide during process design improvements to produce the most desirable products.

A very limited amount of crude and refined products from a few experimental-scale processes is currently available for combustion studies. None are as yet commercially available. Crude shale oil has been produced in greater quantity than coal-derived oils. A 100,000-barrel production run, using the above-ground retort at Anvil Points, Colorado, Paraho facility,* is in progress to supply the U.S. Navy with a sufficient amount of oil to conduct tests of the behavior of refined products as transportation fuels. The Laramie Energy Technology Center (LETC) also operates a 150-ton retort designed to simulate in situ retorting conditions. When runs are made with this equipment, several thousand gallons of crude oil are produced. Although it is unlikely that crude shale oil will be used directly as a fuel, shale oils appear to be good crudes for the production of transportation fuels because of their high H/C ratios, as high as 1.9 (petroleum $\cong$ 2.0).[1] Hydrotreatment prior to refining removes substantial amounts of nitrogen[2] and provides a high-grade refinery feedstock.

Because of the availability of shale oils and their potential importance in providing transportation fuels, these oils were chosen for some preliminary combustion studies to establish the best methods for evaluating the combustion behaviors of synthetic fuels. The oils were supplied by the U.S. Navy through a cooperative agreement with LETC, Development Engineering, Inc., and the USN (Paraho shale oil) and with LETC (Laramie shale oil). A 100-hp combustion facility operated by the Pittsburgh Energy Technology Center (PETC) was used for the experiments. Flue gas was sampled by personnel from Oak Ridge National Laboratory and PETC.

The combustion studies on these two shale oils are divided into two parts: (1) flue gas sampling for organic volatiles and (2) a comparative analysis of combustion behaviors.

FLUE GAS SAMPLING FOR VOLATILE ORGANIC COMPOUNDS

Normal combustion conditions ($\sim$3% excess O_2 in the stack) were chosen to allow comparisons with conventional No. 6 fuel oils. The combustion facility is set up to routinely measure O_2, CO_2, CO, NO/NO_x, SO_2, and total hydrocarbons (gas phase at STP). In addition, a flue gas opacity monitor and instruments to measure flame characteristics are provided with the facility. More details are found in the section describing the combustion behavior of the shale oils.

Among the flue gas constituents, those which pose the greatest threat to human health are the respirable size particles and the unburned or partially burned organic components. A cascade impactor was installed near the top of the stack to collect a representative sample of particles. An analysis of particle sizes was made after one of the tests (Laramie shale oil, test LSF-02).

The collection of organic gas phase components under flue gas conditions is a problem with no straightforward solution. Commercially available sampling trains are cumbersome to install, operate, and dismantle for analysis of the contents. For these studies, the intention was to sample the volatile organics using relatively simple apparatus.

A sampling manifold, constructed for the first test (LSF-02), was designed to provide for the collection of flue gases on beds of solid adsorbent. Both Tenax-GC polymer and XAD-2 resin were used. Deficiencies in this sampling method led to the redesign of the manifold system and the incorporation of a Tenax sampler (Jones sampler)[3] designed at Battelle, Columbus (Figs. 1 and 2). Only the Jones sampler was capable of collecting the less volatile high-molecular-weight compounds since the sampling line

---

*This facility operated by Development Engineering, Inc.

118   B. R. Clark et al.

ORNL-DWG-78-18106

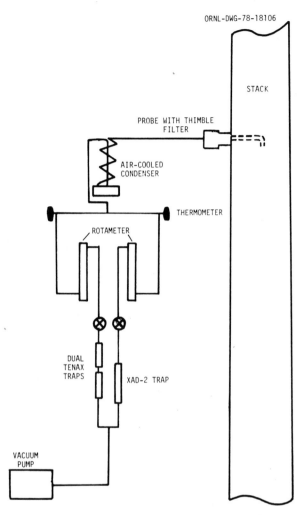

Fig. 1. Simple apparatus for sampling stack gases for low-molecular-weight organic volatile components.

was heated to prevent condensation of gaseous components at flue gas temperatures.

Small Tenax beds (9 mm ID × 5 in. long) were prepared in Pyrex tubing of the proper dimensions to be inserted directly into the injection port of a Perkin-Elmer 3920 gas chromatograph. These beds were thermally desorbed and condensed at the head of the analytical column maintained at −70°C in a cryogenically cooled chamber. XAD-2 resin was packed in cleaned stainless steel tubing ($^3/_8$ in. ID × 10 in. long). The adsorbed contents were removed by Soxhlet extraction with a combination of solvents (pentane, methanol, and acetone). The large bed of Tenax in the Jones sampler (~18 g) was removed and desorbed with pentane in a Soxhlet apparatus. These extracts and the XAD-2 extracts were analyzed by

GC-MS using a DuPont 490 B magnetic sector instrument.

Analyses of Flue Gas Organic Compounds

LSF-02 (Laramie Shale Oil, burned November 15, 1977)

The sampling apparatus used (Fig. 1) limited the types of organic species likely to be collected. The use of a condenser upstream from the sampling manifold would prevent highly water-soluble of high-molecular-weight compounds from reaching the solid adsorbent traps. Collection temperature was about 30°C with flow rates of 300–500 ml/min. Sampling time was ~1h with a few collected over 4–8 min for capillary column GC analysis. Backup Tenax traps were used in case of saturation of the first trap. However, analyses of the backup traps showed no break-through of materials.

Simultaneous flame and flame photometric detection of the thermally desorbed compounds provided a qualitative picture of the types and relative abundances of hydrocarbon and organosulfur compounds. Parallel gas chromatograph/mass spectrograph (GC/MS) data on the components from a separate sample trap did not show a great deal of similarity. Mass spectrograph data were poor because the sensitivity limits were being approached; only aliphatic hydrocarbons were positively identified while no sulfur-containing compounds could be detected.

The XAD-2 trap which collected volatile organics during the entire run was desorbed with methanol, acetone, methylene chloride, and pentane. A variety of compounds were identified by GC/MS (Table 1), especially terpanes, steranes, and other cyclic alkanes. These are in raw retorted shale oil and, apparently, very small amounts remain unburned.

LSF-02 and LSF-03 (Paraho and Laramie Shale Oils, burned April 5–7, 1978)

These tests used a substantially different sampling arrangement (Fig. 2). A single sample probe sampled flue gas for the duration of the test by collecting volatile organics in a Tenax bed (Table 2). The line from the probe was heated up to and including a Gelman filter just at the inlet to the trap. A second probe was used for intermittent sample collection to collect organic volatiles on XAD-2 and small Tenax columns as described previously. Flue gas from this probe was cooled in a condenser but not filtered before passage through the traps.

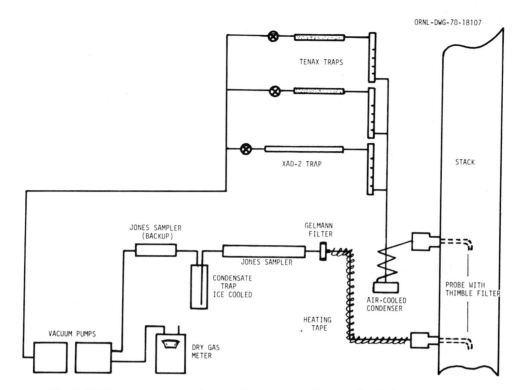

ORNL-DWG-78-18107

Fig. 2. Multipurpose apparatus for sampling stack gases for organic volatile components.

Table 1. Organic flue gas test methods
and components detected
(Test LSF-02)

Tenax, GC-FPD

Dimethylsulfide
Thiophene
Dibutylsulfide
Diethyldisulfide

Tenax, GC/MS

Aliphatic hydrocarbons

XAD-2, GC/MS (200 peaks)

Cyclic terpanes
Alkylphenols (tri-isopropyl-, di-tertbutyl-)
Branched olefin or cyclic hydrocarbon, C_{17}
  (largest GC peak area)
C_{18}-cyclic terpane
N-C_{19}
Phenanthrene/anthracene (small amount)
C_{20}-olefin hydrocarbon
Dialkyl(C_4)phthalate
C_{21}-C_{22} cyclohexyl hydrocarbon
C_{23}-branched alkane
Methyldehydroabietate (small amount)
C_{16}-phthalate
C_{27}-C_{31} alkanes
Steroid

Table 2. Organic flue gas components

Both tests performed with Tenax,
GC-MS (Jones sampler).

Compound
Test LSF-03
Aliphatic hydrocarbons
C_7-carbonyl
Phthalates
Cyclohexyl hydrocarbons
Halocarbons (source?)
Test LSF-04
Aliphatic hydrocarbons, $\sim C_{11}$-C_{30}
Phenol
C_3-benzene
Diethylphthalate
Amyl or isoamyl benzoate
Dibutyl phthalate
Cyclohexyl hydrocarbons
Butyl benzyl phthalate

120   B. R. Clark et al.

Unfortunately, the XAD-2 extracts were not analyzed at the time of this writing. Tenax from the Jones sampler was extracted with pentane and the extract was processed through a modified Rosen separation[4] procedure. Those fractions expected to contain any polycyclic aromatic hydrocarbons were carefully examined by GC/MS. None were found.

Conclusions
1. No appreciable amounts of polycyclic aromatic hydrocarbons were detected.
2. No nitrogen-containing organic compounds were detected.
3. A wide variety of sulfur compounds were detected at trace levels. During startup or upset conditions, high-molecular-weight sulfur-containing heterocyclic compounds were detected (e.g., benzothiophene)
4. Some constituents of raw shale oil were found in the flue gases.

COMBUSTION BEHAVIOR OF SHALE OILS

The combustion tests were run in a 100-HP boiler (Fig. 3). The boiler is a four-pass, dry-back, Cleaver-Brooks firetube boiler capable of generating approxi-

mately 3000 lb/hr of 100-psig saturated steam. The volumetric heat release rate in the unit is in excess of 200,000 Btu/ft^3-h at full load. This unit has been used in combustion test programs on coal-oil mixtures, SRC II and pyrolyzed wood wastes in addition to the shale oil tests. An extensive data base for No. 6 oil firing on this unit is available for comparison.

The boiler is fully instrumented for measuring operating parameters, boiler efficiency, and emissions (Fig. 4). The boiler studies can complement supporting tests (e.g., fuel analysis and characterization). For a particular boiler test, operating data, flame characteristics, and particle characteristics can be monitored. For the shale oil tests, some fuel characterization was attempted along with collecting complete operating data, flame data on temperatures and radiant fluxes, and cascade impactor studies to examine the nature of any particulate emissions.

The test program for the shale oil studies consisted of four tests. Each oil was run at a low excess-air value, 3.3% O$_2$, determined by running Laramie oil at a point where the opacity monitor began to show significant deflection (10% opacity). The low-excess-air test for the Paraho oil was set at the same excess-oxygen value. The high-excess-air tests were run with

Fig. 3. 100-hp PETC boiler.

ORNL–DWG 78-22096

100 HP TEST CAPABILITIES

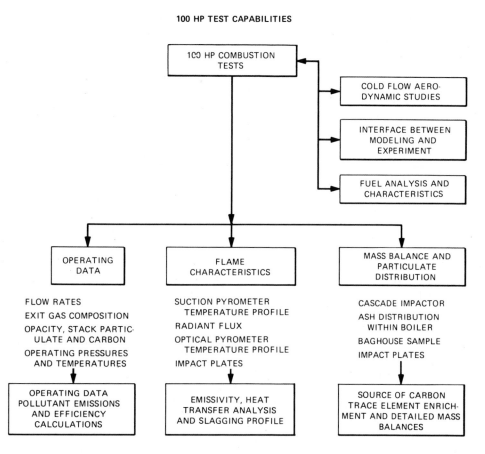

Fig. 4. 100 h.p. test capabilities.

an air rate that gave approximately 5.0% oxygen in the stack.

Each test was run at approximately 3.9×10^6 Btu/h input, based on the fuel heating value, which is full load for this unit. Data were taken every half hour throughout each test, and each test lasted about 8 h. The boiler was set up with a nozzle and secondary air diffuser appropriate for No. 6 oil. The operating conditions for the test were similar to those used when burning No. 2 oil in this same unit. The fuel was characterized before actual testing to select probable operating conditions better, as subsequently confirmed by a short period of firing before the experiments began.

The characterization showed both oils are composed essentially of straight-chain paraffinic hydrocarbons (H/C ratio ~1.65); neither oil exhibits significant aromaticity (Table 3). Viscosity tests on the two oils show the Laramie to be independent of

shear rate with a viscosity of 9 cP at 125° F and 4 cP at 200° F. The paraho oil shows some dependence on shear with a viscosity (at a shear rate of 92 sec^{-1}) of 20 cP at 125° F and 6 cP at 200° F. The specific gravity of

Table 3. Fuel analyses of shale oils

Component (%)	Paraho[a]	Laramie[b]
H	11.4	12.0
C	85.3	84.6
N	2.0	1.56
S	.6	.85
O	.6	.97
Ash	.1	.01
Oils	98.96	100.0
Asphaltenes	0.81	
Benzene insolubles	0.23	

[a] 18411 Btu/lb
[b] 18561 Btu/lb

the Laramie oil was 0.899 at 80°F while the specific gravity of the Paraho oil was 9.923 at the same temperature.

The operating data (Table 4) indicate acceptable combustor operation in all cases, with some reduction in performance for the low-excess-air test on Paraho shale oil. In all four cases, isokinetic stack tests and cascade impactor runs resulted in an immeasurably small capture of particulate matter. The measured NO and NO_x values are higher than those in a comparable No. 6 fuel oil and 20% coal–oil slurry tests (~200 ppm and ~250 ppm) and seem to depend on the stack temperature. These values generally exceed the new source standards for NO_x (0.3 lb/10^6 Btu). The SO_2 depends on the sulfur content of the fuel. The CO values indicate good combustion (CO ≤ 200 ppm) except for test LSF-05, where the value exceeds 800 ppm. The opacity values reflect the approach used in determining the low-excess-air value, 10.0 and 5.5% for those two cases and 0.6 and 1.4% for the high-excess-air tests. Stack temperatures are similar to those produced by conventional fuels. The four efficiencies were calculated using the ASME Heat Loss Method. All four are within 78–80% of the measured values for this boiler on No. 6 oil under a range of conditions.

The flame irradiance was measured with a radiometer equipped with a series of wavelength-sensitive filters (Table 5). The maximum temperature inside the combustor (3090°F) was measured by a two-color optical pyrometer; the combustor exit temperature (2145°F) was measured by a suction

Table 5. Irradiance measurements for shale oil run LSF-02[a]

(Laramie shale oil)

Filter no.	Wavelength range (μm)	Irradiance (W/cm^2)	Emissivity
3	0.60–3.5	66.0	0.106
5	0.23–3.5	93.0	0.149
Open	0.20–3.5	106.4	0.171

[a]Temperature at two-color pyrometer, 3090°F; temperature at suction pyrometer (exit), 2145°F; Average Temperature, 2618°F.

pyrometer. These temperatures are used to calculate an average temperature (2618°F) which is used with the irradiances to calculate the emissivities. The emissivity value over the whole spectral range of the instrument was 0.171.

The emissivity of the flame was also calculated from two-color and infrared pyrometer measurements by sighting on the flame and on zirconia targets in the combustor. The emissivities of the target plates were determined as a function of temperature in an earlier experiment in a clean furnace. Using the infrared pyrometer temperature, which can be adjusted for the emissivity of a target, as the true target temperature, the emissivity of the flame was determined using the relationship:

$$T_{2C} = (\epsilon_f)^{-14} T_t \; ,$$

Table 4. Summary of shale oil tests

	Test			
	LSF-02	LSF-03	LSF-04	LSF-05
Fuel	Laramie	Paraho	Laramie	Paraho
Fuel rate, lb/h	203	208.6	220.7	212.9
Steam rate, lb/h	3019	3044	3035	2837
Atomizing air pressure, psig	20.0	25.0	25.0	25.0
Excess air, %	16.7	25.4	28.4	18.8
CO_2, %	12.9	11.8	11.5	12.8
O_2, %	3.3	4.7	5.2	3.3
CO, ppm	117.0	135.0	223.0	828.0
THC, ppm	5.0	8.0	8.0	15.0
NO/NO_x, ppm		365/380	493/502	583/605
SO_2, ppm	259	188	266	188
Opacity	10.0	.6	1.4	5.5
Stack temp., °F	514	483	522	543
Efficiency	79.6	79.5	78.5	78.4

where T_{2C} is the two-color temperature and T_t is the target temperature from the infrared pyrometer with direct flame impingement. Emissivity values from targets on the lower left and right side of the combustion chamber were computed and averaged. Similarly, the total emissivity along the axis of the combustor through the flame was computed (Table 6). It should be noted that the total emissivity for run LSF-02 computed in Table 6 agrees with that given in Table 5. In summary, the emissivity values for the shale oil tests are comparable to emissivity values obtained with No. 6 oil.

Gas temperatures were measured using a suction pyrometer (Table 7). In all cases, the temperature profiles indicate an asymmetric flame, directed toward the right side of the combustion chamber. This lack of symmetry is in line with results from other tests in this unit and arise from the diffuser design.

Conclusions

1. In general, the shale oils burned like a typical distillate fuel oil requiring little preheat and modest atomizing air pressures for good performance.
2. The boiler efficiencies were similar to those obtained with No. 6 oil and close to the design efficiency (80%) of the boiler for No. 2 fuel oil.
3. Emissivities were similar to those found for No. 6 oil. Similarly, gas temperatures in both the stack and combustion chamber were similar to those found for No. 6 oil.
4. The NO_x emissions were higher than those measured for combustion tests on No. 6 oil and, in fact, exceeded the new source performance

Table 7. Gas temperatures in shale oil combustion (suction pyrometer)

Run	Temperature (°F)	
	Peak	Exit
LSF-02	2352	2281
LSF-03	2138	1800
LSF-04	2337	1954

standards for NO_x, indicating the need for NO_x control techniques. Some success in reducing NO_x emissions through staged combustion (which resulted from diffuser modifications to introduce a collar of air near the combustion chamber wall) has been achieved in this unit for coal-oil slurries with 0.6% nitrogen in the fuel.

REFERENCES

1. J. P. Longwell, "Synthetic Fuels and Combustion," *Progr. Energy Combust. Sc.* **3**, 127 (1977).

2. R. F. Sullivan, "Converting Green River Oil to Transportation Fuels," 11th Oil Shale Symposium, April 12–14, 1978, Golden, Colorado.

3. P. W. Jones et al., "Efficient Collection of Polycyclic Organic Compounds from Combustion " Effluents," *Environ. Sci. Tech.* **10**, 806 (1976).

4. *Technical Manual for Analysis of Organic Materials in Process Streams;* EPA-600/2-76-072, U.S. Environmental Protection Agency, Washington, D.C., March 1976.

Table 6. Flame emissivity of shale oils based on target temperature

Run	Total			R.H.S. target			L.H.S. target		
	T_{2C}^a	T_t^b	ϵ_1	T_{2C}	T_t	ϵ_2	T_{2C}	T_t	ϵ_3
LSF-02	1974	1273	0.173						
LSF-03	1675	1183	0.250	1672	1053	0.157	1755	1133	0.174
LSF-04	1745	1263	0.274	1589	1013	0.165	1797	1123	0.152
LSF-05	1977	1273	0.172	1621	1043	0.171	1755	1143	0.179

[a] T_{2C} = two-color optical pyrometer reading.
[b] T_t = target temperature.

DISCUSSION

Jim Payne Science, Applications, Inc.: When you did your Jones trapping and it was heated, you said you didn't look at the secondary Jones trap to see if you had any polynuclear aromatics in that trap. Did you analyze the condensate trap to see if there were any PAHs in that area? What I'm worried about is a little bit of breakthrough due to high temperatures of the gases passing through the Tenax.

Bruce Clark: That is a possibility. These samples have not yet been analyzed.

SESSION III: BIOLOGICAL EFFECTS STUDIES

Chairman: J. L. Epler
Oak Ridge National Laboratory

SESSION III: BIOLOGICAL EFFECTS STUDIES

SUMMARY

J. L. Epler*

The need for substitute fuels and convenient energy sources has led to renewed efforts in the fields of coal conversion and shale oil. A number of coal liquefaction, coal gasification, and shale oil processes are under development. For this approach to aid in solving energy problems adequately and safely, potential health hazards for workers and the general public should be considered. The overall environmental aspects of any energy technology include pollutant releases, physical disturbances, and socioeconomic disturbances, along with health and safety impacts. This session concentrates on health and safety aspects and illustrates a variety of biological assays coupled to chemical characterizations. The realization that the number of specific compounds and complex effluents and streams in need of biotesting in any technology is overwhelming and would tax the scientific community's whole-animal testing capabilities has led to the development of an array of "screening"—short-term—assays for toxicity, mutagenicity, teratogenicity, and carcinogenicity. The working hypothesis has been that the rapid tests can serve as predictors of biohazards early in the development of the technology. Thus, process changes might be initiated and/or definitive whole-animal testing considered. Conceivably, savings in both time and money on the technology and the biological testing would be gained.

Furthermore, current and/or prospective government regulations consider the biological tests and advocate their use, for example, the Toxic Substances Control Act and the Resource Conservation and Recovery Act. Coal- and shale-derived liquids, aqueous wastes, particulates, and solid wastes can be evaluated as complex mixtures with potential biohazards, common, perhaps to all conversion processes.

The underlying theme of the health effects research summarized in this session becomes obvious—biologists, chemists, and technologists working together to confront the problems of the emerging industries.

Examples of the team approach with biologists and chemists working together are given in the first set of papers from Oak Ridge National Laboratory. The use of fractionated test materials and short-term assays used in a comparative approach is discussed by my group and then contrasted with the whole-animal work of J. M. Holland and his associates. M. D. Waters from the Environmental Protection Agency further explores the use of short-term assays, particularly as carried out in a battery of assays or test matrix, with emphasis is placed on a combined chemical and biological approach. Similarly, the interagency program in health effects is described by D. L. Coffin, of the Environmental Protection Agency, in collaboration with M. R. Guerin and W. H. Griest at Oak Ridge, who discuss the use and function of a materials repository for synthetic fossil fuel test samples. W. Barkley discusses the work of the Kettering Laboratory with toxicity and skin bioassays of various shale oil materials. As a final example of a coupled chemical and biological approach, C. H. Hobbs discusses the work of the group at Lovelace Biomedical and Environmental Research Institute. This program emphasizes the potential inhalation hazards to man from fluidized bed combustion and coal gasification effluents, and again short-term biological tests and chemistry are coupled with animal toxicological tests.

---

*Oak Ridge National Laboratory.

INTEGRATED CHEMICAL AND BIOLOGICAL TESTING OF SYNTHETIC OILS AND EFFLUENTS*

J. L. Epler†

PREFACE

The Health Effects Research Section in the Biology Division of Oak Ridge National Laboratory has expanded its concern with the synthetic fuel technologies greatly over the last few years. The general organization remains much the same as the division overall, i.e., major activities in mutagenesis (including teratogenesis), carcinogenesis, and toxicology. A brief overview of the projects would note significant studies in bacterial mutagenesis, fungal mutagenesis, insect mutagenesis, mammalian cell mutagenesis, mammalian mutagenesis, toxicology, metabolism, carcinogenesis, promotion, cytotoxicity, electron microscopy, neurotoxicity, cytogenetics, cardiovascular disease, and DNA repair. Many of the projects deal not only with "model compounds" that are predicted to occur but also with the actual products or effluents from the various fuel technologies. Complementary chemical characterization and preparation for bioassay are accomplished in a collaborative effort with the Analytical Chemistry Division. The following two parallel and comparative biological studies on similar test materials from the synthetic fuel technologies serve as examples of the coordinated approach carried out in the life sciences.

*Research jointly sponsored by the Environmental Protection Agency (IAG-D5-E681; Interagency Agreement 40-516-75) and the U.S. Department of Energy, under contract W-7405-eng-26 with the Union Carbide Corporation.

†Biology Division, Oak Ridge National Laboratory, Oak Ridge, Tenn. 37830.

SHORT-TERM MUTAGENICITY TESTING*

J. L. Epler†   F. W. Larimer†
T. Ho†   C. E. Nix†
A. W. Hsie†   T. K. Rao†

ABSTRACT

The major goal of the mutagenesis research group in the Biology Division at Oak Ridge National Laboratory is to provide a means of mutagenicity testing of those compounds produced by various existing or proposed methods of energy generation. These compounds include the primary effluents of existing fossil fuel sources such as sulfur dioxide, the oxides of nitrogen, ozone, hydrocarbons, and heavy metals, as well as products of newly proposed methodologies such as coal liquefaction and of auxiliary methodologies such as cooling-tower additives. The work is divided into two phases, one dealing with known compounds expected to occur in the environment through energy production, conversion, or use, and another dealing with actual samples from existing or experimental processes. To approach the problems of dealing with and the testing of large numbers of compounds, we set up a form of the "tier system." Operating units utilizing *Salmonella, E. coli*, yeast, human leukocytes, mammalian cells, and *Drosophila* have been initiated. As a working list we have looked to those compounds expected to be used in fossil fuel production or conversion. Many of these compounds are polycyclic hydrocarbons and require metabolic activation with mammalian extracts. Basic and neutral fractions from crude oils are mutagenic in the microbial systems. Comparative studies with leukocytes, mammalian cells, and *Drosophila* validate these results. The assays represent rapid tests for potential biohazards.

INTRODUCTION

The enormous amount of industrial and technological activity carried out in the modern world creates a large number of chemical pollutants. The developing synthetic fuel industry serves as only one example of a potential source of pollutants that could have significant environmental impacts on man. Considerable research on the health effects of chemicals in the environment is being carried out, but it has become obvious that methods to cut research time and expense are necessary to confront the large number of potentially hazardous substances that man encounters.

In addition to obvious toxic effects from chemical pollutants, the possibility of the compounds producing carcinogenic, mutagenic, or teratogenic effects whose expression may be divorced in time from the actual exposure(s) must be considered. The research effort described here specifically approaches the question of genetic hazard but, as has been pointed out recently,[1] certain microbial genetic assays—the Ames test—show a high correlation between positive results in the mutagenicity testing and the carcinogenicity of the compounds under test. The overall need to subject environmental chemicals to mutagenicity testing has been discussed in the *Committee 17 Report on Environmental Mutagenic Hazards.*[2] The utility of short-term tests for mutagenicity has been summarized by deSerres,[3] and the prospect for the use of these tests in toxicological evaluation has been discussed.[4] Basically, the key recommendation of Committee 17 is that "screening should be initiated as rapidly and as extensively as possible." DeSerres[4] stresses that the data base of knowledge on untested environmental chemicals should be expanded but cautions that we "are not ready, however, to extrapolate directly from data obtained in short-term

---

*Research jointly sponsored by the Environmental Protection Agency (IAG-D5-E681; Interagency Agreement 40-516-75) and the U.S. Department of Energy, under contract W-7405-eng-26 with the Union Carbide Corporation.

†Biology Division, Oak Ridge National Laboratory, Oak Ridge, Tenn. 37830.

tests for mutagenicity directly to man." Tests on other organisms have to be performed to validate and reinforce results from short-term tests. Short-term tests simply point out *potential* mutagenic and carcinogenic chemicals and serve to order priorities for further testing in higher organisms.

Obviously all of the chemical pollutants in question cannot be subjected to whole-animal testing. The expense in time and money would be overwhelming with genetic testing alone. We felt a scheme had to be developed to prescreen the genetically hazardous compounds reliably and allow the investigator to select a smaller sample to be tested thoroughly in other systems. Such a leveled or tiered system of testing (suggested by Bridges[5,6]) would presumably save the scientific world considerable time and expense, while bringing the pertinent information to the industrial community and the general population in an organized manner. If the correlation of genetic damage with potential cancer danger is valid, then added information is received. Conceivably, the choice of compounds that would be subjected to extensive carcinogenic testing in the whole animal would be influenced by the genetic testing.

Perhaps most important in a battery of screening assays is the initial or first level of testing. False negatives here presumably would terminate the testing of any particular compound. False positives would be clarified by further comparative testing. Thus, the initial test should be a high-resolution, sensitive assay, yet rapid and inexpensive. The *Salmonella* histidine reversion system probably fits these criteria best of the currently available short-term assays.

As an initial step in establishing testing procedures, we have investigated the use of the Ames *Salmonella* system with a large number of environmentally important chemicals and effluents, principally those known or predicted to occur in energy use, production, and/or conversion. This report serves as an example of what the test's potential may be in environmental research and states our general scientific approach to mutagenicity testing.

A BATTERY OF TESTS

The *Salmonella* strain series used in testing [obtained through the courtesy of Bruce N. Ames (Table 1)] is composed of histidine mutants that revert after treatment with mutagens to the wild-type state or growth independent of histidine. Both missense mutants and frame-shift mutants comprise the set,

Table 1. Genotype of *Salmonella* strains used for testing mutagens

Additional mutations in		Histidine mutation in strain		
LPS	Repair	hisG46 (missense)	hisC3076 (frame shift)	hisD3052 (frame shift)
+	+	hisG46	hisC3076	hisD3052
rfa	ΔuvrB	TA1535	TA1537	TA1538
rfa	+	TA1975	TA1977	TA1978
rfa	ΔuvrB	TA100[a]		TA98[a]

[a]Plus r factor.

and their reversion characteristics with a potential chemical mutagen indicate the mechanism of action. In addition, the detection scheme yields the highest resolution possible by the inclusion of other mutations: (a) the deep rough mutation, *rfa*, which affects the lipopolysaccharide coat, making the bacteria more permeable, and (b) the deletion of the *uvrB* region, eliminating the excision repair system.[7] The procedures with the strains and their use in mutagenicity testing are discussed in detail in Ames et al.[8]

One other feature of the *Salmonella* system is the ease with which metabolic activation can be incorporated into the assay. Spot tests or quantitative plate tests can be performed in the presence of rat or other mammalian liver homogenates so that the mutagen can be metabolized to its ultimate, active form in the *in vitro* short-term test. In assays requiring activation, standard rat liver microsome preparations[8] from rats induced with Aroclor 1254 (gift of Monsanto Corporation) or sodium phenobarbital are used.

Utilizing the *Salmonella* system developed for mutagenicity (or carcinogenicity?) screening as a first level of testing, we are developing the battery of comparative genetic tests so that a large number of chemical pollutants or, more specifically, environmental mutagens can be evaluated for their potential genetic hazard. The key objective of this approach will be the eventual ability of the investigator to compare and extrapolate from one system to another. One example of this approach we are considering is shown in Table 2. The assays used here are discussed in general in the series edited by Dr. Alexander Hollaender, *Chemical Mutagens: Principles and Methods for Their Detection.*[9] Both gene and chromosomal assays are included, and testing involves a move from the simple microbial tests (*Salmonella, E. coli,* and yeast) to tests with higher

Table 2. An example of a tier system of mutagenicity testing

Gene		Chromosomal	
Organism	Test	Organism	Test
Tier I			
Salmonella	Histidine reversion	Yeast	Gene conversion and mitotic recombination
E. coli	Mutation in WP2 (trp strain)		
Tier II			
Drosophila	Visible, sex-linked recessive lethals	*Drosophila*	X chromosome loss, nondisjunction
Mammalian cells	In culture	Mammalian cells	Chromosomal aberrations
		Mouse	Dominant lethals (?), heritable translocations
Tier III			
Mouse	Specific locus test	Mouse	Heritable translocations
New mammalian tests?			

organisms (*Drosophila,* mammalian cells, human leukocytes) to tests with a whole mammal, presumably the mouse. Thus, establishing the testing priorities on the use by or impact on man, complete testing may be reduced to groups of environmentally important compounds, e.g., selected polycyclic hydrocarbons or various classes of nitrosamines.

SCREENING OF COMPLEX MIXTURES

Recently we have extended the assays, specifically with the *Salmonella* system, to a study of the feasibility of investigating crude industrial products and effluents. By using the genetic results as estimates of biohazards, we can establish priorities for further testing. Our initial efforts argue for the applicability of the testing but only when coupled with analytical work and fractionation of the crude mixtures.[10]

To determine the potential biohazards (mutagenicity/carcinogenicity) of various *crude* and *complex* test materials derived from fossil fuel production or conversion activities rapidly, we have examined the feasibility of using short-term genetic assays to predict, isolate, and identify the chemical hazards. Biological screening assays were coupled to analytical chemistry separation procedures so that the chemical work and priorities for identification could be determined by the bioassay. The wide applicability of the bacterial test system developed by Ames[8] can be demonstrated by the use of the assay as a prescreen for potential genetic hazards of complex environmental effluents or products [e.g., tobacco smoke

condensates,[11] hair dyes,[12] soot from city air,[8] fly ash,[13] and (in our work) oils and aqueous wastes from synthetic fuel technologies[14,15]].

To study the feasibility of applying mutagenicity testing to environmental effluents and crude products from the synthetic fuels technology, we initially attempted to perform screening with the highly sensitive histidine reversion strains of bacteria[8] known to respond to a wide variety of proven mutagens. The working hypothesis was that sensitive detection of potential mutagens in *fractionated* complex mixtures could be used to isolate and identify the biohazard. In addition, the information could be helpful in establishing priorities for further testing, either with other genetic assays or with carcinogenic assays. Finally, the procedures might be useful in monitoring plant processes, effluents, or personnel early in the formation of the engineering and environmental technology that will eventually evolve in the synthetic fuels industry. The approach and preliminary results showed that the coupled chemical-biological scheme is a feasible research mechanism applicable to ascertaining potential human health hazards of a wide variety of environmental exposures, either occupationally or to the population in general.

EXPERIMENTAL RESULTS WITH COMPLEX MIXTURES: CRUDE OILS

As an example of the screening of fractionated complex mixtures, we list in Tables 3 and 4 the results

Table 3. Distribution of mutagenic activity in fractions of Composite Crude-1 and Synfuel A-3[a]

Fraction[b]	Composite Crude-1			Synfuel A-3		
	Relative weight (% of total)	Specific activity (rev/mg)[c]	Weighted activity (rev/mg)	Relative weight (% of total)	Specific activity (rev/mg)	Weighted activity (rev/mg)
NaOH$_I$	1.3	10	<1	26.3	3,430	902
WA$_I$	0.1	10	0	1.3	1,310	17
WA$_E$	0.3	190	1	7.4	2,460	182
SA$_I$	0.2	230	<1	0.04	NT[d]	
SA$_E$	0.4	20	0	0.5	30	0
SA$_w$	0.1	400	1	0.3	1,010	3
B$_{Ia}$	1.3	50	1	10.0	42,400	4,242
B$_{Ib}$	0.3	0	0	0.04	NT	0
B$_E$	0.2	150	<1	3.1	43,300	1,342
B$_w$	0.5	460	2	0.1	2,700	3
Neutral	84.2	277 (350)[e]	233	56.4	1,904 (910)	617
Total	88.9		241	105.5		7,308
Neutral subfraction						
Hexane	71.1	260	186	53.6	420	222
Hexane/benzene	7.8	360	28	10.4	340	35
Benzene/ether	8.9	550	49	41.1	750	309
Methanol	4.7	300	14	8.0	6,600	528
Subtotal	92.5		277	113.1		1,094
Initial sample (g)	24.501			26.568		
Chromatographed (g)	10.226			10.361		

[a]All assays carried out in the presence of crude liver S-9 from rats induced with Aroclor 1254. Strain TA98 metabolically activated with Aroclor-induced preparation.

[b]I = insoluble (fractions a and b), E = ether soluble, W = water soluble, WA = weak acid, SA = strong acid, B = base.

[c]rev/mg = revertants/mg, the number of histidine revertants from *Salmonella* strain TA98 by use of the plate assay with 2×10^8 bacteria per plate. Values are derived from the slope of the induction curve extrapolated to a milligram value.

[d]NT = not tested.

[e]Comparable to "specific activity," but based on the activity of the total neutral fraction rather than the summation of the individual fractions.

Source: Adapted from J. L. Epler et al., "Analytical and Biological Analyses of Test Materials from the Synthetic Fuel Technologies: I. Mutagenicity of Crude Oils Determined by the *Salmonella typhimurium*/Microsomal Activation System," *Mutat. Res.* 57, 265–76 (1978).

from oil samples containing a variety of organic contaminants. The samples consist of the crude oils from experimental coal liquefaction and shale oil processes (supplied by the various Energy Research Centers and industry) along with comparative natural petroleum crudes. A large quantity of oil sample was extracted and subjected to fractionation by the Stedman procedure.[16] Previous publications have shown the usefulness of this technique with crude petroleum oils, synthetic crude oils, shale oils, and aqueous wastes.[10,14,15] Table 3 lists the comparative results from a natural crude and an experimental crude product from coal liquefaction.

Primary fractions were obtained as listed in Table 3. The table shows the analytical weight analysis (column 1) of each fraction and lists the mutagenicity test results both as a specific activity of each fraction (column 2) and as a weighted activity (column 3, product of column 1 and 2). Fractions and/or control compounds to be tested were suspended in dimethyl sulfoxide (sterile, spectrophotometric grade from Schwarz/Mann) at concentrations in the range of 10–20 mg per milliliter of solids. The potential mutagenic fraction was in some cases assayed for general toxicity (bacterial survival) with *Salmonella* strain TA1537 (see Table 1). Normally, the fraction

Table 4. Summary of mutagenicity testing results with synthetic oils and
aqueous samples- –class fractionation scheme[a]

Sample	SA$_W$ fraction[b]		B$_E$ fraction[b]		Neutral fraction		Total weighted activity[d] (rev/mg)
	Relative weight (%)	Specific activity (rev/mg)[c]	Relative weight (%)	Specific activity (rev/mg)	Relative weight (%)	Specific activity (rev/mg)	
Composite crude-1	0.1	400	0.2	150	84.2	277	241
Composite crude-2	0.1	750	0.2	500	84.2	166	147
La.-Miss. crude	0.1	240	0.2	180	80.7	90	76
Shale oil	0.6	160	7.1	952	86.7	112	178
Synfuel A-1	0.3	240	2.0	28,900	73.6	517	4,032
Synfuel A-2	0.4	120	2.0	36,200	69.2	583	4,189
Synfuel A-3	0.3	1,010	3.1	43,300	56.4	1,094	7,308
Synfuel B-1	0.4	0	2.6	1,500	82.3	560	516
Synfuel B-2	1.6	0	1.8	3,800	89.3	465	484

[a]Strain TA98 metabolically activated with Aroclor-induced preparation.

[b]SA$_W$ = strong acid, water soluble; B$_E$ = base, ether soluble.

[c]rev/mg = revertants/mg, the number of histidine revertants from *Salmonella* strain TA98 by use of the plate assay with 2×10^8 bacteria per plate. Values are derived from the slope of the induction curve extrapolated to a milligram value.

[d]Weighted activity of each fraction relative to the starting material is the product of relative weight and specific activity. The sum of these products is given as a measure of the total mutagenic potential of each material.

Sources: Adapted from J. L. Epler et al., "Analytical and Biological Analyses of Test Materials from Synthetic Fuel Technologies: I. Mutagenicity of Crude Oils Determined by the *Salmonella typhimurium*/Microsomal Activation System," *Mutat. Res.* 57, 265–76 (1978); J. L. Epler, T. K. Rao, and M. R. Guerin, "Evaluation of the Feasibility of Mutagenic Testing of Shale Oil Products and Effluents," *Environ. Health Perspect.,* in press.

was tested with the plate assay over at least a 1000-fold concentration range with the two tester strains TA98 and TA100. Revertant colonies were counted after a 48-h incubation. Data were recorded and plotted versus the added concentration, and the slope of the induction curve was determined. It is assumed that the slope of this curve in the linear dose-response range reflects the mutagenic activity (specific activity). Positive or questionable results were retested with a narrower range of concentrations. All studies were carried out with parallel series of plates plus and minus the rat liver enzyme preparation[8] for metabolic activation. Routine controls demonstrating the sterility of samples, enzyme or rat liver S-9 preparations, and reagents were included. Positive controls with known mutagens were also included in order to recheck strain response and enzyme preparations. All solvents used were nonmutagenic in the bacterial test system. Additivity of the individual weighted values was assumed, and a final total mutagenic potential was calculated.

The separation scheme (Swain-Stedman) yields approximate class separations based on the relative acidic-basic properties of the material. The test samples are partitioned between an organic solvent (ethyl ether) and 1 N NaOH to yield an aqueous fraction containing acidic compounds and an organic fraction containing basic and neutral compounds. The organic fraction is extracted with 1 N HCl to yield an aqueous fraction containing basic compounds and a fraction containing neutral organics. Further fractionation produces the primary isolates. Detailed chemical data for oils are given by Rubin et al.[10] and Guerin et al.[17] The neutral portion is dried and chromatographed on a Florisil column as described by Bell et al.[18] Organic solvents are used to solubilize the neutrals and elute subfractions from the column.

Table 3 displays representative data from a synthetic crude oil and a natural crude oil "control." The data are listed as the fraction or subfraction tested with the *Salmonella* histidine-reversion assay, the percentage of material recovered in the various fractions (concentrated and solubilized in dimethylsulfoxide), the mutagenic potential ("specific activity") for each fraction (as revertants per milligram of material in the fraction extrapolated from the linear portion of a dose-response curve), the weighted contribution of each fraction to the starting material, and the sum of all fractions as total mutagenic potential. The total "mutagenic potential" of the synthetic oil is much higher than that of the natural crude (7308 revertants per milligram vs 241 revertants per milligram). Additionally, high-specific-activity components are observed in both the neutral and basic fractions.

Reproducibility of results was shown by comparison of data from similar samples (Table 4). Although discrepancies exist from fraction to fraction, the general trend is apparent, and the sum of activities appears to be roughly reproducible.[14] When the major component, the neutral fraction, is assayable (i.e., toxicity does not eliminate the detection of mutagenicity) as with the synfuel in Table 3, the summation of the subfractionation values of the neutrals shows the approximate additivity of the individual mutagenic determinations.

The summary in Table 4 lists the results from two liquefaction products, a shale oil, and natural crudes. An overview of the results with crude oils points to a number of consistencies: (a) all natural and synthetic crude oils showed some mutagenic potential, (b) the neutral and basic fractions showed activities requiring activation regardless of the source of the sample, and (c) the relative total mutagenic potentials varied over two orders of magnitude. Whether these results reflect a comparative biohazard of processes still under development is not the point in question here. The results simply show that biological testing of crude test materials, in this case by means of genetic reversion assays, can be carried out with the newly developed tester systems, but only when coupled with the appropriate analytical separation schemes. Conceivably, this approach could provide rapid information concerning health effects and the active components involved. Primary candidates for the mutagens (and carcinogens?) responsible for activity in the basic fractions include quinoline, substituted quinolines, alkyl pyridines, acridines, aza-arenes, naphthylamines, and aromatic amines; in the neutral fractions potential hazards may be benzanthracenes, dibenzanthracenes, substituted anthracenes, benzopyrenes, benzofluorenes, pyrene, substituted pyrenes, and chrysenes.

COMPARATIVE MUTAGENESIS OF COMPLEX MIXTURES

To validate and compare the results accumulated in the Ames system using complex test materials from synthetic fuel technologies, we selected specific fractions or subfractions on the basis of their activity in the histidine-reversion assay for further testing in the various assays described in the section above on the battery of tests in the tier approach to testing.

Considering the results with class fractionation procedures and the high mutagenic activity of the basic fractions, Guerin et al.[19] developed a procedure specifically designed for subfractionation of the basic materials, now realized to be major contributors to overall mutagenesis. An elution sequence using alumina and Sephadex LH-20 gel with a combination of solvents isolates 90% of the mutagenic activity from basic compounds into a 0.5 wt% fraction of crude oil.

Preliminary results have been published in the proceedings of the Second International Conference on Environmental Mutagens, Edinburgh, 1977[20] and in Guerin et al. [17,19] Qualitative comparisons may be seen in Table 5. The selected fractions or subfractions utilized were basic and neutral isolates from synthetic crude oils from coal liquefaction processes [Synfuel A and B as described by Epler et al.[14]; crude oil products of experimental coal liquefaction processes]. In Drosophila[21] and in the mammalian cell gene mutation assay,[22] detection of genetic damage has been a function of newly developed fractionation schemes (e.g., the use of column chromatography[23]) that result in higher-specific-activity (more highly purified) mutagenic subfractions. In general, the results validate the initial screening carried out in the Salmonella assay, but they have not yet been used to test materials exhaustively that are negative in the Ames system.

CONCLUSIONS—PRECAUTIONS

In these initial studies, the purpose has not been to compare the biohazards resulting from different materials or processes. The results show that biological testing—within the limits of the specific system used—can be carried out with complex organic materials, but only when coupled with the appropriate extraction and analytical separation schemes. An extrapolation to relative biohazard at this point would be, at the least, premature. The primary function of such combined chemical and biological studies may be to aid in isolating and identifying the specific classes or components involved. A number of precautions should be observed in the use of these assays.

The detection or perhaps the generation of mutagenic activity may well be a function of the chemical fractionation scheme utilized. The inability

Table 5. Comparative mutagenicity of crude oil fractions[a]

Test material	Test system					
	Salmonella typhimurium, his⁻→his⁺	*E. coli,* arg⁻→arg⁺ or gal⁻→gal⁺	*Saccharomyces cerevisiae,* his⁻→his⁺ or CAN^S→can^R	*Drosophila melanogaster* sex-linked recessive lethals	CHO cells, 6-thioguanine resistance	Human leukocytes, chromosome aberrations
Crude synfuel[b]	+[c]	NT	NT	NT	NT	NT
Basic fraction	+	+	+	+	+	NT
Neutral fraction	+	+	+	−	NT	+?

[a]Liquid crude oils from coal liquefaction processes.

[b]Crude synfuels are generally too toxic to test in most systems.

[c]+ = mutagenic; − = nonmutagenic; NT = not tested.

to recover specific chemical classes or the formation of extraneous products by the treatment could well produce erroneous results.

Along with the obvious bias that could accompany the choice of samples and their solubility or the time and method of storage, a number of biological discrepancies can also enter into the determinations. For example, concomitant bacterial toxicity can nullify any genetic damage assay that might be carried out. The dose-response relationship may not be linear, and some other method for a quantitative comparison may be mandatory. The choice of inducer for the liver enzymes involved could be wrong for selected compounds or mixtures. Furthermore, induction of metabolic enzymes of rat liver includes both activating and deactivating enzymes for potential mutagens. Results with mixtures requiring activation can be complex and different from those with pure organic compounds. Mutagenicity studies should include not only proper metabolic activation systems but also appropriate quantitation of the metabolic enzymes (determined by titration studies) in the assays. Mutagenic analyses of complex mixtures of organic constituents activated with crude and complex enzyme homogenates require careful examination and cautious interpretation. The choice of strain in a reversion assay could be inappropriate for selected active components of a mixture; therefore, a battery of tests should be considered, including an assay for forward mutation.

The applicability of the generally used *Salmonella* test to other genetic end points and the validation of the apparent correlation between mutagenicity and carcinogenicity still need further work through sufficient fundamental research. The question of a correlation between *mutagenic potency* in the Ames

assay and *carcinogenic potency* should be treated with caution. Furthermore, the short-term assays chronically show negative results with certain substances which are known biohazards (e.g., heavy metals and certain classes of organics). Similarly, compounds involved in or requiring co-carcinogenic phenomena would presumably go undetected. Recent studies point to synergistic effects between compounds that may further complicate quantitative interpretation of results with complex mixtures.

As a prescreen to aid investigators in establishing their priorities, the short-term testing appears to be a valid approach with complex mixtures. Overinterpretation at this stage of research, especially with respect to relative hazard or negative results, should be avoided. The most effective use of the short-term assays may well be as a tool to isolate and identify mutagens in complex mixtures. Further extension and validation of short-term assays and their relationship to genetic and carcinogenic potential in the whole animal should be carried out. Considerable basic research will be necessary before extrapolation to man is appropriate.

ACKNOWLEDGMENTS

This research was jointly sponsored by the Environmental Protection Agency (IAG-D5-E681; Interagency Agreement 40-516-75) and the Division of Biomedical and Environmental Research, U.S. Department of Energy, under contract W-7405-eng-26 with the Union Carbide Corporation.

The authors thank the staff at the Pittsburgh Energy Research Center and the Laramie Energy Research Center for their cooperation in providing samples, the Analytical Chemistry Division staff coordinated by M. R. Guerin for their efforts in providing the fractionated test materials, the Environmental Mutagen Information Center for their aid, and their colleague Dr. A. W. Hsie for permission to cite his unpublished work.

REFERENCES

1. J. McCann et al., "Detection of Carcinogens as Mutagens in the *Salmonella*/Microsome Test: Assay of 300 Chemicals, Part I," *Proc. Nat. Acad. Sci. USA* **72**, 5135–39 (1975).

2. J. W. Drake et al., "Environmental Mutagenic Hazards," *Science* **197**, 503–14 (1975).

3. F. J. de Serres, "The Utility of Short-Term Tests for Mutagenicity," *Mutat. Res.* **38**, 1–2 (1976).

4. F. J. de Serres, "Prospects for a Revolution in the Methods of Toxicological Evaluation," *Mutat. Res.* **38**, 165–76 (1976).

5. B. A. Bridges, "Some General Principles of Mutagenicity Screening and a Possible Framework for Testing Procedures," *Environ. Health Perspect.* **6**, 221–27 (1973).

6. B. A. Bridges, "The Three-Tier Approach to Mutagenicity Screening and the Concept of Radiation Equivalent Dose," *Mutat. Res.* **26**, 335–40 (1974).

7. B. N. Ames, F. D. Lee, and W. E. Durston, "An Improved Bacterial Test System for the Detection and Classification of Mutagens and Carcinogens," *Proc. Nat. Acad. Sci. USA* **70**, 782–86 (1973).

8. B. N. Ames, J. McCann, and E. Yamasaki, "Methods for Detecting Carcinogens and Mutagens with the *Salmonella*/Mammalian-Microsome Mutagenicity Test," *Mutat. Res.* **31**, 347–64 (1975).

9. A. Hollaender, ed., *Chemical Mutagens: Principles and Methods for Their Detection,* vols. 1–5, Plenum Press, New York, 1971–1979.

10. I. B. Rubin et al., "Fractionation of Synthetic Crude Oils from Coal for Biological Testing," *Environ. Res.* **12**, 358–65 (1976).

11. L. D. Kier, E. Yamasaki, and B. N. Ames, "Detection of Mutagenic Activity in Cigarette Smoke Condensates," *Proc. Nat. Acad. Sci. USA* **71**, 4159–63 (1974).

12. B. N. Ames, H. O. Kammen, and E. Yamasaki, "Hair Dyes Are Mutagenic: Identification of a Variety of Mutagenic Ingredients," *Proc. Nat. Acad. Sci. USA* **72**, 2423–27 (1975).

13. C. E. Chrisp, G. L. Fisher, and J. E. Lammert, "Mutagenicity of Filtrates from Respirable Coal Fly Ash," *Science* **199**, 73–75 (1978).

14. J. L. Epler et al., "Analytical and Biological Analyses of Test Materials from the Synthetic Fuel Technologies: I. Mutagenicity of Crude Oils Determined by the *Salmonella typhimurium*/Microsomal Activation System." *Mutat. Res.* **57**, 265–76 (1978).

15. J. L. Epler, T. K. Rao, and M. R. Guerin, "Evaluation of Feasibility of Mutagenic Testing of Shale Oil Products and Effluents." *Environ. Health Perspect.,* in press.

16. A. P. Swain, J. E. Cooper, and R. L. Stedman, "Large Scale Fractionation of Cigarette Smoke Condensate for Chemical and Biologic Investigations," *Cancer Res.* **29**, 579–83 (1969).

17. M. R. Guerin et al., "Polycyclic Aromatic Hydrocarbons from Fossil Fuel Conversion Processes," in *Carcinogenesis, Vol. 3: Polynuclear Aromatic Hydrocarbons,* ed. P. W. Jones and R. J. Freudenthal, pp. 21–33, Raven Press, New York, 1978.

18. J. H. Bell, S. Ireland, and A. W. Spears, "Identification of Aromatic Ketones in Cigarette Smoke Condensate," *Anal. Chem.* **41**, 310–13 (1969).

19. M. R. Guerin et al., "Separation of Mutagenic Components in Synthetic Crudes," presented to the Division of Environmental Chemistry, American Chemical Society, Anaheim, California, March 12–17, 1978.

20. J. L. Epler et al., "Comparative Mutagenesis of Test Material from the Synthetic Fuel Technologies," in *Progress in Genetic Toxicology,* ed. D. Scott, B. A. Bridges, and F. H. Sobels, pp. 275–84, Elsevier/North-Holland, New York, 1977.

21. C. E. Nix and B. S. Brewen, "The Role of *Drosophila* in Chemical Mutagenesis Testing," presented at the Symposium on Application of Short-Term Bioassays in the Fractionation and Analysis of Complex Environmental Mixtures, Williamsburg, Viginia, February 21–23, 1978, EPA-600/9-78-027, September 1978.

22. A. W. Hsie et al., "Mutagenicity of Carcinogens: Study of 101 Individual Agents and 3 Subfractions of a Crude Synthetic Oil in a Quantitative Mammalian Cell Gene Mutation System," presented at the Symposium on Application of Short-Term Bioassays in the Fractionation and Analysis of Complex Environmental Mixtures, Williamsburg, Virginia, February 21–23, 1978, EPA-600/9-78-027, September 1978.

23. A. R. Jones, M. R. Guerin, and B. R. Clark "Preparative-Scale Liquid Chromatographic Fractionation of Crude Oils Derived from Coal and Shale," *Anal. Chem.* **49**, 1766–71 (1977).

CARCINOGENICITY OF SYNCRUDES RELATIVE TO NATURAL PETROLEUM AS ASSESSED BY REPETITIVE MOUSE SKIN APPLICATION*

J. M. Holland†

M. S. Whitaker   J. W. Wesley

ABSTRACT

The relative carcinogenicities of coal- and shale-derived liquid crudes were compared with a composite blend of natural petroleum using discontinuous exposure of mouse skin. All of the syncrudes tested as carcinogenic but the natural crude composite was negative, following three times weekly application of 50% w/v solutions for 22 weeks followed by a 22 week observation period. In addition to eliciting progressive squamous carcinomas, the syncrudes were also capable of inducing persistent ulcerative dermatitis. This inflammatory or necrotizing potential appeared to be inversely proportional to the carcinogenicity of the material. A measure of the relative solubility of the materials in mouse skin was obtained by quantitation of native fluorescence in frozen sections of skin. There appeared to be a general, although nonquantitative, association between fluorescence intensity in sebaceous glands and carcinogenicity in epidermal cells; however, it will be necessary to examine a greater number of samples to establish such a correlation.

The feasibility of using athymic mice to compare the relative *in vivo* susceptibility of intact human and mouse skin to carcinogenic hydrocarbons has been evaluated. Although the approach is technically feasible, the small portion of grafts which survive indefinitely is evidence that many technical improvements will be necessary before this approach can be exploited.

INTRODUCTION

The production of liquid hydrocarbons from coal or oil shale is inevitable. The only questions are when and by what methodology. Historical evidence suggests that crude shale oil[1] and coal-derived liquids[2] are more potent skin irritants and carcinogens than natural petroleum. To determine whether similar materials, produced by contemporary technologies, have the same potentials for harm, it will be necessary to compare their relative carcinogenicities quantitatively using experimental animals. Application of the animal bioassay data to the question of human occupational exposure will require measurement of the degree to which mouse skin differs from human skin in its responsiveness to topically applied whole crudes. Although the availability of genetically defined athymic nude mice[3] may facilitate this comparison, many technical problems remain unsolved. This presentation relates findings of exploratory experiments and describes one approach to determining the bio-availability of syncrudes in the intact skin.

MATERIALS AND METHODS

The materials used in these experiments were obtained from various sources, whose cooperation is gratefully acknowledged. The samples included two coal liquids: *Coal liquid A* was a centrifuged oil produced by a synthoil process under development and was provided by the Pittsburgh Energy Research Center. *Coal liquid B* was a heavy oil produced by the COED process from Western Kentucky Coal and was provided by the FMC Corporation. In addition to the two prototype coal liquids a *shale oil crude*

*Research sponsored by the Division of Biomedical and Environmental Science, U.S. Department of Energy, under contract W-7405-eng-26 with Union Carbide Corporation.

†Biology Division, Oak Ridge National Laboratory, Oak Ridge, Tenn. 37830.

produced in the 150-ton simulated *in situ* retort run was provided by the Laramie Energy Research Center. To place the data in perspective a *composite natural petroleum* was included that consisted of the following natural crudes, by volume: 20% Wilmington, California; 20% South Swan Hills, Alberta, Canada; 20% Prudhoe Bay, Alaska; 20% Gach Sach, Iran; 10% Louisiana-Mississippi Sweet; and 10% Arabian light. The composite natural crude was provided by the Laramie Energy Research Center.

A mixed solvent consisting of 70% acetone, 30% cyclohexane (AC), by volume, was found that would satisfactorily suspend (with sonication) or dissolve all of the materials. For the skin fluorescence studies, solutions of each of the materials were placed in preweighed glass vials, and the AC solvent evaporated to constant weight with a gentle stream of air under ambient conditions. The weight of "nonvolatile" solids was determined and an appropriate volume of solvent added to yield solutions containing 20 mg/ml of total solids. Skin carcinogenesis studies were done with the native materials (without adjusting for total solids) by the addition of the AC solvent to a final solute concentration of 50% (w/v). The resulting dilute crudes were sonicated briefly to achieve dispersion. Fresh solutions were prepared weekly.

For the fluorescence studies 10–12-week-old male C3Hf/Bd mice were shaved, and those determined not to be in active hair growth were selected. Fifty μl of the 20 mg/ml solution was applied to the dorsal skin. In separate experiments this amount of AC solvent covered an area of approximately 5 cm^2. At 1, 4, and 24 hours and 8 and 14 days after exposure, three mice treated with each material were killed by CO_2 inhalation. Exposed skin was excised, frozen in liquid nitrogen, and sectioned by cryostat. Sections were mounted in a nonfluorescing medium, and fluorescence intensity was determined using a procedure described elsewhere in greater detail.[4]

Dermatologic effects were evaluated by Monday, Wednesday, and Friday applications of 50 μl of the 50% solutions to the shaved dorsal skin of 30 C3Hf/Bd mice. Separate groups were treated with each material commencing at 10–12 weeks of age. Exposure continued for 22 weeks, after which time the mice were held for an additional 22 weeks without additional manipulation other than periodic shaving. During both the exposure period and the follow-up, the time of first skin tumor observation and character of non-neoplastic skin changes were recorded. Progression of individual lesions was monitored and their locations, relative sizes, and characters recorded on scale drawings. Mice dying throughout the study or killed at its termination were necropsied. Tissues were subjected to histologic examination to establish (1) the identity and specificity of neoplasms arising in the treated skin; (2) the fraction of the mice with tumors, metastatic either to regional lymph nodes or distant sites; and (3) the detection of systemic injury arising as a consequence of percutaneous absorption of the applied materials.

RESULTS

Equivalent amounts of the materials applied at the same dose rate resulted in widely differing skin responses. The data summarizing the clinical and anatomic responses are given in Table 1. Of interest is the lethality of the squamous epidermal tumors induced, only a portion of which could be attributed to systemic metastasis. Of 30 mice exposed to shale oil, 14 developed progressive epidermal neoplasms, and 10 of these died before the end of the study, four with histologically confirmed metastasis. From the first observation of a tumor, average time to death was similar for all materials.

Table 1. *In Vivo* effects of continuous topical application of various crude petroleums

Compound	Skin tumor incidence[a] (%)	Tumor latency (days ± S.E.)	Deaths	% Survival		Average time to death with tumor (days ± S.E.)	Number of animals with ulcerative dermatitis	
				Tumor bearer	Nontumor bearer		At 22 weeks	At 44 weeks
Shale oil	47	154 ± 9	11	28	94	117 ± 10	9	2
Coal liquid A	63	149 ± 8	6	68	100	103 ± 11	0	0
Coal liquid B	37	191 ± 14	1	91	100	90	20	5
Composite crude	0	0	0		100		0	0

[a]Squamous epidermal tumors present either at death or the end of the experiments.

In addition to inducing autonomous and progressive epidermal neoplasms, the materials were also capable of inducing dermatitis with varying degrees of efficiency. In its mildest form, skin irritation was reflected by transient epilation and well demarcated depigmentation of the exposed skin, immediately followed by stimulation of the air follicle and a subsequent persistent hypertrichosis out of phase with the systemic hair cycle. These changes were manifest in mice exposed to coal liquid A and the composite natural crude. More severe irritation occurred with shale oil and coal liquid B. With these materials, hair growth stimulation progressed into a frank epidermal ulceration, permanent epilation, and cicatrization. Table 1 also indicates the number of mice, out of the starting group of 30, that developed ulcerative dermatitis following exposure to each material. The incidence is given at the end of exposure (22 weeks) and again at the end of the clinical follow-up (44 weeks). Histologic studies of the ulcerative lesion revealed no evidence of occult malignancy. The central, denuded area consisted of a granulation tissue covered by a thin, insipissated, crust. Adenexal structures were absent from the dermis. The absence of hair follicles suggests that failure of the lesions to heal was the result of loss of epithelial stem cells which normally would participate in reepithelialization. The denuded dermis has responded by fibrosis. Healing of the damaged area, if and when it eventually occurs, is dependent upon contraction of connective tissues which brings the wound margins into opposition.

It is likely that compositions of these materials would influence their solubility in skin and consequently their biological activity. To obtain a measure of the skin penetrability, distribution, and persistence of these materials, we have used their natural fluorescence to follow tissue distribution. Our working hypothesis is that biological activity should be positively correlated with the magnitude and persistence of fluorescence and possibly also with its localization.

The results of these measurements are given in Table 2. Several differences are readily apparent. Coal liquid A had greater initial fluorescence than any of the other materials and was also observed to persist for at least 14 days following a single application of approximately 200 $\mu g/cm^2$ total solids. At the same concentration, clearance of both shale oil and the composite crude appeared to be more rapid than either coal liquid. Although there appears to be a positive association among total fluorescence, persistence of fluorescence, and carcinogenicity, a quantitative correlation cannot be inferred on the basis of these limited data. For example, shale oil and natural petroleum exhibited similar fluorescence intensities and skin clearances, but shale oil was much more carcinogenic in mouse skin. The most probable explanation for these discrepancies is that fluorescence is not totally correlated with carcinogenicity. Obviously, if fluorescence is to be used as a method for monitoring the work place[5], it will be necessary to determine the degree to which it is correlated and thus predictive of carcinogenicity for each specific composite exposure.

In order to compare human and mouse dose response relationships for known carcinogens, it has been necessary to compare experimental data with epidemiologic data, a comparison that is never easy and usually retrospective. To avoid these limitations we propose determining the species response variable directly. Our approach will be to graft normal human skin onto athymic nude mice. After healing is complete, the grafted human skin will be challenged with pure carcinogenic hydrocarbons as well as syncrudes in order to compare the relative sensitivity

Table 2. The intensity of fluorescence in sebaceous glands at various times after topical application of petroleum liquids[a]

Time after application	Average fluorescence intensity[b]			
	Coal liquid A	Coal liquid B	Shale oil	Natural oil
1 h	540(123)	172(49)	70(21)	137(44)
4 h	422(101)	185(61)	55(12)	140(88)
24 h	454(114)	190(24)	109(21)	135(40)
8 d	112(21)	43(6)	29(1)	27(1)
14 d	42(11)	23(1)	25(1)	25(1)

[a] After application of approximately 200 $\mu g/cm$
[b] In $10^3 s^{-1}$. The parenthetical figure is the $\pm 95\%$ confidence limit.

of human skin to that of similarly grafted mouse skin *in vivo*.

Two exploratory experiments have been done to determine the feasibility of this approach. Groups of specific pathogen-free athymic mice were grafted with approximately 2-cm elliptical pieces of adult human skin. In one experiment full thickness skin was used. The only preparation was removal of the subcutaneous fat. In the second experiment skin thickness was reduced to approximately 1 mm using an electro-keratome. The mice were anesthetized with 0.1% pentobarbital, graft beds prepared, and human skin implanted under aseptic conditions. Bandages were removed at 11 days and the status of the grafts monitored periodically. After an initial slough of the original donor tissue, reepithelialization was observed. Mouse and human skin could be easily distinguished in that the mouse skin was albino, and the human skin was pigmented. Over the succeeding weeks the majority of human skin grafts underwent progressive contracture and eventually were lost. Disappearance of the grafts was not accompanied by clinical evidence of an active or inflammatory rejection process. Our current hypothesis to account for loss of the grafts is that connective tissue, induced during the initial healing process, contracts and decreases the blood supply to the graft which, deprived of nutrition, slowly recedes. An assessment of the persistence and integrity of the grafts from each of the initial studies is given in Table 3. It should be noted that in spite of many known and unknown technical problems, a small percentage of the grafts in each of the experiments has remained indefinitely.

DISCUSSION

The present paper demonstrates that syncrudes differ greatly from one another as well as from natural petroleums, both as skin irritants and carcinogens, *in vivo*. The index of relative carcinogenicity[6] in order of decreasing potency was 42, 30, 19, and 0 for coal liquid A, shale oil crude, coal liquid B, and natural petroleums, respectively. These are considered to be minimum estimates since they are based upon a single, high dose. It is possible that at the 50% concentration used to induce tumors, expression of neoplasms could have been prevented or delayed in one or more of the test groups by toxic constituents of the mixtures.

The observed lethality of the tumors was unexpected. In previous experiments in which various hydrocarbon and non-hydrocarbon carcinogens have been applied to the same mice under identical conditions, the epidermal tumors induced, while locally malignant, seldom contributed to death.[7] An important difference between this and previous experiments was that in the present experiment exposure to the skin carcinogen was discontinuous. Although there are a number of possible explanations for the observed lethality, the hypothesis we favor is that discontinuous or short term exposure, coupled with an extended clinical follow-up, encourages more rapid clinical progression of the incipient neoplasms, possibly because toxic or inhibitory constituents (which may or may not include the active carcinogens) of the mixtures may kill or inhibit the growth of neoplastic epidermal cells. If future experiments verify this observation,

Table 3. Survival of full-thickness human skin grafts on specific pathogen-free athymic nude mice

Experiment	Time of observation (d)	Effective number of mice	Number of animals with grafts		
			Retained	Questionable	Rejected
1	11	44	44		
	22	44		43	1
	49	44	12	26	6
	131	44	8		36
	225	16	7(2)[a]		9
2	11	50	50		
	94	50	20(4)[a]		30

[a]The parenthetical numbers indicate the animals which retained human epidermis as well as dermis.

then discontinuous exposure may be the preferred means to assess the carcinogenicity of complex mixtures.

Fluorescence of the materials in skin has proven to be a convenient and relatively precise method to detect the presence as well as demonstrate the distribution and persistence of the various petroleum crudes in mouse skin. Many refinements of the approach used are possible and are being actively pursued. We are interested in calibrating the method using known reference carcinogens such as benzo[a]pyrene. It may also be possible to use selective, narrow-bandpass filters to achieve greater specificity.

Addressing the question of species-dependent differences in target tissue susceptibility by using athymic nude mice grafted with human skin would appear feasible. However, after somewhat disappointing long-term graft retentions, we anticipate the many technical issues that remain to be resolved before meaningful data can be obtained.

REFERENCES

1. A. Scott, "The Occupation Dermatoses of the Paraffin Workers of the Scottish Shale Oil Industry, with a Description of the System Adopted and the Results Obtained at the Periodic Examinations of These Workers," in *Eighth Scientific Report of the Imperial Cancer Fund*, pp. 85–142, Taylor and Frances, London, 1923.

2. R. J. Sexton, "The Hazards to Health in the Hydrogenation of Coal, IV: The Control Program and the Clinical Effects," *Arch. Environ. Health* 1, 208–31 (1960).

3. E. M. Pantelouris, "Absence of the Thymus in a Mouse Mutant," *Nature* 217, 370–71 (1968).

4. J. M. Holland, M. S. Whitaker and J. W. Wesley, "Correlation of Fluorescence Intensity and Carcinogenic Potency of Synthetic and Natural Petroleums in Mouse Skin," *Am. Ind. Hyg. Assoc. J.,* in press, June 1979.

5. N. H. Ketcham and R. W. Norton, "The Hazards to Health in the Hydrogenation of Coal, III: The Industrial Hygiene Studies," *Arch. Environ, Health. Health* 1, 194–207 (1960).

6. J. Iball, "The Relative Potency of Carcinogenic Compounds," *Am. J. Cancer* 35, 188–90 (1939).

7. J. M. Holland, D. G. Gosslee and N. G. Williams, *Epidermal Carcinogenicity of Bis(2,3-Epoxycyclopentyl)Ether, 2-2-Bis (p-Glycidyloxyphenyl) Propane, and m-Phenylenediamine in C3H*

and C57BL/6 Inbred Male and Female Mice, Cancer Res. 39, 1714–21 (1979).

DISCUSSION

David Coffin, Environmental Protection Agency-Research Triangle Park, North Carolina: I just wanted to say that this is an excellent presentation, and I think it is good to see an integrated approach to these things. However, I had one comment in regard to the basic fraction and its importance in this general scheme. This has to do with work we did back in the middle sixties with material from the air in which we collected large amounts of airborne particulates in New York City and extracted them. We found the basic fraction in this material was, on a weight-to-weight basis, by far the most reactive as measured by action in a system that we called photodynamic bioassay in paramecia (an indirect measure of carcinogenesis—at least we thought at that time it was) and by injection into neonate or newborn mice. We found that the basic fraction was by far the most reactive in airborne material. However, it represented only about 2% of the organic extract, which, of course, in the total picture is not very great. These tumors that we elicited in these newborn mice were largely in the liver, which suggested, I think, that we were dealing with some of the heterocyclic compounds. I thought that this was of interest in view of the report here that the basic fraction in coal liquids is so important. But the only question I had is what proportion of the total organic extract or the positive compounds does the basic fraction represent in coal?

Jim Epler, Oak Ridge National Laboratory: I suspect that as a general answer to that, the basic fractions would amount to somewhere in the range of 2 to 10% on a relative weight basis. Additionally, we have taken the basic material and run other fractionation involving additional steps of column chromatography and have been able to isolate the bulk of the mutagenic activity in about half a percent of the total starting material. In other words, an increase in specific activity, if you will, of almost an order of magnitude.

Mike Guerin, Oak Ridge National Laboratory: Dave, the basic fraction only constituted anything from 2% to 10% of some of these materials, but in terms of its total contribution of mutagenicity over the materials we've looked at, it ranges anything from 15% of the mutagenicity that is caused by this fraction to, in one case, 85%, even though it's only in that case about 7% of the material.

142   J. M. Holland, M. S. Whitaker, and J. W. Wesley

Harvey Drucker, Pacific Northwest Laboratories: On the skin-painting studies, are the dose-response curves for the various syncrudes about the same, or do they differ?

Mike Holland, Oak Ridge National Laboratory: Dr. Drucker, I think that we really don't have sufficient data to answer you directly. Two-year studies that were begun slightly less than two years ago are about to reach their fruition, and I think the next time we have one of these get-togethers that information will be available. Our expectation is that if we titer out toxicity, we titer out interfering side reactions and that those response relationships will look comparable. In other words, the slopes will be parallel. I think that is the answer that we expect to see, but I want to hedge.

A PHASED APPROACH TO THE BIOSCREENING OF EMISSIONS AND EFFLUENTS FROM ENERGY TECHNOLOGIES*

M. D. Waters [†]

ABSTRACT

Short-term bioassays are being applied together with chemical analytical procedures in the detection and evaluation of potentially hazardous emissions and effluents from conventional and developmental energy technologies. The reliability of these screening tests is improved when they are combined with other short-term bioassays and employed in concert as batteries of tests. Short-term test batteries for genotoxic effects should evaluate point mutations, chromosomal aberrations, primary damage to DNA, oncogenic transformation *in vitro*, and toxicity related to each of these effects. Tests such as these may be combined effectively into a text matrix which permits a stepwise or phased evaluation of energy-related effluents. Such an approach involving the iterative application of biological and chemical procedures facilitates a cost-effective utilization of limited resources.

INTRODUCTION

The National Energy Plan has affirmed that "the U.S. and the world are at the early stage of an energy transition." It is indeed this transition and the future reliance on multiple sources of energy that has provided the problem and the challenge of "bioscreening" for health and ecological effects.

During the next ten years the major contribution to energy-related emissions and effluents will be of a conventional nature. We will see increased combustion of coal by conventional methods together with the gradually increased utilization of alternative procedures involving fossil fuel conversion.[1] Our bioassay methods must, therefore, be geared to the evaluation of conventional sources as well as to the newly evolving energy conversion technologies. We have made significant progress in both areas.

Short-term bioassays are being applied effectively in the detection and evaluation of potentially hazardous emissions and effluents from conventional and developmental energy sources.[2] Biological screening tests such as the Ames *Salmonella*/microsome assay[3] have demonstrated their utility (1) as indicators of potential long-term health effects such as mutagenesis and carcinogenesis; (2) as a means to direct the fractionation and identification of hazardous biological agents in complex mixtures; (3) as a measure of relative biological activity to be correlated with changes in process conditions; and (4) to establish priorities for further confirmatory short-term bioassays, testing in whole animals, and more definitive chemical analysis and monitoring.

Clearly, however, these tests do not circumvent the need for conventional toxicological, clinical, and epidemiological evaluation. Likewise it is not possible to divorce our vested concern for human health effects from a more basic concern for the welfare of the ecological environment.

Although our concern is with health as well as ecological effects, this paper will examine the applicability of short-term bioassays in studies on the potential health effects of energy-related technologies. This examination will be carried out in the context of the total environmental assessment problem, i.e., the total impact of new technology on man and his environment. This paper suggests a phased approach to the bioscreening of emissions and effluents from energy technologies, which will be realized through the stepwise application of a matrix of currently available short-term bioassays.

---

*Research supported by the Environmental Protection Agency, Office of Energy, Minerals, and Industry.

†Biochemistry Branch, Environmental Toxicology Division, Health Effects Research Laboratory, Research Triangle Park, North Carolina.

143

ENVIRONMENTAL ASSESSMENT

The introduction of emissions and effluents into the environment from energy-related technologies creates a multifaceted toxicological problem. The problem is a function of the environmental media, i.e. air, water, and food and of the routes of exposure. Indeed, mechanisms of toxic effects may be highly media- or route-specific. Most exposures to hazardous agents in the environment are multiexposures; that is, there are very few instances in which a singular substance or a single exposure is solely responsible for adverse health or environmental effects. Each of these exposure factors must be taken into account in assessing the potential health effects of energy-related technologies.

Man is but a part of a total ecosystem. His well being is ultimately dependent upon the well being of the system as a whole. Hence, it is critical to evaluate health effects in man in the light of relevant and significant perturbations of the ecosystem, especially those which directly influence exposure media including food, water, and air.

Environmental effects may be slow to manifest themselves. Thus, our current methods of disposal and recycling of toxic wastes and byproducts may ultimately compound the health and ecological effects of substances introduced into the environment. We need to have detailed knowledge of the transport and fate of environmental toxicants. Careful consideration of these factors is as important as measuring the toxicity of the industrial discharge waste stream and must ultimately be included in the assessment of the environmental impact of new technologies.

We have considered the components of the environmental assessment problem in the broadest sense. The development of a rapid, effective, and inexpensive means for evaluating the potential health hazards associated with emissions and effluents from energy-related technologies is a critical step in this process. Short-term bioassays provide a means to this end but do not offer an independent solution to the assessment problem.

Dual Role of Chemical and Biological Analysis

It is clear that chemical and biological analyses have a dual role in the process of environmental assessment. Each of these disciplinary approaches has its advantages and limitations. Chemical analysis, while indicative, cannot provide sufficient data for complete evaluation of potential pollutant effects, because the biological activity of complex samples cannot be consistently predicted. Although bioassays can indicate the biological activity of a given sample, they cannot specify which components of a crude sample are responsible for the observed toxicity.

In considering the application of analytical chemical methodology, it is important to realize that the presence or absence of a known toxic component within a complex sample neither indicates nor precludes a relationship between that component and the biological activity of the sample. To avoid the possibility of overlooking unanticipated biologically active components, chemical analysis should not be restricted to determination of preselected known toxic compounds or suspected hazardous components of a complex sample. Indeed, it may be technically overwhelming to analyze for all of the known or suspected hazardous components in a large number of complex samples. For this reason and for cost effectiveness, it can be argued that a phased approach involving stepwise application of biological and chemical methodology is appropriate in evaluating the potential health hazard of complex mixtures. Such a phased approach would involve the iterative application of short-term bioassays to determine which samples are biologically active, together with chemical fractionation and analysis to ascertain which agents are responsible for the observed effects.

The theme of the recent Williamsburg symposium[4] sponsored by the EPA Office of Energy, Minerals, and Industry through the Biochemistry Branch of EPA's Health Effects Research Laboratory, Research Triangle Park, North Carolina, was that short-term bioassay techniques can be used effectively to assess the biological activity of complex samples and their components. Assays for toxicity, mutagenesis, oncogenic transformation, and related effects are being applied to an array of complex samples including ambient air, water, and food, automotive emissions, industrial emissions, coal and its combustion products, and natural and synthetic oil products.

In a number of cases it has been necessary to fractionate such complex mixtures chemically in order to separate the toxic from the mutagenic or potentially carcinogenic species. In such cases, an iterative process involving alternate application of biological and chemical procedures proved most effective in isolating the biologically active species.

The work of Epler, Guerin, and their associates at Oak Ridge National Laboratory is the best example of this collaborative approach as it pertains to synthetic fuels technology.

An Environmental Source Assessment

The Industrial Environmental Research Laboratory of the Environmental Protection Agency at Research Triangle Park, North Carolina (IERL-RTP), has delineated a three-phased approach to performing an environmental source assessment, i.e., the evaluation of feed and waste streams of industrial processes in order to determine the need for control technology.* Each of the three phases involves a separate sampling and analytical procedure.

According to the plan outlined by IERL-RTP,[5-7]

The first level (1) provides preliminary environmental assessment data, (2) identifies problem areas, and (3) generates the data needed for the prioritization of energy and industrial processes, streams within a process, and components within a stream for further consideration in the overall assessment. The Level 2 sampling and analysis effort is designed to provide additional information that will confirm and expand the information gathered in Level 1. Level 1 results serve to focus Level 2 efforts. The Level 2 results provide a more detailed characterization of biological effects of the toxic streams, define control technology needs, and may, in some cases, give the probable or exact cause of a given problem. Level 3 utilizes Level 2 or better sampling and analysis methodology in order to monitor the specific problems identified in Level 2 so that the toxic or inhibitory components in a stream can be determined exactly as a function of time and process variation for control device development. Chronic, sublethal effects are also monitored in Level 3.

To meet the environmental source assessment requirement of comprehensiveness, the IERL-RTP phased approach provides for physical, chemical, and biological tests. Physical and chemical characterization of environmental emissions is critical to the definition of need for and design of control technology. However, the final objective of the Industrial Environmental Research Laboratory's environmental assessment is the control of industrial emissions to meet environmental or ambient goals that limit the release of substances that cause harmful biological (health and ecological) effects. Consequently, the testing of industrial feed and waste streams for biological effects is needed to complement the physical and chemical data and ensure that the assessment is comprehensive.

This effort is cited as an example of a large-scale phased application of short-term bioassays and chemical analytical techniques to the analysis of complex environmental samples. It represents an important beginning in an area of research and application that will require continued intensive collaboration among engineers, chemists and biologists if success is to be achieved. The results of Level 1 pilot studies on samples from textile wastewater, fluidized-bed combustion, and coal gasification processes are still being analyzed; however, a number of conclusions can be drawn from the results obtained thus far:

1. In general, chemical analytical techniques are quantitatively more sensitive than are the health-effects bioassays.

2. Biological activity, especially genetic activity, may be masked by toxicity and may require chemical fractionation before it is demonstrable.

3. In some cases, tests for potential ecological effects may be more sensitive and more critical than tests for potential health effects.

4. The prediction of relative toxicity on the basis of chemical analysis alone is subject to error.

5. Results of biological and chemical tests are complementary—considered together the two types of tests provide useful information not obtainable when considered separately.

A MATRIX OF SHORT-TERM BIOASSAYS

There now exists a matrix of short-term health effects bioassays which can be applied in a phased or stepwise manner in the biological evaluation of energy-related samples. As with the IERL-RTP example, any program aimed at identifying and reducing release of hazardous emissions or effluents requires the utilization of inexpensive short-term bioassays to assign priorities to samples for further evaluation by conventional toxicological procedures. However, some short-term bioassays are known to be insensitive to specific chemicals or classes of chemicals. Indeed, no single bioassay is adequate to monitor all types of chemical and biological activity. The problem of potential false negatives may be alleviated by the use of a required core battery of tests and by the identification, prior to biological testing, of those samples which contain chemicals structurally or physicochemically similar to known false negative agents or classes of agents. Such agents or

---

*L. Johnson and J. A. Dorsey, "Environmental Protection Agency Level 1 Chemistry: Status and Experiences," in this proceedings.

fractions of complex samples might then be selected for higher level testing without the need for preliminary screening.

In the areas of mutagenicity and carcinogenicity testing, several variations of the tiered or phased approach have been discussed in the literature.[8-10] However, there is considerable agreement on the essential point of emphasis at each level of evaluation and on the need to employ a battery of tests to detect various genotoxic effects, i.e., those effects involving damage to the genetic material.

For purposes of illustration and example we have organized a number of the existing bioassays into a three-phased matrix based primarily on the endpoint measured, cost and complexity. Although the short-term bioassays that will be mentioned are not to be considered exclusive, they do represent state-of-the-art methodology. Figure 1 depicts the kinds of bioassays to be employed in the biological evaluation of energy-related technologies. A battery of tests is suggested at each level of evaluation. The batteries are designed to perform a specific function in each of the three phases of the text matrix. The emphasis in the Phase 1 test battery is on the *detection* of acute toxicity using mammalian cells in culture and intact animals and genotoxic effects including point mutation and primary DNA damage in microbial species, and chromosomal alterations in mammalian cells in culture. The Phase 2 battery is designed to *verify* the results from Phase 1 tests by employing higher level toxicity tests involving mammalian cells

in culture and intact mammals and genotoxicity assays using plants, insects, and mammals. Genotoxicity assays at Phase 2 are separated into tests for mutagenicity *per se* and specific tests for carcinogenic potential. Phase 3 testing is devoted to quantitative *risk assessment* using conventional toxicological methods. For the purpose of defining a probable negative result for genotoxicity, the "core" battery of short-term tests is most important.

Core Battery of Tests for Genotoxic Effects

Because no single test is capable of indicating all of the various types of biological activity which may be relevant to the processes of mutagenesis and carcinogenesis, it is generally held that a battery of short-term tests should be performed. The battery approach is intended to reduce "false negatives" to a minimum and thus assure reasonable protection of human health. Batteries of tests have been proposed in the development of EPA's Pesticide Guidelines for Mutagenicity Testing and in the Consumer Product Safety Commission's "Principles and Procedures for Evaluating the Toxicity of Household Substances."[11] These documents reflect the thinking expressed in the Committee 17 Report on "Environmental Mutagenic Hazards"[12] and in the report of the working group of the DHEW Subcommittee on Environmental Mutagenesis "Mutagenic Properties of Chemicals: Risk to Future Generations."[13] Indeed, there is considerable agreement that a core battery of tests for mutagenic

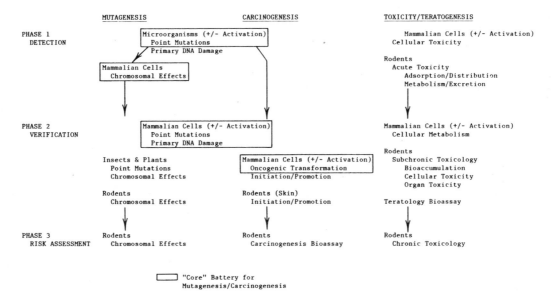

Fig. 1. A phased approach to bioscreening for environmental health effects.

and carcinogenic effects should include, as minimum, tests for point mutation in microorganisms and mammalian cells in culture; a test for chromosomal alterations, preferably an *in vivo* test; a test for primary damage to DNA using mammalian, preferably human, cells in culture and a test for oncogenic transformation *in vitro*. Such a battery of tests might be considered to represent the "core" or the most essential of the genotoxicity tests in the phased evaluation process. Redundancy in the test battery is considered desirable until a more complete data base of test results has been assembled. To aid in interpretation, it is also necessary to ascertain the influence of cellular toxicity in these tests. The following is a description of the kinds of biological activity detectable in short-term tests comprising the "core" battery.

Biological Activity Detected in the Core Battery

Point mutations

Point mutations are alterations which affect single genes. These alterations include base pair substitutions and frameshift mutations as well as other small deletions and insertions. Applicable test systems include both forward and reverse mutation assays in bacteria, yeast, and mammalian cells in culture. Most of these assays employ an exogenous source of metabolic activation provided by a mammalian liver microsomal preparation. It has become apparent that a majority of the genotoxins are procarcinogens or promutagens which must be converted into their reactive forms before their effects can be evaluated. The metabolic conversion is believed to be mediated by oxidative enzymes and to involve the formation of reactive electrophilic metabolites which bind covalently to DNA. Gene mutation assays which incorporate whole-animal metabolic activation (e.g., urine screening) are very desirable since it is not possible to ensure metabolic fidelity in entirely *in vitro* systems. One must employ intact animals to demonstrate the heritability of mutational effects.

Chromosomal alterations

Chromosomal alterations include the loss or gain of entire chromosomes, chromosome breaks, nondisjunctions and translocations. Short-term tests for these abnormalities involve searching for chromosomal aberrations in somatic and germinal cells usually from insects and mammals. Chromosomal aberrations observed in germinal tissues of

intact animals provide important evidence of the accessibility of the test chemical to the reproductive organs.

Primary damage and repair in informational macromolecules

Damage and repair bioassays do not measure mutation directly but do measure the direct damage to DNA and other macromolecules by chemical agents and its subsequent repair. Bioassays to detect macromolecular damage and repair are available using bacteria, yeast, mammalian cells and whole animals. Except for the whole-animal bioassays these systems generally employ an exogenous source of metabolic activation.

Oncogenic transformation *in vitro*

Oncogenic transformation is the process whereby normal cells grown in culture are converted into malignant cells after treatment with a carcinogen. The demonstration of malignancy (tumor formation) can be observed by injecting the transformed cells into whole animals, although this is not an obligatory requisite for oncogenic transformation. A number of mammalian oncogenic transformation bioassays utilizing cells derived from different rodent species are currently available. Some of these cell systems have the endogenous capability to activate procarcinogens, and with others exogenous microsomal activation has been used successfully.

Toxicity *in vitro*

An initial requirement in mammalian cell mutagenesis and oncogenesis bioassays is the determination of the lethal toxicity of each test agent. This information may be used to establish the range of concentrations to be employed in the mutagenesis or oncogenesis assays and to quantify the observed mutation or transformation frequency in terms of the number of cells surviving the treatment.

Utility of Genotoxicity Tests as Predictive Tools

For these genotoxicity tests, two critical issues are the validity of the tests and the interpretation of the results.

Carcinogenesis

At the present time, microbial mutagenesis test systems are most widely used to prescreen substances for potential oncogenicity. Tests for gene or point mutations in microorganisms, as for example those involving the *Salmonella typhimurium*/microsome system, have been found to be highly predictive of oncogenic potential.[14] Indeed, most chemical mutagens which have been adequately tested have been found to be oncogenic in whole-animal bioassays. It is well established that most but not all oncogens are mutagens when appropriate metabolic activation is provided in the short-term tests. Research on test systems which permit sequential evaluation of mutagenesis and oncogenic transformation has enhanced our understanding of the relationship between these two phenomena.

Oncogenic transformation *in vitro* is considered to be directly relevant to the process of tumor formation in the intact animal. Few, if any, false positives are detected using this methodology.[15] However, because of the laborious nature of cell transformation assays it may not be feasible, because of time or fiscal constraints or the availability of facilities, to put a very large number of samples through such testing procedures immediately. It is for reasons such as these that oncogenic transformation assays are generally considered to be higher level tests.

When more fully developed, human-cell bioassays for oncogenic transformation may afford a final short-term test for substances found positive in phylogenetically lower organisms. This will be true especially if epithelial cell systems can be developed which retain their metabolic activation capability. Those chemicals which produce positive responses in human cell systems might be given highest priority for evaluation in conventional whole-animal oncogenesis bioassays.

Mutagenesis

The fundamental concern in mutagenesis testing is the risk to future generations. Alterations of the genetic material in germinal cells are rarely expressed in the exposed individuals. These alterations may not become apparent for several generations, but they contribute to an increased genetic burden in the exposed population. The observation and quantitation of these mutational effects in germinal tissues requires the use of intact animals, e.g., the sex-linked recessive lethal test in the insect *Drosophila*.[16] It would be highly desirable to include such a test in the core battery if significant human exposure is anticipated to a suspect mutagen.

The usefulness of cells other than germinal cells as a predictive tool is judged to be high for certain kinds of genetic alterations. Microbial cells are widely used

to detect point mutagens. Many of these systems have been "genetically engineered" to enhance their sensitivity as detection systems. Mammalian and human cells in culture can provide more relevant information on the ability of the substance to induce both point mutations and chromosomal alterations. It is important to evaluate both types of potential genetic activity. Chromosomal alterations are best evaluated in intact animals, but inexpensive whole-animal tests for point mutagens are lacking.

Ancillary effects

Tests which detect primary damage to informational cellular macromolecules (e.g., DNA) have been found to show moderate to high correlation with mutagenic and oncogenic potential as indicated by animal bioassays.[17,18] Microbial tests in this category are extremely rapid and inexpensive and offer the possibility of examining various manifestations of macromolecular damage. Mammalian and human cell-culture bioassays for primary DNA damage offer the possibility of detecting macromolecular damage in tests which have demonstrated promising correlations with mutagenic and oncogenic potential as evaluated using experimental animals.[19] Several of these systems permit concomitant measurement of primary DNA damage, point mutation, chromosomal effects and cellular toxicity.

Strategy for the Employment of a Phased Approach

As indicated previously, a phased or stepwise mode of application has been favored as a cost-effective approach to the bioscreening of large numbers of complex environmental samples and their components.

A strategy for the employment of short-term bioassays based on biological activity, cost, and complexity may be delineated as follows:

In Phase 1, tests representing each kind of biological activity would be performed. The extent of redundancy in testing within a category of biological activity would be dictated by a number of factors including production volume, anticipated human exposure, known hazards of feed stocks, etc. If any of these tests prove positive, the appropriate follow-up tests would be pursued in Phase 2 and, if required, in Phase 3 depending upon the degree of associated risk. If those tests were to prove negative, no further testing would be performed unless there were over-riding considerations, such as mentioned previously. In such cases a "core" battery of tests for genotoxicity would be completed with negative results before short-term testing would cease. Extensive health risk could entail further long-term testing to define a negative result. This approach would facilitate a cost-effective utilization of limited testing resources and would at the same time provide protection for human health in proportion to the anticipated risk involved.

Specific Bioassays

Specific tests may be organized within the test matrix as described in the following sections. It should be emphasized that these bioassays are cited as examples and should not be considered exclusive.

Phase 1 bioassays

A battery of short-term tests for this phase is illustrated in Fig. 2. As mentioned previously, the emphasis at this level of testing is on detection of mutagens, potential carcinogens, and acutely toxic chemicals in a battery of *in vitro* and *in vivo* tests. The results obtained from Phase 1 tests are used to assign

POINT MUTATIONS

° SALMONELLA/MICROSOME (AMES) REVERSE MUTATION, PROTOTROPHY TO HISTIDINE
° ESCHERICHIA COLI-WP2/MICROSOME REVERSE MUTATION, PROTOTROPHY TO TRYPTOPHANE
° SACCHAROMYCES CEREVISIAE REVERSE AND FORWARD MUTATION

CHROMOSOMAL EFFECTS

° IN VITRO CYTOGENTICS
    CHINESE HAMSTER OVARY CELLS
    WI-38 HUMAN FIBROBLASTS

PRIMARY DNA DAMAGE

° ESCHERICHIA COLI POL A⁻ REPAIR DEFICIENT STRAINS
° BACILLUS SUBTILIS, REC⁻ REPAIR DEFICIENT STRAINS
° SACCHAROMYCES CEREVISIAE GENE CONVERSION AND MITOTIC RECOMBINATION

CYTOTOXICITY

° RABBIT ALVEOLAR MACROPHAGE (FOR PARTICULATES)
° CHINESE HAMSTER OVARY CELLS
° WI-38 HUMAN LUNG FIBROBLASTS

RODENT ACUTE TOXICITY

Fig. 2. Phase-1–short-term bioassays for mutagenesis-carcino-genesis-toxicity.

priorities for further testing in appropriate confirmatory bioassays at Phase 2. The *in vitro* end points which are considered, based upon expense, complexity, and the current level of development of bioassay systems, are point mutations, chromosomal alterations, primary DNA damage, and cellular toxicity. Redundancy is desirable to the extent which is economically feasible. All bioassays are performed with and without mammalian metabolic activation systems where appropriate. Conventional rodent acute toxicity tests are considered essential at Phase 1 in view of the limitations of cytotoxicity screening tests. The latter tests cannot represent intact animals but provide useful preliminary information about the relative cellular toxicity of selected samples (e.g. airborne particulate materials).[20] Rodent acute toxicity tests can, in addition, provide a source of body fluids and tissues to be examined for the presence of active mutagens and carcinogens by the use of short-term genotoxicity bioassays.

Phase 2 bioassays

As mentioned previously, the Phase 2 tests illustrated in Fig. 3 are designed to verify the results obtained in Phase 1. The test systems are

POINT MUTATIONS

° MAMMALIAN CELLS IN CULTURE (CHO, L5178Y, V79)
° INSECTS-DROSOPHILA
° PLANT-TRADESCANTIA AND MAIZE

CHROMOSOMAL EFFECTS

° IN VIVO CYTOGENETICS - LEUCOCYTE CULTURE AND BONE MARROW CELLS

PRIMARY DNA DAMAGE

° UNSCHEDULED DNA SYNTHESIS (WI-38)
° SISTER-CHROMATID EXCHANGE FORMATION (IN VITRO AND IN VIVO)

NEOPLASTIC TRANSFORMATION

° SYRIAN HAMSTER EMBRYO CELLS
° MOUSE FIBROBLAST CELL LINES (C3H10T1/2 AND BALB/c 3T3)

CELLULAR METABOLISM

° PRIMARY LIVER CELLS

(RODENT SUBCHRONIC TOXICOLOGY)

(TERATOLOGY)

Fig. 3. Phase-2–short-term bioassays for mutagenesis-carcino-genesis-toxicity.

selected to provide confirmatory information on point mutations, chromosomal alterations, primary DNA damage and repair, and cellular oncogenic transformation. The latter test provides more explicit information on the carcinogenic potential of a sample. The test organisms are mammalian cells in culture, supplemented with exogenous metabolic activation, plants, insects, and intact mammals. Conventional teratology studies should be performed in this phase. Phase 2 systems are considered to provide more relevant and definitive information in the continuing process of health hazard evaluation, especially where intact organisms are employed.

Phase 3 bioassays

Phase 3 testing involves the use of conventional whole-animal methods. The emphasis here is on

quantitative risk assessment. Experimentation with intact mammals is needed to provide information on the presence, concentration, and biological activity of toxins in the target tissues. It is here that established whole-animal mutagenesis and carcinogenesis bioassays are performed. In addition, information on pharmacokinetics involving absorption, distribution, metabolic transformation, and excretion cannot be obtained without studies using intact mammals.

Resource implications

The resource implications of a mutagenicity-carcinogenicity bioscreening program are shown in Table 1. It is evident that most of the short-term bioassays designed to detect point mutation,

Table 1. Resource implications of carcinogenicity-mutagenicity bioscreening program

Test	Cost[a] ($)	Study time[b]	Quantity of material required (g)
Gene (point) mutations			
Bacteria (Ames plate test)	350–600	2–4 weeks	2
Bacteria (liquid suspension)	1,000–2,000	2–4 weeks	2
Eukaryotic microorganisms (yeast)	200–500	2–4 weeks	2
Insects (Drosophila, recessive lethal)	6,000–7,500	4–6 months	10
Mammalian somatic cells in culture (mouse lymphoma)	2,500–4,800	1–2 months	2
Mouse specific locus	20,000+	1 year	25
Chromosomal mutations			
In vitro cytogenetics	1,000–2,000	2–4 weeks	2
In vivo cytogenetics	3,000–6,500	6–8 weeks	20
Insects, heritable chromosomal effects	3,000–6,500	4–6 weeks	10
(Drosophila) nondisjunction	3,000	1–3 months	
Dominant lethal in rodents	6,000–10,000	3 months	20–25
Heritable translocation in rodents	40,000–67,000	12–18 months	25
Primary DNA damage			
DNA repair in bacteria	200–500	2–4 weeks	2
Unscheduled DNA synthesis	350–2,000	4–6 weeks	2–5
Mitotic recombination and/or gene conversion in yeast	200–500	4–6 weeks	2–5
Sister chromatid exchange	1,000–1,200	4–6 weeks	2–5
Oncogenic transformation in vitro			
Chemically induced transformation	6,500–7,500	10–12 weeks	2–5

[a]Cost of these tests has varied and can be expected to vary until test requirements are stabilized.
[b]This time period covers the experimental time and report preparation.

chromosomal alterations *in vitro* and primary DNA damage and repair are relatively rapid, inexpensive, and require small amounts of test material. Together with cytotoxicity bioassays for selected applications and rodent acute toxicity tests, these bioassays constitute effective screens for toxic and genotoxic effects of energy-related emissions and effluents.

CONCLUSIONS

Several conclusions may be reached on the basis of the foregoing discussion.

1. Short-term bioassays may be applied effectively to energy-related samples as indicators of potential health effects.

2. These tests facilitate rapid detection and prioritization of samples for further evaluation by confirmatory biological and chemical methodologies.

3. The bioassays are useful as a means to direct the chemical fractionation and identification of hazardous components within complex mixtures.

4. Chemical fractionation may be required to demonstrate genetic or genotoxic activity in complex mixtures which contain interfering toxic components.

5. The results of chemical and biological analysis are complementary and together provide useful information not obtainable when the two approaches are applied separately.

6. A phased approach involving the iterative and stepwise application of biological and chemical procedures permits a cost-effective utilization of limited resources.

ACKNOWLEDGMENTS

The author acknowledges the extensive participation of his colleagues in this work: Dr. Joellen Huisingh, Dr. Stephen Nesnow, Dr. Shahbeg Sandhu, Mr. Larry Claxton, and the staff of the Biochemistry Branch, Health Effects Research Laboratory, and the Industrial Environmental Research Laboratory, Environmental Protection Agency, Research Triangle Park, North Carolina.

REFERENCES

1. S. J. Gage, "Control Technology Bridges to the Future," *Energy/Environment II.* R&D Decision Series, EPA-600/9-77-012, November 1977 pp. 15–26.

2. S. Nesnow, M. D. Waters, and H. V. Malling, "Detection and Evaluation of Potential Health Effects Associated with Hazardous Agents from Alternate Sources of Energy," *Energy/Environment II.* R&D Decision Series, EPA-600-77-012, November 1977, pp. 301–6.

3. B. N. Ames, F. D. Lee, and W. E. Durston, "An Improved Bacterial Test System for the Detection and Classification of Mutagens and Carcinogens," *Proc. Nat. Acad. Sci. USA* 71, 782–86 (1973).

4. Symposium on Application of Short-term Bioassays in the Fractionation and Analysis of Complex Environmental Mixtures, sponsored by EPA Office of Energy, Minerals, and Industry, Williamsburg, Virginia. February 21–23, 1978, proceedings in press.

5. J. A. Dorsey et al., *Enviomental Assessment Sampling and Analysis: Phased Approach and Techniques for Level I*, EPA-600/2-77-115, June 1977.

6. J. W. Hamersma, S. L. Reynolds, and R. F. Maddalone, *IERL-RTP Procedures Manual: Level I Environmental Assessment,* EPA-600/2-76-160a, June 1976.

7. K. M. Duke, M. E. Davis, and A. J. Dennis, *IERL-RTP Procedures Manual: Level I Environmental Assessment Biological Tests for Pilot Studies,* EPA600/7-77-43, April 1977, p. 2.

8. W. G. Flamm, "A Tier System Approach to Mutagen Testing." *Mutat. Res.* 26, 329–33 (1974).

9. B. A. Bridges, "Use of a Three-Tier Protocol for Evaluation of Long-Term Toxic Hazards, Particularly Mutagenicity and Carcinogenicity," in *Screening Tests in Chemical Carcinogenesis,* WHO/IARC Pub. No. 12, ed. R. Montesaro, H. Bartsch, and L. Tomatis, Lyon, France, 1976, pp. 549–68.

10. B. J. Dean, "A Predictive Testing Scheme for Carcinogenicity and Mutagenicity of Industrial Chemicals," *Mutat. Res.* 41, 83–88 (1976).

11. "Principles and Procedures for Evaluating the Toxicity of Household Substances," prepared for the Consumer Product Safety Commission by the Committee for the Revision of NAS Publication 1138, National Academy of Sciences, Washington, D.C., June 1977, pp. 86–98.

12. J. W. Drake, Chairman, "Environmental Mutagenic Hazards," prepared by Committee 17 of the Environmental Mutagen Society, *Science* 187:503–14 (1975).

13. W. G. Flamm, "Approaches to Determining the Mutagenic Properties of Chemicals: Risk to Future Generations," prepared for the DHEW Committee to Coordinate Toxicology and Related

Programs by working group of the Subcommittee on Environmental Mutagenesis, April 1977.

14. J. McCann and B. N. Ames, "Detection of Carcinogens as Mutagens in the Salmonella/ Microsome Test: Assay of 300 Chemicals: Discussion." *Proc. Nat. Acad. Sci. USA* **73**:950–54 (1976).

15. R. J. Pienta, J. A. Poiley, and W. B. Lebherz, "Morphological Transformation of Early Passage Golden Syrian Hamster Embryo Cells Derived from Cryopreserved Primary Cultures as a Reliable *in vitro* Bioassay for Identifying Diverse Carcinogens," *Int. J. Cancer* **19**, 642–55 (1977).

16. F. E. Würgler, F. H. Sobels, and E. Govel, "Drosophila as Assay System for Detecting Genetic Changes," in *Handbook of Mutagenicity Test Procedures,* ed. B. J. Kilbey, M. Legator, W. Nichols and C. Ramel, Elsevier Sci. Pub. Co., Amsterdam, 1977, pp. 335–73.

17. T. Sugimura et al., "Overlapping of Carcinogens and Mutagens," *Fundamentals in Cancer Prevention,* ed. P. N. Magee et al., Univ. Park Press, Baltimore, 1976, pp. 191–215.

18. Energy Resources Co. Inc., "Short-Term Toxicological Bioassays and Their Applicability to EPA Regulatory Decision Making for Pesticides and Toxic Substances," report prepared for the Office of Planning and Evaluation, EPA Contract No. 68-01-4383, 1977.

19. R. H. C. San and H. F. Stich, "DNA Repair Synthesis of Cultured Human Cells as a Rapid Bioassay for Chemical Carcinogens," *Int. J. Cancer* **16**, 284–91 (1975).

20. M. D. Waters, J. L. Huisingh, and N. E. Garrett, "The Cellular Toxicity of Complex Environmental Mixtures," Symposium on Application of Short-Term Bioassays in the Fractionation and Analysis of Complex Environmental Mixtures, Williamsburg, Virginia, 1978, proceedings in press.

THE INTERAGENCY PROGRAM IN HEALTH EFFECTS OF SYNTHETIC FOSSIL FUELS TECHNOLOGIES: OPERATION OF A MATERIALS REPOSITORY

David L. Coffin*
Michael R. Guerin†
Wayne H. Griest†

THE INTERAGENCY PROGRAM FOR SYNTHETIC FUELS TOXICOLOGIC EVALUATION

It is generally agreed that, in the long run, the energy needs of the nation cannot be met solely by petroleum products. Furthermore, it is doubtful that the technologies of atomic energy, solar energy, or geothermal energy can be brought on line quickly enough to prevent a gap in the continuum of our energy supplies. In fact, these three sources are possibly unsuited to providing energy for a vehicular transportation as we know it today. Therefore, to take up the slack, there must be an extraction of liquid hydrocarbons from nonpetroleum sources available on this continent (i.e., oil shale and tar sands).

To use coal and oil shale as a source of energy for broad spectrum uses such as home heating and vehicular transportation, technologies of conversion to liquid and gas must be used on an unprecedented scale. There is considerable precedent for the use of coal as a source of gaseous and liquid fuel; for example, producer gas has been used for home heating and cooking, and synthetic gasoline was produced from coal by Nazi Germany. These technologies are now principally of historical interest since they have been subordinated to the widespread use of petroleum as energy sources. Because of this, coal and oil shale extraction technology has not stayed abreast of the present state of knowledge in other areas. Thus, new technologies reflecting the present chemical knowledge are being developed on a crash basis to meet the enormous energy needs of the nation.

Potentially, the technologies involved present risks to human health and environmental hazards in various phases of materials extraction, conversion, and use.[1] For example, the residual material from retorting of oil shale contains biologically active substances which may pose environmental hazards through contamination of groundwater. The retort oil contains carcinogens to which industrial workers are potentially at risk, as they may be to any effluent during refining of the oil.

There are many technological problems associated with these developing industries. To quote from the report of the Conference on Health Problems of Energy Technologies, Pinehurst, North Carolina, January 7–9, 1976, sponsored by NIEHS, EPA, ERDA, and NIOSH:

> The unique aspect is how toxicologic methods will be applied to monitor the processes and products of a developing industry so that a feed-back mechanism may be instituted in order to modify production methods, secure protective measures for industrial workers and protection of handlers during distribution and use of the product or possibly turn off the technology entirely if alternate means could be substituted having less severe health and ecological consequences. Therefore, it seems that the toxicologic purview of the developing energy technologies offers an opportunity for toxicologists to have an impact on health and ecological consequences before huge capital investments make modifications exceedingly difficult.

---

*U.S. Environmental Protection Agency, Research Triangle Park, North Carolina.
†Oak Ridge National Laboratory, Oak Ridge, Tennessee.

Establishment of a Synfuel
Materials Repository

To use toxicological evaluations of these emerging technologies rapidly for considerations of environmental impact, occupational health, and product safety, an interagency systems approach has been established. Central to this approach has been the establishment of a chemical or materials repository through an interagency agreement under which the Environmental Protection Agency furnished operating funds and the Department of Energy furnished personnel and facilities.

This repository provides a center where relevant substances can be collected, coded, stored, and subdivided for distribution to participating scientists, thus providing a reference source from which scientists may draw identified material together with what relevant background chemical data is available. In this way the toxicity of a product (or its component fractions) may be evaluated by specialists working in separate laboratories. Furthermore, the repository stores reference material that may be drawn on for verification of data by the same scientists or for new investigations by scientists not hitherto using the repository. The term "repository" is used because some samples are expected to undergo long-term study and will require long-term, carefully controlled (and recorded) storage.

However, the intent is to maximize the distribution of samples rather than store them. Further, the staff is to provide whatever assistance it can (within available resources) in recommending appropriate samples, preparing samples for bioassay, and chemically characterizing them. The repository is to be a bioassay chemistry facility. The staff focuses on the biology-chemistry interface and relies on process developers to identify highest priority sample types. Lastly, the research results are sent back to those who provide the samples.

The repository will not provide samples for process- or site-specific assessment or replace existing sample collections. Process- and site-specific assessments can only be made using sample suites acquired in carefully integrated sampling and analysis efforts with fully defined process operating parameters. Sample collections maintained by government- and privately-sponsored process development groups are required to support those specific process development efforts.

The repository will support the methods development and basic research required to make reliable process- and site-specific environmental assessments.

This research and development generally requires a much more thorough study of samples than can be done by already costly environmental assessments. The potential for acquiring a wide variety of chemical and biological data on the same samples promises a more efficient use of national toxicology resources by correlating a wide variety of studies (e.g., mutagenicity with chemistry and with carcinogenicity).

Samples and User Guidelines

The repository contains samples from coal and shale conversion processes and from coal combustion. A few crude and refined petroleum products are also available for comparative study. The samples are primarily of the product, the process stream, and the solid residue. Samples of aqueous and airborne environmental releases are generally not available. Small quantities of respirable-size-classified coal combustion and shale-related particles are available which have been prepared from bulk samples. These were not taken by ambient environmental sampling. Small quantities of aqueous process effluent samples are also available but are of questionable usefulness because of the instability of aqueous samples in storage.

With few exceptions, the samples do not constitute self-consistent sets and cannot be tied to specific process conditions. Most samples are of value only as research materials to test and develop methods or to derive biological and chemical data for further use. Until recently, most samples have not been sufficiently valuable to warrant carefully controlled and recorded storage conditions and stability determinations. Materials are now being acquired which do deserve this care, so facilities are being built to give it. Highest priority samples (most important to process developers and/or expected to be extensively studied as generic research materials) are to be stored in glass at 4°C under an inert gas blanket and protected from exposure to light. The chemical integrity of the samples is to be monitored by periodic chemical and physical testing if funding is available.

Almost without exception, sample suppliers are eager to give their materials widespread distribution. Most ask only the courtesy of being kept informed of the results of work involving their samples. An occasional summary of progress and preprints and reprints of abstracts and reports is usually sufficient.

Requests for samples can often be handled verbally. This is particularly the case for requests for generically identified samples (e.g., "coal-derived liquid A") required in small quantities. Depending on the supplier, background information can also often be released with the samples. Brief summaries of others currently examining the sample and of results available are being prepared to be given with the samples.

The repository staff cannot guarantee that all requests will be honored. Each request must be approved by the sample supplier. While this is generally handled verbally by the repository staff, access to some samples may require written summaries of intended work and an agreement that a serious attempt will be made to give the supplier an opportunity to comment on open literature publications resulting from work with the samples. In general, the larger the quantity required and the more valuable (e.g., one-of-a-kind or one extensively used by others) the sample is, the more difficult it is to get immediate release.

Requests for samples may be addressed to the Bio/Organic Analysis Section, Analytical Chemistry Division, Building 4500S, Room E-160, Oak Ridge National Laboratory, Oak Ridge, TN 37830, or to the attention of Dr. W. H. Griest ([615] 574-4869) or the Section Office (574-4862).

Example Experimental Activities

A primary experimental activity involving the repository has been joint contributions (Division of Pollutant Characterization and Safety, Office of Health and Environmental Research, and Department of Energy) to the bioassay chemistry of synfuels materials. Contributions have included the supply of samples, the development and comparison of class fractionation methods,[2,3] chemical characterization of the samples,[4,5] identification of chemicals contributing to bioactivity,[6] and the use of fractionation methods in biotesting.[7] The same materials are being examined for other biological and physical-chemical properties as part of other methods development studies. Experineces with these materials have been used to design the next collaborative study now that materials of greater relevance are available.

A second example of repository input is the preparation and characterization of particulate matter for bioassay. Activities have included assistance in sampling particulate matter associated with shale retorting, small-scale centrifugal-size classification of the particulate matter, preparation of large quantities of the respirable-size fraction through subcontract, and the comparative characterization of the particulates with those collected in the vicinity of the operation by high volume sampling. Capabilities

within the Analytical Chemistry Division have been drawn upon for physical characterization (size distribution and morphology), major and trace element analysis, and adsorbed organic content determination. Similar services have been allocated to the sampling and characterization of particles from coal combustion. A goal of these studies is to determine whether resuspended particles constitute a useful bioassay research material.

Anticipated Direction

Facilities are being readied for more carefully controlled storage and handling, and protocols are being developed for assessing and recording the composition of the samples as a function of storage parameters. Costs of sample storage and characterization prohibit handling large numbers of materials. It is likely that a small number (up to fifty) of sample types will be handled with greater control in the future. Sufficient quantities of those deemed by process developers as most important for generic research may be available as reference research materials. Resources are being sought to provide quantitative analytical services in support of high priority biological research programs.

REFERENCES

1. Ronald G. Oldham et al., *Assessment, Selection, and Development of Procedures for Determining the Environmental Acceptability of Synthetic Fuel Plants Based on Coal,* U.S. Department of Commerce, NTIS FE-1795-3 (Pt. 2), May 1977.

2. A. R. Jones, M. R. Guerin, and B. R. Clark, "Preparative-Scale Liquid Chromatographic Fractionation of Crude Oils Derived from Coal and Shale," *Anal. Chem.* **49**, 1766–1771 (1977).

3. I. B. Rubin et al., "Fractionation of Synthetic Crude Oils from Coal for Biological Testing," *Environ. Res.* **12**, 358–365 (1976).

4. H. Kubota, W. H. Griest, and M. R. Guerin, "Determination of Carcinogens in Tobacco Smoke and Coal-Derived Samples—Trace Polynuclear Aromatic Hydrocarbons," in *Trace Substances in Environmental Health-IX, 1975,* a symposium, ed. D. D. Hemphill, University of Missouri, Columbia.

5. M. R. Guerin et al., "Polycyclic Aromatic Hydrocarbons from Fossil Fuel Conversion Processes," in *Carcinogenesis, Vol. 3: Polynuclear Aromatic Hydrocarbons,* ed. P. W. Jones and R. I. Freudenthal, Raven Press, New York, 1978.

6. M. R. Guerin, J. L. Epler, and C. W. Gehrs, *Short-Term Bioassay and Bioassay Chemistry of Materials Related to Synthetic Fossil Fuels,* Report ORNL/TM-6390, Oak Ridge National Laboratory, Oak Ridge, Tenn., July 1978.

7. J. L. Epler et al., "Analytical and Biological Analyses of Test Materials from the Synthetic Fuel Technologies. I. Mutagenicity of Crude Oils Determined by the *Salmonella Typhimurium*/Microsomal Activation System," *Muta. Res.* 57, 265–278 (1978).

DISCUSSION

Larry Jenkins, Naval Medical Research Institute: I would like to add just a little bit to what Dr. Coffin has presented. As soon as Sohio Refinery in Toledo finishes the Paraho crude refining, we will be doing mammalian multi-species, long-term, continuous inhalation with lifetime follow-up. I think we are up to 58-slide pathology on each rat and mouse right now. We are carrying out those studies on the shale-derived JP 5, which is Navy standard jet fuel, and on shale-derived DFM or marine diesel fuel. Concomitantly with these studies, although perhaps time-phased a month or so behind, the Air Force will carry out in the same laboratory (WPAFB) a long-term inhalation exposure with lifetime follow-up on JP 4, which is Air Force standard jet fuel. To my knowledge, this is one of the first, if not the first, look at end-use fuels that will be coming out of the synthetic fuel programs. We, in my laboratory, for example, are getting over 200 barrels of the fuels for inhalation studies. We will have some excess; if anybody wants a gallon of shale JP 5, we may be able to arrange that. The results of these studies will be published as soon as they are completed. If anyone has any questions about these studies, or any comments, we would be most pleased to hear them now before we start rather than a year from now when we are halfway into it. Thank you.

Frank Schweighardt, Department of Energy: My question will actually be going from the front back to Mike. Sitting back and listening to all the presentations from yesterday and also the details today of the biological work, you have planned these things out very well. My question sort of goes to Mike: What care is being taken and thought to sampling the plants, sample handling, and sample storage of these products? What care is being taken because you can't just keep these things sitting out in the open, particularly if you have barrels of this stuff?

Mike Guerin, Oak Ridge National Laboratory: With the materials that Dave is talking about in terms of very carefully controlled studies (the shale-derived

materials, for example, and some of the new coal-derived materials), the storage conditions are going to be generally the standard type that are used any place: cold and dark. We hope to be able to monitor them in terms of general chemical composition and character. The materials that are in the repository now have always been used primarily as research materials. The selection of the four materials that we talked about earlier today in terms of biological activity wasn't because they were in any way relevant to the process but rather, based on the chemistry, based on the biology, we felt they could be used to tell us something about mutagenicity. In a case like that it isn't so terribly important that they be handled in such a way that they are tied to a process, when they weren't tied to the process to begin with; it was just whatever was available. Clearly this is a problem, and clearly it's a problem in any kind of a long-term study, particularly mouse skin carcinogenesis, but we have one of two options. One option is to say we've got to know all of what's happening, therefore we can't do anything; or we take what we have and begin to work with it, monitor it in the meantime, and find out what kind of difficulties we have. We have chosen the second option—at least to try to do something with what we have. Again, an important point I think to make is that the materials that are available are coal-derived or shale-derived samples that we have had for some time. They have been used by various investigators and are not tied to the process. They are simply research materials. They have not (not all, anyway) been handled in a very carefully controlled manner, nor have they been monitored for physical and chemical change, but these that would be used for an extensive study will be.

Frank Schweighardt, Department of Energy: Well, just for one other point on the same subject—I was interested because that's pretty much my area in the nitrogen basis of coal-derived materials—those nitrogen-base compounds in particular will change their solubility in the isolated fraction. That was one of my examples there when you were showing that the base components from a coal liquefaction product; these components will change (it wasn't defined where that came from, that is, if it was pentane soluble, if we are going to the classical form of oils). If you let that sit around, that will change to a pentane-insoluble–benzene-soluble component after awhile, and your nitrogen-base components start—for lack of a better word—polymerizing. They will change solubility, and finally they become almost intractable after a period of time. That's why I was very concerned with the nature of those base components.

THE TOXICOLOGY AND CARCINOGENIC INVESTIGATION OF SHALE OIL AND SHALE OIL PRODUCTS*

W. Barkley†
D. Warshawsky†
R. R. Suskind†
E. Bingham‡

ABSTRACT

The toxicity of several shale oils, produced by various retort methods, and spent shale samples is reported. A spent shale sample was studied for its teratogenic effects on rabbits. The benzo(a)pyrene content of these samples and other by-products was determined. The mouse skin bioassay technique was utilized to evaluate the carcinogenic potency of various shale oils and spent shale samples.

INTRODUCTION

Studies of workers involved in the production and use of shale oil strongly suggest that the major health concern of this industry is the carcinogenicity of shale oil. It is well known from the reports of Bell[1] (1876) and Scott[2] (1922) that prolonged exposures to shale oil can produce skin cancer in humans. These reports, as well as reports of others, dictate that an assessment of the potential health effects of shale oil and shale oil products should be made as new technologies are developed for the expanding industry.

As great emphasis is being placed on developing domestic energy sources, it is most likely that oil shale will soon be commercially developed. The feasibility of extracting and processing oil shale has been known for some time, and this industry would have developed sooner had it not been impeded by economic and political problems. The development of the oil shale industry in Estonia and Brazil gives added impetus to the possibility that oil shale industry will soon be in full operation.

There are additional health and environmental concerns associated with oil shale extraction and processing. Blasting and mining will produce dust, particulate organic matter (POM), CO, NO_x, SO_2, hydrocarbons, silica, and metal salts. Crushing and screening the ore produces more dust, silica, and POM. In addition to the above compounds, retort operations can produce polycyclic organic compounds, H_2S, NH_3, and volatiles. Some of these compounds, as well as arsenic and other metals, are produced in the upgrading process and disposal of the solid waste products is a major environmental concern. Many of these compounds may contaminate not only the atmosphere but water as well. Since water is a necessary commodity in the oil shale industry, many of these compounds may be found in the waste and runoff waters.

Although there may be many potentially hazardous chemicals present in the work area and the

*This work was partially supported by Center Grant ES-00159 from the Institute of Environmental Health Science and by a contract from the American Petroleum Institute.

†University of Cincinnati Medical Center, Department of Environmental Health, Kettering Laboratory, 3223 Eden Avenue, Cincinnati, Ohio 45267.

‡Occupational Safety and Health Administration, U.S. Department of Labor, Washington, D.C. 20210.

surrounding environment, the major health concern appears to be the carcinogenicity associated with shale oil use and production.

ACUTE AND CHRONIC TOXICITY

Weaver and Gibson[3] (1978) reported on the acute and chronic studies on four crude shale oils, three raw shales and four spent shales (Table 1). The results do not show that the materials are acutely toxic.

Weaver and Gibson[3] also reported on chronic inhalation studies of two raw shales and two spent shales. In the study rats and monkeys were subjected to the inhalation of the respirable dusts at concentrations of 10 and 30 mg/m^3. After one year of exposure no adverse effect was seen in the animals.

TERATOGENICITY

Disposal of solid wastes produced in processing and extracting of oil shale presents a major environmental problem. The potential hazard to workers handling spent shale, which may contain significant quantities of silica as well as polycyclic aromatic hydrocarbons, should be of primary concern. Secondarily, the possibility of carcinogens leaching into the general environment represents a concern. Since spent shale may be used in revegetation of the disturbed land, it is possible that products leached from spent shale may, upon ingestion of plants (or water) containing the contaminants, create a teratogenic hazard.

A study[4] was undertaken to evaluate the potential embryotoxic and/or teratogenic effects of spent shale in New Zealand white rabbits.

Method

Pregnant New Zealand white rabbits, 3–5 kg, were administered spent shale in distilled water by oral intubation in doses of 250 mg/kg and 500 mg/kg on the 8th and 12th days of gestation. Control animals received carbon at the same dosage and frequency. Another group received only the vehicle, water. The does were sacrificed (chloroform) on the 28th day of gestation. At sacrifice the embryos or implants were examined for anomalies.

Results

The results are summarized in Table 2. Control rabbits given water showed no anomalies except for rudimentary ribs, and four of the 41 implants were dead or resorbed. The animals given carbon or spent shale did not differ from the control group in the number of implants dead or resorbed. As can be seen, there was a slight increase in the number of anomalies. The data showed that spent shale does not cause significantly more abnormalities than does inert carbon. There was not a display of a dose-dependent response since higher dosages caused fewer resorptions than lower doses.

CARCINOGENIC STUDIES

In our carcinogenic animal studies we have reported the benzo(a)pyrene (BaP) content of test materials. This practice was predicated on the early belief that BaP was the carcinogen present in coal tar and therefore might also be present in shale oil. The isolation and identification of BaP from coal tar by

Table 1. Acute studies of shale oil and shale oil products[a]

Sample	Test	Results
Crude Shale Oils (4)	Oral LD50 (Rat)	8–10 g/kg
Crude Shale Oils (4)	Dermal LD50 (Rabbit)	5 ml/kg
Crude Shale Oils (4)	Eye Irritation (Rabbit)	minimal, reversible
Raw Shales (3)	Eye Irritation (Rabbit)	negative
Spent Shales (4)	Eye Irritation (Rabbit)	irritating, reversible
Crude Shale Oils (4)	Dermal Irritation (Rabbit)	0.5 g in 72 hr abraided-unabraided
Crude Shale Oils (4)	Sensitization (Guinea Pig)	negative
Raw Shales (3)	Sensitization (Guinea Pig)	negative
Spent Shales (4)	Sensitization (Guinea Pig)	negative

[a]Reported by N. K. Weaver and R. L. Gibson, Amer. Indust. Hyg. Conf. (1978).

Table 2. The number and types of abnormality found in rabbit embryos
after oral doses of spent shale

Agent and dose (mg/kg)	Number of litters	Number of implants	Number dead or resorbed	Location of abnormality				
				Ribs	Skull	Vertebra	Limbs	Others
Control	4	41	4	4				
Carbon 250	8	67	10	8	1			2
Carbon 500	4	50	6	3	1	4	1	1
Spent Shale 250	10	84	9	7	2	8	1	3
Spent Shale 500	15	129	9	17	13	7	1	1

Hieger[5] (1930) and Cook (1933)[6] led many investigators to study the carcinogenic constituents of shale oil. The search for the carcinogen in Scottish shale eluded many investigators until 1943, when Berenblum and Schoental[7] identified BaP in shale oil, but they also observed a fraction of shale oil to be carcinogenic that did not contain detectable quantities of BaP. Later, Hueper and Cahnmann[8] (1958) and Bogovsky[9] (1962), respectively, reported BaP-free American and Estonian shale oil to be carcinogenic in animal mouse skin painting studies.

Method

Young adult C3H/HeJ male mice were treated twice weekly with 50 mg of the test material. The material was applied to the interscapular area of the shaven backs with a microliter pipette or a calibrated dropper. In the case of solid materials, such as raw and spent shale, the materials were suspended in white mineral oil in 1:2 ratio (by weight). Mice were treated for 80 weeks or until the appearance of a papilloma. If a papilloma progressed and was diagnosed grossly as a carcinoma, the mouse was killed and autopsied. However, if the papilloma regressed the treatments were resumed.

Results

Table 3 shows the results of skin painting with two crude shale oils from two different processes. The two oils produced very little difference in tumor incidences and average latent period.

Another study determined the BaP content found in a raw shale oil, upgraded shale oil, processed shale, processed water, native grass, leached water, raw shale, and native soil. The results of these determinations can be seen in Table 4. BaP concentrations found in all samples were considered low. The results of topical applications of raw shale oil and upgraded oil to mouse skin are presented in Table 5. It may be seen that shale oil induced tumors in 86% of the effective number of mice with an average time for appearance of about 30 weeks, whereas the upgraded oil induced tumors in 13% of the effective number with an average latent period of 49 weeks. A solution of 0.05% BaP in toluene resulted in 91% of the effective number of mice developing tumors with an average latent period of 46 weeks. No tumors were observed in the mice treated with toluene alone after eighty weeks of topical application.

Four shale oils and raw and spent shales that Weaver and Gibson[3] studied are currently on test at

Table 3. Carcinogenic potency of two raw shale oils

Sample	Strain of mice	Dosage	Number of mice	Final[a] effective number	BaP (%)	Number of mice developing tumors		Average time of appearance of papillomas (weeks)
						Malignant	Benign	
Shale oil #1, Heat transfer process	C3H	50 mg, twice weekly	15	12	<0.00001	8	3	43
Shale oil #2, Retort combustion process	C3H	50 mg, twice weekly	15	12	<0.00001	8	2	43

[a]The final effective number is the number of mice alive at the time of appearance of the median tumor plus those mice that may have died with tumors.

Table 4. Benzo(a)pyrene content
of shale oil products

Sample	BaP concentration
Raw shale oil	<0.00005%
Upgraded oil	0.0006%
Processed shale	<0.000005%
Processed water	<1 ppb
Native grass	0.000065%
Leached water	<1 ppb
Raw shale	<0.000005%
Native soil	<0.000005%

Kettering Laboratory for their carcinogenic potential. After 60 weeks of topical applications of the materials to the backs of mice, all the animals treated with the shale oils have developed tumors or have died. The incomplete results of this study are presented in Table 6. No tumors have been observed in the mice receiving raw and spent shale suspended in white mineral oil.

Analysis*

The raw and spent shale samples contain only a few parts per billion of BaP (<1 ppb to 24 ppb). The shale oil retort samples, on the other hand, contain 1000 to 5000 ppb of BaP and 3000 to 20,000 ppb of mono- and dimethylated BaP. The mono- and dimethylated BaP were analyzed together and reported as one value. The biological activities of the oil samples do not correlate with the amount of benzo(a)pyrene in the samples. As an example, the ratios of benzo(a)pyrene to benz(a)anthracene in samples RO_2 and RO_4 are 0.32 and 3.5, respectively, while the ratios of methylated compounds are 1.6 to 0.69 in samples RO_2 and RO_4, respectively.

In addition to benzo(a)pyrene and benz(a)anthracene, ten other polycyclic aromatic hydrocarbons are identified (Table 7). One of these, acridine, is found only in sample RO_1. It should be noted that pyrene, fluoranthene, benzo(e)pyrene, and benzo (g,h,i)perylene have cocarcinogenic activities[10] in addition to aliphatic compounds[11] that are found in these complex mixtures. A number of heterocyclic aromatics are not found in these samples: carbazole, 11 H-benz(a)carbazole, benz(c)acridine, 7 H-benzo(a)-carbazole, dibenz(a,h)acridine, and 7 H-dibenz(c,g)-carbazole.

These results show that it is important to look at the mixtures as wholes and to consider cocarcinogens and inhibitors that are present in the samples. Lastly, it has to be noted that BaP may not be a good indicator for carcinogenicity of shale oil.

Discussion

Animal studies using shale oil and shale oil products have shown that they are relatively low in acute toxicity. There is, however, great concern about the long term effects of these products. One such concern is the carcinogenic properties of shale oil. It has been clearly demonstrated that raw shale oils, produced by a number of retort methods, are carcinogenic for the mouse's skin. The content of benzo(a)pyrene in these oils was quite low and could not account for the carcinogenic activity. It is likely that other carcinogens and perhaps cocarcinogens are in raw shale oil.

In the Estonian oil shale industry, Bogovsky and Jons[12] report no increase cancer incidence among the shale oil workers. Unlike the Scottish shale oil

*These data were supplied by the American Petroleum Institute.

Table 5. Carcinogenic potency of raw and upgraded shale oil

Sample	Strain of mice	Dosage	Number of mice	Final[a] effective number	BaP (%)	Number of mice developing tumors		Average time of appearance of papillomas (weeks)
						Malignant	Benign	
Raw shale oil	C3H	50 mg, twice weekly	50	45	<0.00005	21	18	30
Upgraded shale oil	C3H	50 mg, twice weekly	50	39	0.0006	3	2	49
Positive control (0.05% BaP in toluene)	C3H	50 mg, twice weekly	100	92	0.05	75	9	46
Negative control (toluene only)	C3H	50 mg, twice weekly	100	91	0	0	0	

[a]The final effective number is the number of mice alive at the time of appearance of the median tumor plus those mice that may have died with tumors.

Table 6. Carcinogenic potency of shale oils and shale oil products[a]

Sample	Strain of mice	Dosage	Number of mice	Final effective number	BaP (%)	Number of mice developing tumors		Average time of appearance of papillomas (weeks)
						Malignant	Benign	
Raw oil #1	C3H	50 mg, twice weekly	50		0.00018	11	19	28.6
Raw oil #2	C3H	50 mg, twice weekly	50		0.00018	30	1	19.1
Raw oil #3	C3H	50 mg, twice weekly	50		0.00012	27	10	26.3
Raw oil #4	C3H	50 mg, twice weekly	50		0.00042	27	11	24.9
Raw shale 101	C3H	50 mg, twice weekly	50		<0.00001	0	0	
Raw shale 102	C3H	50 mg, twice weekly	50		<0.00001	0	0	
Raw shale 103	C3H	50 mg, twice weekly	50		<0.00001	0	0	
Spent shale 201	C3H	50 mg, twice weekly	50		<0.00001	0	0	
Spent shale 202	C3H	50 mg, twice weekly	50		<0.00001	0	0	
Spent shale 203	C3H	50 mg, twice weekly	50		<0.00001	0	0	
Spent shale 204	C3H	50 mg, twice weekly	50		<0.00001	0	0	
Control – no treatment	C3H	50 mg, twice weekly	50			0	0	
Control – mineral oil only	C3H	50 mg, twice weekly	50			0	0	
0.05% BaP in mineral oil	C3H	50 mg, twice weekly	50		.05	40	6	37.8
0.15% BaP in mineral oil	C3H	50 mg, twice weekly	30		.15	29	1	27.4

[a]Incomplete: 54 weeks duration.

Table 7. Polycyclic aromatics present in shale oil

Carcinogen

benzo(a)pyrene (strong)
benz(a)anthracene (weak)

Borderline Carcinogen

chrysene

Noncarcinogen

pyrene[a]
fluoranthene[a]
triphenylene
benzo(e)pyrene[a]
perylene
anthanthrene
benzo(g,h,i)perylene[a]
coronene
acridene

[a]Cocarcinogenic activity.

industry, the Estonian industry has good hygiene practices, which are cited as one of the reasons for this lack of a cancer problem. Although after more than 20 years of operation there is no increase in the cancer incidence, more time and study are necessary before the potential hazards of this industry can be truly evaluated.

REFERENCES

1. B. Bell, "Paraffin epithelioma of the scrotum," *Edin. Med. J.* **22,** 135–37 (1876).

2. A. Scott, "On the occupational cancer of the paraffin and oil workers of the Scottish shale industry," *Br. Med. J.* **2,** 1108–09 (1922).

3. N. K. Weaver and R. L. Gibson, "The U.S. oil shale industry: a health perspective," presented at American Industrial Hygiene Conference, May 7–12, 1978. Los Angeles, California.

4. E. Bingham, M. Zerwas, and W. Barkley, unpublished data.

5. I. Hieger, "Spectra of cancer-producing tars and oils and of related substances," *Biochem. J.* **24** 505 (1930).

6. J. W. Cook, C. L. Hwett, and I. Hieger, "The isolation of a cancer producing hydrocarbon from coal tar," *J. Chem. Soc.* 1, 395 (1933).

7. I. Berenblum and R. Schoental, "Carcinogenic constituents of coal tar," *Br. J. Exper. Path.* **24,** 232–39 (1943).

8. W. C. Hueper and H. J. Cahnmann, "Carcinogenic bioassay of benzo(a)pyrene-free fractions of American shale oils," *Arch. Path.* **65,** 608–14 (1958).

9. P. Bogovsky, "On the carcinogenic effect of some 3,4-benzopyrene-free and 3,4-benzopyrene-containing fractions of Estonian shale-oil," *Acta Unio. Int. Contra Cancrum* 18, 37–39 (1962).

10. B. L. Van Duuren and B. M. Goldschmidt, "Carcinogenic and tumor-promoting agents in tobacco carcinogenesis," *J. Natl. Cancer Inst.* **56,** (6) 1237–42 (1976).

11. E. Bingham and H. L. Falk, "Environmental carcinogens," *Arch. Environ. Health* **19,** 779–83 (1969).

12. P. Bogovsky and H. J. Jons, "Toxicological and carcinogenic studies of oil shale dust and shale oil," presented at Workshop on Health Effects of Coal and Oil Shale Mining, Conversion, and Utilization, Jan. 27–29, 1975. Department of Environmental Health, Kettering Laboratory, University of Cincinnati, Cincinnati, Ohio.

DISCUSSION

Mike Holland, Oak Ridge National Laboratory: This is a comment and a question for Dr. Barkley. First of all, I want to congratulate you for an excellent presentation—full of useful data. First of all, in our work we have observed that how the bioassay is carried out could influence latency, the timed appearance of the tumor. Since this is a very critical endpoint for establishing relative potency among different full crudes, I think this is an issue that all of us should be aware of. What I am implying by this is that some of the toxic, potentially inhibitory constituents of these complex mixtures can delay or retard the expression of neoplasms in skin when the materials are applied repetitively. While it is varied in the data that I presented earlier and would require some additional exposition, a short-term or pulsed exposure may actually be a preferred method for demonstrating the true carcinogenicity of these materials. Now to the question, in the experiments that you have reported, Dr. Barkley, have you noted evidence of local inflammatory reactions, dermal toxic sequelae, that might be the tip of the iceberg that I am alluding to?

William Barkley: Yes, we have noted toxic effects. In fact, on these samples we did a pilot study where we painted everyday. In a week, all the animals had lost their fur and were in very bad condition. So that is one of the reasons why we don't do three times a week, because there are toxic effects from repetitive doses. As far as the sample changing, we have explicit instructions from our sponsor, the American Petroleum Institute, that these samples should be kept under nitrogen, in the dark, and in the cold when not in use. So we try to minimize what changes might occur there, but we do have some toxic effects and the animals do show some toxicity, especially to the skin.

INHALATION TOXICOLOGY OF PRIMARY EFFLUENTS
FROM FOSSIL FUEL CONVERSION AND USE*

Charles H. Hobbs†   Roger O. McClellan†

Charles R. Clark†   Rogene F. Henderson†
Larry C. Griffis†   Joseph O. Hill†
Robert E. Royer†

ABSTRACT

Studies have been started to better define inhalation hazards to man from potential effluents from fluidized bed combustion (FBC) or low-Btu gasification of coal. Samples of potential airborne effluents have been collected from an experimental FBC and an experimental gasifier located at the Morgantown Energy Technology Center. Aerosols from these units have been characterized as to their chemical and physical properties. A series of *in vivo* and *in vitro* toxicological tests on the potential effluents have been started. Some extracts of particles collected from the FBC were mutagenic in the *Salmonella*/mammalian microsome mutagenicity (Ames) test, as were fractions of extracts of samples of the gasifier. However, the mutagenicity of FBC fly-ash extracts appears to be no greater and perhaps less than that reported for fly ash from a conventional coal combustor. Further *in vitro* and *in vivo* testing is needed to establish the significance of these results. Studies on the deposition and retention of FBC fly ash following inhalation by experimental animals have been started as have tests for acute pulmonary injury. These studies will be followed soon by chronic inhalation studies in experimental animals.

INTRODUCTION

With the increased use of coal to meet our energy needs, it is essential that the potential hazards to man from the extraction, combustion, and conversion of coal be established. The fluidized bed combustion of coal and low-Btu gasification of coal are two emerging technologies which have the potential for significant use in the near future.

The Inhalation Toxicology Research Institute (ITRI) has established programs to better define the hazard to man from inhaling effluents from these two developing technologies and to compare their potential hazard to man with that from the conventional combustion of coal. A major component of these programs was the establishment, in April of 1977, of a collaborative effort with the Morgantown Energy Technology Center (METC). In collaboration with METC personnel, ITRI has established a field sampling program in which the potential airborne effluents from an experimental

fluidized bed coal combustor and an experimental low-Btu coal gasifier located at METC are being characterized. The effort consists of characterizing the process streams of the units as to the chemical and physical characteristics and toxicological properties of potential effluents. Major emphasis has been placed on characterizing the aerosols less than 10 μm in aerodynamic diameter and the vapor phase hydrocarbons that may be released as fugitive emissions, resulting in exposure of workers, or emissions that could be released from stacks resulting in exposure of the general population. Some samples that could be released as solid wastes have also been obtained.

*Research done under U.S. Department of Energy Contract Number EY-76-C-04-1013 and in animal facilities fully accredited by the American Association for Accreditation of Laboratory Animal Care.

†Inhalation Toxicology Research Institute, Lovelace Biomedical and Environmental Research Institute, Inc., P.O. Box 5890, Albuquerque, New Mexico 87115.

These units at METC are experimental, and control devices have not been perfected. In fact, a significant effort at METC is related to developing methods of fine particle control, both for technological and environmental reasons. Thus, the results from these studies are expected to aid in determining the need for control devices as the units are scaled to full size and to guide the technologies of fluidized bed combustion and low-Btu gasification in their development in an environmentally acceptable fashion.

It is expected that the sampling efforts will be expanded to include samples from other and perhaps larger FBC and low-Btu gasifiers as they become available. Comparable samples must also be obtained from conventional coal combustion to make appropriate toxicological comparisons and eventual risk assessments for man from the different technologies.

EXPERIMENTAL APPROACH

It is important to sample the process streams of the units to establish the physical and chemical characteristics of the source of the exposure to man as it is released to the atmosphere (Fig. 1). This aids in establishing the exposure atmosphere which the population at risk will inhale. Following inhalation of a material, it is essential to establish its fate in the body to establish the dose to the critical tissues. The goal is to be able to predict by appropriate testing the biological response of man to the inhaled material.

The purpose of these inhalation toxicology studies is to aid in the orderly development of the technology so that human health is protected and other societal costs are minimized (Fig. 2). The initial phase of these studies has been analyses of samples from the METC, FBC, and low-Btu coal gasifier to establish what the effect of various coal types being used as feedstocks and the particular operating conditions being used

have on the physical, chemical, and toxicological properties of potential effluents.

This paper summarizes the results of the physical and chemical characterization of potential effluents from these units as related to their potential toxicological effect and to report on the results of the toxicological testing done to date. A multi-tiered approach to toxicological testing is being used. Much of the work done to date has been in in vitro systems to establish the mutagenic and cytotoxic properties of many samples obtained from the units under various operating conditions using various types of coal as feedstocks. The results of these studies will be used to select materials for studies in experimental animals to determine the fate of the inhaled material and its biological effects including carcinogenesis and potential for producing chronic respiratory disease.

Detailed descriptions of the FBC and low-Btu gasifier, schematic diagrams of the units, location of the sampling ports, a complete description of the process stream sampling system, and a more detailed presentation of the results of the physical and chemical characterization of the potential effluents from the FBC and low-Btu gasifier are presented earlier in these proceedings in the paper by Newton et al. and in a longer report on the FBC,[1]

MATERIALS AND METHODS

Briefly, the METC FBC is an 18-inch atmospheric unit which has been under development at METC since 1967. Various fuels including coals, lignite, and culm (anthracite waste) have been burned in the combustor. Bed materials have also been varied to enhance SO_2 capture. The unit is equipped with primary and secondary cyclone and bag filters as control devices. The material from the primary cyclone is reinjected into the FBC for further combustion. Samples have been obtained from four

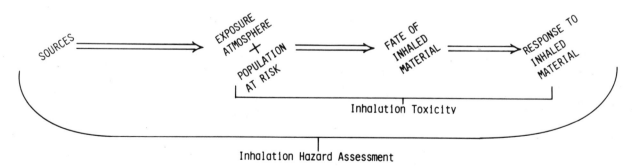

Fig. 1. Key links in inhalation toxicity assessment.

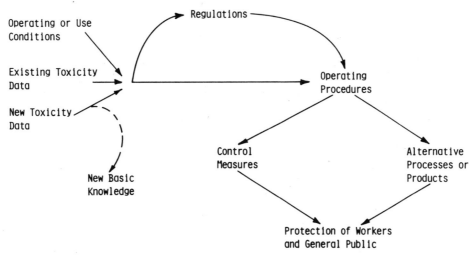

A. Protection of man
B. Minimize societal costs
- Health costs
- Control measures
- Alternative processes or products

Fig. 2. The purpose of inhalation toxicology studies is to protect the health of man and minimize other societal costs.

locations in the process stream of the FBC: before the first cyclone, between the first and second cyclones, between the second cyclone and the bag filters, and after the bag filters. The FBC has been sampled under various operating conditions (Table 1), with different coal types as fuel, and with different bed materials (Table 2). The aerosol from the process stream is diluted and cooled before its collection by the various sampling devices. The sampling devices being used as well as the quantity of material that can be collected by each device is given in Table 3. Samples of coal, bed material, cyclone ash, bag filter ash, and bottom ash have been collected.

Table 1. Typical FBC operating conditions during sampling

Fuel	Montana Rosebud subbituminous
Bed material	Geer limestone
Fluidizing air flow	2600 scfh
Fluidizing air pressure	3 psi
Injector air flow	960 scfh
Total air flow	4980 scfh
Fluidizing velocity	3.0 ft/s

The METC coal gasifier is a pressurized version of the McDowell-Wellman atmospheric stirred-bed gasifier. The unit is equipped with a gas cleanup system through which one-fourth of the total gas flow passes to remove vapors of tars and oils from the gas.

Table 2. Summary of coal feed stocks, bed temperatures, and bed materials used in the FBC during the ITRI field sampling trips

Coal	Bed temperature (°F)	Bed material
Montana Rosebud	1450 1650	Limestone
Texas lignite (bulk)	1550	Glass sand
Texas lignite (refuse)	1600	Glass sand
Texas lignite (refuse)	1600	Limestone
Texas lignite (refuse)	1450	Glass sand
Texas lignite	1450	Limestone
Texas lignite (washed)	1600	Glass sand
Western Kentucky	~ 1500	Limestone

At the present time, both the cleaned and uncleaned gas are burned in twin flares. Thus, it has not been possible to sample an exhaust aerosol. However, if the low-Btu gas is depressurized, diluted, and cooled (as would happen if leaks occurred), an aerosol is formed. To obtain samples of this aerosol, samples were withdrawn from both the cleaned and the uncleaned gas process streams. Samples have also been taken of liquids from the disengagement chamber of the gasifier to assess the effect of this cleanup stage. Due to a more limited operation of the gasifier and the difficulties associated with sampling

Table 3. Samplers used on effluent streams from METC FBC

Sampling device	Flow rate (liters/min)	Total sample weight	Comments
Lovelace Multi-Jet Cascade Impactor	25	400 mg	Aerodynamic size fractions 0.6 to $\cong$ 10 μm
Sierra Radial Slit Jet Cascade Impactor	21	500 mg	Aerodynamic size fractions 0.6 to $\cong$ 10 μm
Mercer Cascade Impactor	0.5	1.0 mg	Aerodynamic size fractions 0.3 to 5.0 μm
Lovelace Aerosol Particle Separator	0.3	~10 mg	Aerodynamic size fraction ($\cong$ 26) 0.5 to 3.0 μm. Grids for transmission electron microscopy as a function of aerodynamic size
Filter (47 mm)	25	~1 g	On Selas Ag membrane to help determine mass loading
Point-to-plane electrostatic precipitator	0.5	~1 ng	Samples for transmission electron microscopy for geometric size analysis and scanning electron microscopy for morphological and elemental analyses
Tenax sampler	2.0	~10 mg	Samples for determination of vapor-phase polycyclic aromatic hydrocarbons above C_5
Concentric electrostatic precipitator	100–125	~10 g	Device is used as a sample chamber exhaust cleanup. Samples are used for major elemental analysis and particulate hydrocarbon analysis

from the high-temperature and high-pressure process stream of the gasifier (Table 4), fewer samples have been taken and analyzed from the gasifier than from the FBC.

Table 4. Typical low-Btu gasifier operating conditions during sampling

Using Arkwright subbituminous coal

	Location 1[a]	Location 2[b]
Air intake, scfh	74,000	46,250
Steam intake, lb/h	870	910
Coal feed rate, lb/h	1,450	1,525
Gas pressure, psi	130	150
Gas output, scfh	160,000	92,000

[a]Clean gas stream.
[b]Raw gas.

Inorganic analysis of the samples from the aerosol sampling system used for both the FBC and gasifier was done by spark source mass spectrometry (SSMS),* using rhodium and indium as internal standards. The methods for organic analysis of the particle-associated and vapor phase hydrocarbons have been presented elsewhere.[1,2] Particles were ultrasonically extracted with methylene chloride while Tenax material was Soxhlet extracted with pentane. Gas chromatography, gas chromatography/mass spectrometry, and high-pressure liquid chromatography were used to characterize the samples. Some samples were further fractionated by a modification of the method of Jones et al.[3] and by standard techniques.[4]

The mutagenicity testing of the particle or Tenax (a polymeric absorbant used to trap vapor phase hydrocarbons) extracts was done[5] with strains TA1535, TA100, TA1537, TA1538 and TA98 of

*All SSMS analyses were done by CDM/Accu Labs, Wheatridge, Colo.

*Salmonella typhimurium.** Microsomes for enzyme activation were prepared from the livers of rats treated with Aroclor 1254 (100 mg/kg IP). For mutagenicity testing, samples of particles were extracted ultrasonically with methylene chloride, but Tenax samples were Soxhlet extracted with pentane. The extracts were evaporated to dryness at room temperature and reconstituted in DMSO for the Ames test. Some FBC bag filter ash was extracted with horse serum, saline, or cyclohexane by the methods reported by Chrisp et al.[6] to compare extraction techniques. The cytotoxicity of fly ash from the FBC and coal-gas aerosols was evaluated in rabbit alveolar macrophages by a modification of the methods of Waters et al.[7]

Fly ash collected from the bag filter of the FBC was resuspended as an aerosol for the inhalation exposure of rats or Syrian hamsters. A fluidized bed aerosol generator recently developed at ITRI was used to generate the aerosol.[8] Thirty days before exposure of the rats, some fly ash was neutron activated for 40 h with a flux of 9×10^{13} neutrons cm^{-2} s^{-1} by methods previously reported.[9] The deposition, retention, and distribution of the fly ash following inhalation was determined by whole-body counting and serial sacrifice of the animals followed by liquid scintillation or Ge(Li) detector counting of tissues to obtain the distribution of the material. Syrian hamsters were also exposed in modified Rochester chambers to graded levels of ash that had not been activated for 8 h. These animals were also serially sacrificed and the levels of selected enzymes determined for airway fluid and lungs.[10]

RESULTS AND DISCUSSION

The aerosols released from the stack of the FBC have aerodynamic properties (mass median aerodynamic diameters 2–3 μm) that will allow their wide dispersal and, if inhaled, result in the deposition of particles deep in the lungs of man. The particles from the FBC are irregular in shape as compared with the spherical, often hollow, particles emitted from pulverized coal power plants.[6] The FBC ash shows less size-dependent concentration of elements than has been reported for ash from conventional coal combustion.[11] The levels of hydrocarbons potentially emitted from the FBC either associated with particles or in the vapor phase are difficult to compare at this time with hydrocarbon emissions from conventional

combustors, as few measurements from the latter have been reported. There is the need to compare the hydrocarbon emissions from conventional and FBC combustion of coal using comparable sampling, and it does appear that even when sampling is done following about 2 min of diluting and cooling from the FBC, more of the hydrocarbon is in the vapor phase than is associated with particles. There is a need to determine the mechanisms of interaction between hydrocarbons and particles before and after they are released from the stack from coal combustion processes.

Following cooling and diluting, the raw and clean gas process streams of the gasifier respirable aerosols were collected by the sampling system. The majority of the aerosol particles appeared to be tars and oils and were chemically labile, changing from a liquid to a "varnish" overnight. The coal gas aerosol showed size dependence for some elemental concentrations. Organic analyses of vapor phase and particle-associated hydrocarbons show that complex mixtures are present in these potential effluents. Aerosols such as these could be released from gasifiers either as fugitive emissions (affecting the general population) or from pipe or valve ruptures (primarily exposing workers). More work is needed to establish the nature of the material that could be released and its biological significance.

Mutagenicity testing of potential effluents collected from the FBC at METC is in progress. Initially an attempt was made to compare the mutagenicity of the METC FBC fly-ash samples with that of fly ash from a conventional power plant.[6] When FBC baghouse fly-ash samples from Montana Rosebud coal were extracted with horse serum, saline solution, or cyclohexane,[6] none of the samples tested demonstrated mutagenic activity. However, when a sample (~125 mg) of METC baghouse fly ash was extracted with methylene chloride, mutagenic activity was observed in the extract. Cyclohexane extracts of ash showed no mutagenic activity, but subsequent methylene chloride extracts of the same ash showed mutagenic activity when tested (Table 5). To date, this particular fly-ash sample has the highest mutagenic activity per gram of ash of any of the FBC fly-ash samples tested. This particular sample was obtained from METC before the ITRI field sampling trips, and thus its history is not well characterized, compared with the other FBC fly-ash samples available for testing. The amount of methylene chloride–extractable hydrocarbons and the mutagenic activity were both much higher from this fly ash

---

*Obtained from B. N. Ames.

Table 5. Mutagenicity of methylene chloride and cyclohexane extracts of Montana Rosebud fly ash

Sample	Fly ash extracted per plate (mg)	Number of revertants in *Salmonella typhimurium* (without/with S9[a])				
		TA1535	TA100	TA1537	TA1538	TA98
Spontaneous revertants (control)		9/7	152/141	10/13	27/40	9/37
Montana Rosebud[b] (methylene chloride extract)	500		*319/433*[c]		*83/485*	
	250		*269/391*		*81/142*	
	125	8/13	*216/304*	45/36	*91/87*	*96/82*
Montana Rosebud[b] (cyclohexane extract)	500		112/131		12/29	
	250		107/117		17/36	
	125	7/5	122/117	8/8	11/25	12/27
Montana Rosebud[b] (methylene chloride extraction after cyclohexane extraction)	500		*316/626*		*98/526*	
	250		*269/435*		*125/218*	
	125	7/7	*152/259*	52/23	*70/55*	*105/116*

[a]S9 = rat liver homogenate, source of metabolizing enzymes.
[b]Collected in bag filter of METC FBC.
[c]Italicized values indicate mutagenic response in this test system.

than from any of the other samples. Both frame shift and base pair substitution mutations have been observed from extracts of this fly-ash sample. The mutagenic activity was increased in all cultures by the addition of enzymes from rat liver homogenates. This indicates the presence of promutagens that require metabolic activation and also suggests that a mixture of mutagenic compounds are present.

To determine if operating conditions can influence the mutagenic activity observed in the effluents from the FBC, a number of samples have been collected under known conditions. The influence of the coal type in the combustor was the first variable tested. Methylene chloride extracts of other FBC baghouse fly-ash samples were tested (Table 6). Extracts of ash collected in the baghouse under controlled conditions when the FBC was burning Montana Rosebud and Western Kentucky coal showed no mutagenic activity. The only other sample tested that has shown mutagenic activity was taken when the FBC was burning Texas lignite coal.

The next variables tested were the operating temperature of the bed and the bed materials. The coal burned in these runs was Texas lignite, which had been shown to be positive in the Ames test. When bulk Texas lignite was burned at 1650°F, the ash yielded extracts with mutagenic activity (Table 7).

Bulk material burned at 1400–1550°F showed a slight response or no mutagenic activity. When the Texas lignite was washed before burning at 1600°F, no mutagenic activity was observed. Thus, pretreatment of the coal may change the potential hazard of the effluent. The resulting refuse from the washing was burned at 1600°F, and the ash extract produced cytotoxic effects masking any potential mutagenic response. Burning the extract of the refuse at lower temperatures (1400–1550°F) produced no mutagenic or cytotoxic effects.

The mutagenic activity of the hydrocarbons from the vapor phase collected on Tenax was also studied. The vapor phase hydrocarbons showed little mutagenic activity when tested with the frame shift bacterial strain TA1538 (Table 8). In addition to the baghouse fly ash, samples of size-selected fly ash have been screened for mutagenic activity. Due to the limitations of the sampling devices, the amount of size-selected ash available for extraction was small compared to the amount of baghouse ash. It has generally been necessary to expose each plate to an extract from at least 125 mg of fly ash before a mutagenic response has been observed (Table 6). The recent addition of a high-volume sampling system in the stack of the FBC at METC makes it possible to collect larger quantities of size-selected fly ash. These

Table 6. Mutagenicity testing of various coal fly-ash extracts
collected from METC fluidized bed combustor

Fly ash collected in bag filter of METC FBC and extracted with methylene chloride

	Fly ash extracted (mg)	Total hydrocarbon per plate (µg)	Number of revertants in TA1538	
			Without S9[a]	With S9
Spontaneous revertants (control)			14	59
Fluidized bed combustion				
Montana Rosebud	125	54	*146[b]*	*239*
(bulk sample for	250	108	*295*	*660*
technique development unknown operating conditions)	500	215	*396*	*1081*
Montana Rosebud	125	0.7	8	15
(controlled conditions)	250	1.3	5	11
	500	2.6	7	17
	1000	5.2	10	23
Western Kentucky	125	5	9	33
	250	10	9	28
	500	20	11	24
Texas lignite	125	17	17	61
	250	35	26	112
	500	69	*43*	*246*

[a]S9 = rat liver homogenate, a source of metabolizing enzymes.
[b]Italicized values indicate mutagenic response in this test system.

Table 7. Mutagenic activity from bag filtered fly ash produced under different operating conditions

Fly ash extracted with methylene chloride

Sample	Bed temperature (°F)	Bed material	Ash extracted (mg)	Total hydrocarbon (µg)	Number of revertants in TA1538		Toxicity
					Without S-9[a]	With S-9	
2-acetylaminofluorene					*230[b]*	*>4000*	−
Control					15	35	−
Solvent blank					8	31	−
Fluidized bed combustion							
Texas lignite	1650	Limestone	2000	63	*1065*		−
(bulk)			1000	32	*414*		−
			500	16	*180*	*423*	−
Texas lignite	1400	Limestone	900	61	20	20	+
(bulk)	1550		900	18	*129*	*246*	+
Texas lignite	1600	SiO₂	2000	38			++++
(refuse)			1000	19			+++
			500	10			++
Texas lignite	1600	Limestone	2000	43	18		+++
(refuse)		+	1000	22	17		++
		SiO₂	500	11	21	38	−
Texas lignite	1450	SiO₂	2000	57	20	18	−
(refuse)			1000	29	14	19	−
			500	14	12	19	−
Texas lignite	1450	Limestone	2000	45	10	31	−
(refuse)		+	1000	23	11	20	−
		SiO₂	500	11	6	14	+
Texas lignite	1600	SiO₂	2000	55	19	22	−
(washed)			1000	28	19	14	−
			500	14	13	15	−

[a]S-9 = rat liver homogenate, a source of metabolizing enzymes.
[b]Italicized values indicate mutagenic response in this system.

Table 8. Mutagenic survey of Tenax vapor phase hydrocarbon samples

Fly ash collected in bag filter of METC FBC and extracted with methylene chloride
No evidence of cytotoxicity (i.e., pinpoint colonies) on any plates

Coal source	Sample location[a]	Hydrocarbon per plate (µg)	Number of revertants in TA1538	
			Without S-9[b]	With S-9
Texas lignite (bulk)	1	58	11	9
Texas lignite (bulk)	1	24	11	37
Texas lignite (refuse)	1	1165	13	45
Texas lignite (refuse)	2	40	8	32
Texas lignite (refuse)	4	19	7	39
Texas lignite (refuse)	1	57	12	35
Texas lignite (washed)	2	48	7	20
Texas lignite (washed)	1	30	9	34
Montana Rosebud (bag ash)		18 (1000 mg ash extracted)	35	53
Spontaneous			9	35

[a]Sample locations: 1 and 2, exhaust stack after all control devices (effluent released to atmosphere); 4, above freeboard of combustor and before control cyclones.

[b]S-9 = rat liver homogenate, a source of metabolizing enzymes.

samples are more representative of the ash discharged into the environment and similar to those tested for a conventional power plant.[6] Using this new sampling system, one set of size-selected samples from the combustion of bituminous Western Kentucky coal has been tested. No mutagenic activity was observed in extracts of this ash even when quantities of ash extracted and hydrocarbons present were equivalent to those from the Montana Rosebud baghouse ash that showed mutagenic activity (Table 9). The Western Kentucky baghouse ash obtained at the same time as the size-selected samples also did not show mutagenic activity (Table 6).

Direct comparisons of this mutagenicity data from FBC fly ash and that reported elsewhere[6] can be made only with certain qualifications. In the two studies, methods of sample collection, type of fuel used, and operating conditions differed markedly. Additional testing of samples from the FBC and of samples collected from a conventional power plant and treated in an identical fashion are needed to resolve these differences.

On the basis of the information that is available at this time, the following conclusions concerning the mutagenicity of extracts of fly ash collected from the FBC at METC and that reported for fly ash from a conventional coal combustor[6] may be made. Mutagens extractable from a sample of conventional fly ash by horse serum, saline solution, or cyclohexane are not present in the samples of FBC fly ash that have been tested in a similar fashion. However, larger samples of FBC baghouse ash from certain coal types yielded methylene chloride extracts that demonstrated mutagenic activity. The sample extracts (Table 9) most comparable to those previously reported[6] showed no mutagenic activity, even when large quantities of ash were extracted. Thus, it is our tentative conclusion that the fly ash collected from the atmospheric FBC at METC does not have a greater mutagenic activity associated with it than that reported for conventional fly ash and, in fact, it may be less. Limited testing of the vapor phase hydrocarbons extracted from the Tenax samplers has not demonstrated mutagenic activity in the few samples tested to date. Since there is a large amount of hydrocarbon in the vapor phase, additional emphasis will be given to mutagenicity screening of the vapor phase hydrocarbon in future testing.

Some of the potential effluents from the METC low-Btu coal gasifier have also been characterized for mutagenic potential in the Ames test. Although tar that has been collected from the disengagement chamber would most probably never be released to the environment, occupational exposures of personnel to similar materials could occur. The disengagement chamber tar is, furthermore, material that can

Table 9. Mutagenicity testing of size-selected
Western Kentucky FBC fly-ash extracts

Fly ash extracted with methylene chloride

Sample	Effective cut-off diameter (μm)	Total hydrocarbon per plate[a] (μg)	Number of revertants in TA1538	
			Without S-9[b]	With S-9
Spontaneous revertants (control)			15	48
High volume Sierra impactor stage				
1	8.27	46	18	25
2	3.59	89	12	28
3	2.25	46	15	29
4	1.15	32	15	31
5	0.67	80	19	32
6	<0.67	86	14	30

[a]Represents extraction of 125 mg fly ash.
[b]S-9 = rat liver homogenate, source of metabolizing enzymes.

be compared with materials that could be released as aerosols or vapors from the gasifier. Thus, a sample of disengagement tar has been fractionated and tested (Table 10). The results indicated that fraction 5 contained a large proportion of the mutagenic activity. Thus, fraction 5 was further fractionated into ether insolubles, ether-soluble acids, ether-soluble bases, ether-soluble neutral compounds, ether-soluble amphoterics, and compounds soluble in ether but much more soluble in water. These subfractions were then tested (Table 11). The ether insolubles, ether-soluble neutral compounds, and, in particular, the ether-soluble bases were significantly mutagenic. Preliminary analysis of the basic subfraction by gas chromatography/mass spectrometry has shown this subfraction to be primarily alkyl-substituted, two-ring nitrogen heterocycles; three-ring and alkylated, three-ring nitrogen heterocycles; and some unsubstituted four-ring nitrogen heterocycles. Testing methylene chloride extracts of the coal gas aerosols collected on filters indicates that, while the filter sample extracts appear to be less active than the tar sample fractions, the activity was also concentrated in fractions 4, 5 and 6.

Table 10. Results of Ames tests of disengagement chamber tar and fractions tar

Sample	Number of revertants in Salmonella typhimurium[a] (without/with S-9)[b]			
	TA98	TA100	TA1537	TA1538
Control (DMSO-saline/DMSO-liver)	23/45	102/109	5/11	14/34
Whole tar	21/357[c]	150/209	9/98	17/336
Fraction 1	34/144	120/276	6/26	11/35
Fraction 2	45/251	143/242	6/37	10/272
Fraction 3	117/195	166/192	33/22	35/53
Fraction 4	48/386	117/218	6/22	21/279
Fraction 5	26/1,432	96/412	2/110	11/1,377
Fraction 6	51/1,647	159/319	18/61	24/357

[a]Average of two plates, 250 μg per plate.
[b]S-9 = rat liver homogenate, source of metabolizing enzymes.
[c]Italicized values indicate mutagenic response in this test system.

Table 11. Results of Ames tests of subfractions
of tar fraction 5

Sample	Number of revertants[a] in *Salmonella typhimurium* (without/with S-9)[b]	
	TA98	TA100
Control (DMSO-saline/DMSO-liver)	19/46	150/137
Ether insolubles	25/*341*[c]	163/256
Ether soluble neutrals	45/*180*	151/*360*
Ether soluble acids	42/52	173/182
Ether soluble bases	24/*824*	164/*3,000*
Amphoterics	22/60	156/160
Water and ether solubles	19/46	150/138

[a]Average of two plates.
[b]S-9 = rat liver homogenate source of metabolizing enzymes.
[c]Italicized values indicate mutagenic responses in this test system.

The finding of mutagenic potential in these materials from an FBC and a gasifier should not be extrapolated to a direct carcinogenic potential hazard to man. The samples were obtained from experimental units being operated under various conditions and without optimized control devices. In fact, some of the materials tested, such as the disengagement-chamber tar, will probably not be released to the environment nor (with proper control procedures) be associated with significant exposures to workers. Thus, these materials would not be associated with a carcinogenic hazard to man. Also the results of this *in vitro* test must be confirmed in mammalian systems and in whole-animal tests before defining the hazard to man.

The cytotoxicity of size-selected samples of FBC fly ash was tested using rabbit alveolar macrophage cell cultures exposed for 20 h to 1 mg/ml of fly ash (Table 12). Under these conditions, the most toxic fly ash were the particles in the 2-μm-diam range. Similar

Table 12. Cytotoxicity of size-selected fly-ash samples from the METC FBC burning Western Kentucky coal

Impactor stage	Effective cut-off diam (μm)	Cytotoxicity[a] (%)
1	8.27	1
2	3.59	9
3	2.25	30
5	0.67	6

[a]At 1.0 mg/ml. Data are means of three determinations.

results have been obtained with other samples of fly ash taken when the FBC was burning other coal types (i.e., Montana Rosebud). Approximately 10 μg/ml of Cd^{+2} or V^{+5} would be required in the media to produce the same toxic effect ($\sim$30% killing). The observed concentration of these and other metals in the ash,[1] assuming 100% solubility at this concentration of fly ash in the media, indicates that less than 0.1 μg/ml of Cd^{+2} or V^{+5} would have been present. Furthermore when culture media preincubated with the fly-ash samples were tested for cytotoxicity, no reductions in macrophage viability were observed. Therefore, the limited cytotoxicity of the ash was probably not related to the metal content of the ash. The cytotoxicity may be related to the particle size of the ash, since smaller and larger particles from the same coal type had reduced cytotoxicity. The size dependence of the cytotoxicity corresponds roughly with the size dependence for phagocytosis of particles by macrophages.[12] Cytotoxicity testing of gasifier aerosols is in progress at this time.

The *in vitro* mutagenicity and cytotoxicity tests will be of considerable value in determining which variables should be selected for further study. However, the human hazard of the material will depend on deposition and retention in the respiratory tract and the resulting dose to critical tissues. As a first step in determining this dose, studies on the deposition and retention of neutron-activated FBC fly ash are being made. In these studies, ash collected from the bag filter of the FBC has been used because of the quantities needed for inhalation exposures of animals. Although it would be desirable to conduct future exposures to material collected from the stacks of FBC, conventional combustors, or gasifiers, present technology does not permit collection of sufficient quantities of material. Efforts are under way at ITRI and other laboratories to develop means for collecting sufficient material from stacks. Until these are available, surrogate material such as this FBC bag filter ash will be used. It is extremely important that the comparability of this ash with that of ash collected before or after emission from the stack be established by physical and chemical characterization as well as by *in vitro* toxicological tests. This evaluation is part of our early evaluation of the samples that have been collected and are being considered for use in inhalation studies.

Whole-body retention was measured by whole-body liquid scintillation counting of neutron-activated FBC bag filter fly ash following a brief nose-only inhalation exposure to an aerosol of fly ash with a mass median aerodynamic diameter (MMAD) of about 3.8 μm and a geometric standard deviation of 2.9 (Fig. 3). This MMAD is considerably larger than that observed for aerosols collected from the stack of the FBC just before emission to the atmosphere (Table 3). Ways to reduce this particle size before introducing the aerosol into the exposure

chamber are being studied, as are the comparative physical, chemical, and *in vitro* toxicological characterizations of aerosols generated in this fashion from FBC bag filter fly ash and that of fly ash collected just before emission from the METC FBC stack.

The relative activity of several elements in the lung were compared as a function of time after exposure, and it appears that Sc and Fe remain associated with the ash particles while Cs, Eu, Mn, and Co appear to be preferentially dissolved. It will be of considerable interest to compare the differential solubility of FBC fly ash with that of conventional fly ash due to the apparent differences in the concentration of certain trace elements discussed earlier in this report.

To determine the potential of FBC fly ash for producing relatively acute damage to the lung *in vivo* and to compare it to the toxicity of other pulmonary toxicants such as cadmium and chromium,[13-15] Syrian hamsters were exposed by inhalation to FBC bag filter fly ash for 6 h. The concentrations of fly ash to which animals were exposed were either 60 or 120 mg/m³ of air. Control animals were exposed to filtered air for 6 h. Animals were sacrificed at 1, 2, 7, and 21 d following exposure. Final analysis of the results of this study are not available at this time. However, preliminary results indicate that this fly ash

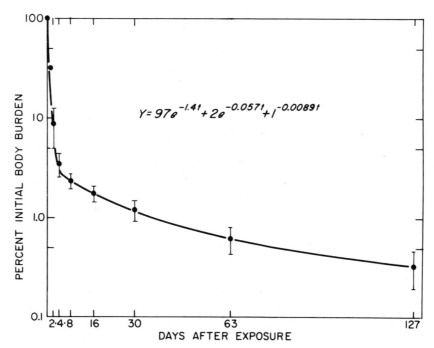

$$Y = 97e^{-1.41t} + 2e^{-0.0571t} + 1e^{-0.0089t}$$

Fig. 3. Whole-body clearance of radioactivity in rats following a single inhalation exposure to neutron-activated FBC fly ash. The data are means ± one standard deviation, corrected for physical decay.

sample induced little, if any, pulmonary damage following this relatively short exposure. These studies are being pursued with more chronic inhalation studies. Studies will also be made of animals exposed to aerosols of ash from conventional coal combustors for comparison.

No studies in which animals have been exposed to simulated or collected samples of coal-gas aerosols from the gasifier have been started. However, methods for the exposure of animals to aerosols of polycyclic aromatic hydrocarbons (PAH) have been developed.[16,17] Animals have been exposed to aerosols of pyrene and the tissue distribution determined by techniques recently developed. Due to the implicated importance of PAH in pulmonary carcinogenesis and the positive results of mutagenesis assays of potential effluents from conventional and advanced (FBC) combustion processes and conversion of coal as well as from other sources such as diesel exhaust, it is essential that particle-hydrocarbon interactions and biological responses be better defined. Studies are planned to assess the hazard to man from the inhalation of hydrocarbons either as vapor or perhaps, more importantly, in association with inhaled small particles.

There are three reports available in the literature on the chronic inhalation of fly ash from conventional coal combustion in experimental animals.[18-20] These studies have shown that $1-1\frac{1}{2}$-y exposures to 0.5 mg/m^3 of fly ash have not resulted in biological effects other than accumulation of ash in lung cells and lymph nodes and some nodules of aggregated macrophages. However, these studies have not used functional indices such as forced expiratory flow-volume and volume-time which are recommended for detecting lung disease from environmental pollutants[21] and shown to be of use in human epidemiological studies.[22] Techniques for measuring functional indices in rodents have recently been developed at ITRI[23] and will be used with rodents chronically exposed to fine particles. Reported studies have not, however, sufficiently established the carcinogenic potential of aerosols that have now been found to have mutagenic and thus, by association, carcinogenis potential for man. Observations of animals and their life-span exposure to relevant atmospheres and better understanding of such factors as actual doses of toxicants to the critical tissues must be achieved before the risk to man from developing fossil fuel technologies such as the FBC combustion or low-Btu gasification of coal can be estimated.

ACKNOWLEDGMENTS

The authors thank METC personnel who supplied FBC and low-Btu gasifier design and operating conditions, especially J. J. Kovach for coordination of efforts, W. E. Wallace, G. D. Case, L. C. Headly, J. Wilson, R. Rice, and U. Grimm for constructive criticism and review. We also acknowledge the efforts of the ITRI sampling team, R. Tamura, D. Horinek, R. Peele, T. Stephens, and E. Barr. Finally we thank our scientific colleagues at ITRI for review and support.

REFERENCES

1. R. L. Carpenter et al., *Characterization of Aerosols Produced by an Experimental Fluidized Bed Coal Combustor Operated with Sub-Bituminous Coal,* Lovelace Inhalation Toxicology Research Institute Report LF-57 Albuquerque, N.M. 1978.

2. R. L. Hanson et al., "Characterization of Potential Organic Emissions from a Low-Btu Gasifier for Coal Conversion," in *Proceedings of Third International Symposium on Polynuclear Aromatic Hydrocarbons,* Battelle Columbus Laboratories (October 25–27, 1978), in press.

3. A. R. Jones, M. R. Guerin, and B. R. Clark, "Preparative-Scale Liquid Chromatographic Fractionation of Crude Oils Derived from Coal and Shale," *Anal. Chem.* **49,** 1766–71 (1977).

4. R. L. Shriner, R. C. Fuson, and D. Y. Curtin, *The Systematic Identification of Organic Compounds,* Fifth Edition, John Wiley and Sons, New York, 1964, pp. 101–02.

5. B. N. Ames, J. McCann, and E. Yamasaki, "Methods for Detecting Carcinogens and Mutagens with the *Salmonella*/Mammalian-Microsome Mutagenicity Test," *Mutat. Res.* **31,** 347–63 (1975).

6. C. E. Chrisp, G. L. Fisher, and J. E. Lammert, "Mutagenicity of Filtrates from Respirable Coal Fly Ash," *Science* **199,** 73–75 (1978).

7. M. D. Waters et al., "Metal Toxicity for Rabbit Alveolar Macrophages In Vitro," *Environ. Res.* **9,** 32–47 (1975).

8. R. L. Carpenter and K. Yerkes, "On the Nature of Fluid Beds as Dry Powder Aerosol Generators," *J. Aerosol Sci.,* submitted for publication.

9. O. G. Raabe et al., "An Improved Apparatus for Acute Inhalation Exposure of Rodents to Radioactive Aerosols," *Toxicol. Appl. Pharmacol.* **26,** 264–73 (1973).

10. R. F. Henderson, E. G. Damon, and T. R. Henderson, "Early Damage Indicators in the Lung. I. Lactate Dehydrogenase Activity in the Airways," *Toxicol. Appl. Pharmacol.* **44**, 291–97 (1978).

11. D. F. S. Natusch, J. R. Wallace, and C. A. Evans, "Toxic Trace Elements: Preferential Concentration in Respirable Particles," *Science* **183**, 202–04 (1974).

12. F. F. Hahn, G. J. Newton, and P. L. Bryant, "In Vitro Phagocytosis of Respirable-Sized Monodisperse Particles by Alveolar Macrophages," in *Pulmonary Macrophage and Epithelial Cells,* ed. C. L. Sanders et al., ERDA Symposium Series 43, Oak Ridge, ERDA Technical Information Center, 1977, pp. 424–35.

13. R. F. Henderson and A. H. Rebar, "Early Damage Indicators in the Lung III. Biochemical Response of the Lung to Lavage with Metal Salts," *Toxicol. Appl. Pharmacol.,* submitted for publication.

14. R. F. Henderson et al., "Early Damage Indicators in the Lung IV. Biochemical and Cytological Response of the Lung to Inhaled $CdCl_2$," *Toxicol. Appl. Pharmacol.,* submitted for publication.

15. R. F. Henderson et al., "Early Damage Indicators in the Lung V. Biochemical and Cytological Response of the Lung to Inhaled $CrCl_3$," *Toxicol. Appl. Pharmacol.,* submitted for publication.

16. K. W. Tu, and G. M. Kanapilly, "A High-Capacity Condensation Aerosol Generation System," *Environ. Sci. Technol.,* submitted for publication.

17. K. W. Tu and G. M. Kanapilly, "Generation and Characterization of Condensation Aerosols of Vanadium Pentoxide and Pyrene," *Amer. Indus. Hyg. Assn. J.,* submitted for publication.

18. H. MacFarland et al., "Chronic Exposure of Cynamolgus Monkeys to Fly Ash," in *Inhaled Particles II, Vol. 1,* ed. W. Walton, Gresham Press, Surrey, 1970, pp. 313–26.

19. Y. Alarie et al., "Long-Term Continuous Exposure to Sulfur Dioxide and Fly Ash Mixtures," *Arch. Environ. Health* **27**, 251–53, (1973).

20. Y. Alarie et al., "Long-Term Exposure to Sulfur Dioxide, Sulfuric Acid Mist, Fly Ash and Their Mixtures," *Arch. Environ. Health* **30**, 254–62 (1975).

21. A. Bouhys, "Environment and Lung Disease," in *Breathing,* Grune & Stratton, New York, 1974, pp. 261–89.

22. R. Knudson, B. Burrows, and M. Lebowitz, "The Maximal Expiration Flow-Volume Curve: Its Use in the Detection of Ventilatory Abnormalities in a Population Study," *Am. Rev. Resp. Dis.* **114**, 871–79 (1976).

23. J. L. Mauderly et al., "Respiratory Measurements of Unsedated Small Laboratory Animals Using Nonrebreathing Valves," *Lab. Anim. Sci.,* submitted for publication.

DISCUSSION

Jim Epler, Oak Ridge National Laboratory: Who does your *Salmonella* mutagenesis work for you? Have you established a group?

Charles Hobbs: We don't have a really large group; it's a very small group. Dr. Rich Clark at the University of California at Davis is primarily responsible for overseeing that work.

Thomas Worthy, METPATH, Inc.: I was wondering whether you had done isozyme patterns particularly on something like LDH where the patterns fluctuate from tissue to tissue. At least in humans there are some indications that altered patterns are present in various carcinomas.

Charles Hobbs: We have done some isozyme work. I think a lot of it is initially related to determining that the LDH that we were looking at actually was of the lung-type LDH, and not an influx of serum or blood-type LDH. This was confirmed in an earlier paper. We really haven't tried to correlate LDH patterns with any types of carcinoma. I think there has been some interest shown by the papers that have been published, perhaps, in looking at some of these end-airway fluids as a diagnostic technique.

SESSION IV: ENVIRONMENTAL TRANSPORT AND ECOLOGICAL EFFECTS

Chairman: R. A. Lewis
U.S. Department of Energy
Cochairman: C. W. Gehrs
Oak Ridge National Laboratory

SESSION IV: ENVIRONMENTAL TRANSPORT AND ECOLOGICAL EFFECTS

SUMMARY

C. W. Gehrs*

Environmental research related to synthetic fuel processes is concerned with evaluating the potential ecological effects arising from effluent and product releases and with determining the ultimate fate of released materials in the environment. The first concern includes assessing potential changes in productivity (such as decreased crops and forest growth) and gene pools (of environmental species). Evaluation of the ultimate fate of materials provides the data for estimating critical environmental pathways whereby contaminants may reach man. As such, it is necessary input for hazard assessment activities. The vectors of interest include atmospheric, solid, and liquid releases.

The six papers included in this session were selected to provide the most recent data in each of the areas of research (atmospheric, solid, and liquid releases). The large quantities of solids produced by synthetic fuel processes, coupled with the potentially stringent requirements of conforming to the Resource Conservation and Recovery Act, make the disposal of solid residues perhaps the most important determinant of economic feasibility for synthetic fuel processes. The first two papers deal with these materials. D. S. Shriner and his group discuss a laboratory-scale experiment in which both chemical and biological screening data for three solid residues from a Lurgi facility are compared. In the second paper, R. K. Skogerboe presents information relative to characterization of spent oil shale. This study was conducted in a field situation in which large-scale lysimeters were used to determine effects of evironmental parameters on controlling fluxes of components of leachate.

Evaluation of the effects of atmospheric releases of synthetic fuel processes is one of the more difficult tasks to undertake at the technological scale of operation currently in existence. This arises from the reactivity of the gases and the inability to "bottle them" up and return them to the laboratory for ecological evaluation. J. L. Shinn and J. R. Kercher of Lawrence Livermore Laboratory present information about photosynthetic and growth response of vegetation to several gases. Through use of modeling capabilities, they have been able to evaluate complex gases superimposed on ambient air quality.

Aqueous effluents and fluxes related to product or process spills and leachates of solid residues suggest that aquatic environments will sustain the greatest impact from synthetic fuel processes. Three questions are addressed by aquatic scientists: (1) what are the hazards, in the near field, associated with operating a particular facility or process, (2) what are the far field effects to the aquatic environment associated with chronic low level releases of effluents, and (3) what are the critical pathways whereby contaminants move through the environment and reach man. The fourth and fifth papers of this session address the first question, while the final paper addresses the last question. H. L. Bergman has been working on in situ coal and oil shale processes for several years. He presents results of his group's efforts to identify several test systems. J. B. States and his colleagues at Pacific Northwest Laboratory have been working with the staff of the Fort Lewis, Washington, SRC pilot facility to identify the toxic components of the effluents to enable design and operations of an optimal waste treatment facility.

---

*Oak Ridge National Laboratory.

The final paper in this session by S. E. Herbes and colleagues presents data from a field study of aqueous transport of polycyclic aromatic hydrocarbons around a coking facility in Pennsylvania. This study corroborated the results of laboratory studies in identifying the critical environmental parameters affecting removal or enhancement of PAH in waters. Ultimately these results will enable development of source terms for use by health assessment scientists in evaluating hazards associated with releases from advanced fossil energy systems.

PHYSICAL, CHEMICAL, AND ECOLOGICAL CHARACTERIZATION OF SOLID WASTES FROM A LURGI GASIFICATION FACILITY*†

D. S. Shiner‡   B. R. Parkhurst‡
H. S. Arora‡   C. W. Gehrs‡
N. T. Edwards‡   T. Tamura‡

ABSTRACT

Coal gasification results in large quantities of solid residuals that require disposal. The method of disposal required for such waste material will depend not only on physical and chemical properties of the ash, but also on the potential biological activity and environmental mobility of leachate components of the waste material. The results of an integrated evaluation of physical, chemical, and biological data from unquenched Lurgi gasifier bottom ash derived from three U.S. coals suggest a low potential for short-term, acute hazard to aquatic or terrestrial ecosystems if proper landfill disposal procedures are used. Of ten heavy metals analyzed in leachates of the wastes, only three—Fe, Mn, and Zn—appeared in concentrations greater than ten times drinking water standards, suggestive of potential chronic hazard to biota in receiving waters. Ecological toxicity tests confirmed toxicity levels of potential concern should one of these leachate fractions occur beyond a landfill boundary.

INTRODUCTION

Coal conversion and, specifically, coal gasification processes are expected to be prominent energy options by 1990. Among the most significant issues facing efforts to utilize our coal resources fully are those environmental and land-use issues associated with the appropriate disposal of solid wastes.

Methods of disposal or by-product use of coal gasification solid wastes will depend strongly on physical and chemical characteristics of the ash. Closely allied to the physical and chemical properties, and perhaps even more critical to disposal or by-product utilization planning are the potential toxicological properties of the wastes. Should the disposal of such wastes result in the transport of toxic trace contaminants either directly or through ecological pathways to man, special handling of those wastes would be required.

A problem most likely to arise is contamination of surface and groundwaters by solid wastes leached from landfills. For most trace contaminants expected to be leachable from a landfill, a number of primarily physical steps such as dilution, dispersion, and attenuation in the soil media will regulate the movement of the contaminant from the landfill site. Dispersion would be expected to be a function of soil structure and parent material properties at the site. Dilution would result from intrusion of the leachate plume into the surface and groundwater. Attenuation would be a function of the physical, chemical, and mineralogical properties of the landfill liner and soil materials at the site.

A first step to assess the requirements reasonably for environmentally safe disposal of coal gasification solid wastes is to evaluate the potential toxicity of extractable components of that waste material. This permits us to establish the magnitude of potential

*Research sponsored by the Office of Environmental Research, U.S. Department of Energy, under contract W-7405-eng-26 with Union Carbide Corporation.

†Publication 1301, Environmental Sciences Division, Oak Ridge National Laboratory.

‡Environmental Sciences Division, Oak Ridge National Laboratory, Oak Ridge, Tennessee 37830.

disposal problems by evaluating the toxic potential of the waste relative to that of known hazardous materials. By analyzing a range of potential dilution factors for surface-to-groundwater and ground-water-to-surface transport of a toxic constituent, we can relate relevant short-term biological tests to actual physical and chemical attributes of a waste material. This paper discusses the results of experiments designed to integrate the physical, chemical, and biological characterization of coal gasification solid wastes.

The particular material selected was chosen for a number of reasons: (1) availability, (2) existing characterization data from which comparisons can be made, and (3) representation of a major coal type.

A number of coal gasification processes are presently under consideration for demonstration plant tests and eventual commercial scale operation in the United States. The Lurgi process was originally developed in the 1930s for noncaking and non-agglomerating coals. The system employs fixed-bed, pressurized, nonslagging gasifiers, which operate either with air or with oxygen and steam. More than 50 individual Lurgi gasifiers are in operation throughout the world.

Short-run trials were conducted by the Office of Coal Research and the American Gas Association at Westfield, Scotland, in 1973-74, to demonstrate the ability of a modified Lurgi process to handle caking and agglomerating U.S. coals. Large quantities of Illinois 5, Illinois 6, Montana Rosebud, and Pittsburg 8 coals were shipped to Scotland for the test.[1] The materials used in this study were samples of ash resulting from the gasification of Illinois 5, Illinois 6, and Montana Rosebud coals in a modified Lurgi gasifier in Westfield, Scotland. There are several previously published reports on the gasification runs, and also a recent report of some preliminary environmental toxicity data for leachates of this material.[2-4]

We present data first on the waste material itself, then on leachates of the wastes, and finally on toxicity tests conducted with those leachate materials. The physical and chemical characterization data emphasize information not previously reported for these waste materials.

A DESCRIPTION OF THE LURGI GASIFIER SOLID WASTES

Limited data are available on environmental characterization of wastes derived from these tests.

Griffin et al.[5] reported that the Lurgi ash from Illinois 6 coal consisted predominantly of noncrystalline, glassy material. Quartz, mullite, hematite, and plagioclase feldspar were the crystalline phases present. The major objective of this section is to characterize the physical and chemical properties of the above solid wastes further.

Materials and Methods

Unquenched Lurgi ash residues of three U.S. coals were received from the central laboratories of Peabody Company, Freeburg, Illinois. Salient physical characteristics, bulk density, particle density, and sieve analysis were determined by routine analytical methods. Portions of the ash were separated by sieve into three fractions for ashing studies: >5-mm siliceous ash, <5-mm siliceous ash, and predominantly carbonaceous material. Samples of approximately 20 g from these fractions were heated in crucibles in a muffle furnace at several predetermined temperatures. A given temperature was maintained for 24 h, after which the crucibles were placed in a dessicator and weighed. When a constant weight (±2%) was attained, the temperature was raised to the next higher level.

The pH was determined with a Beckman pH meter. A 1-to-1 solid-to-deionized water ratio was used. A Yellow Springs Instruments conductivity bridge was used to determine electrical conductivity. Elemental composition was obtained by neutron activation analysis (NAA). Total sulfur was determined with an automatic LECO titrator (Laboratory Equipment Co., St. Joseph, Michigan). Tetrabromoethane (TBE), specific gravity 2.9, was used to obtain two density fractions—specific gravity greater than and less than 2.9—by a double-tube heavy-liquid separation method.[6]

Results and Discussion

Two size fractions of each coal sample were gasified during the Westfield trials. The average chemical composition of feed coal and wastes obtained from these runs differed slightly for a given feed coal (Table 1). Illinois 5 and 6 wastes received for present studies were derived from the gasification of 6- to 32-mm-sized coals. However, we have not been able to associate our Rosebud sample with a given run. These trials were termed "successful" by Woodall-Duckham Ltd.,[3] the operating engineers on the project, and we assume, therefore, that residues obtained from these trials represent wastes that might

Table 1. Salient properties of feed coals and Lurgi waste in gasification trials of U.S. coals in Westfield, Scotland

Each coal listed is divided into two size ranges, in mm

| Constituents | Weight percent composition of coals | | | | | |
| | Rosebud | | Illinois 5 | | Illinois 6 | |
	6–32	2–10	6–32	ROM[a]	6–32	ROM[a]
Coal						
Ash, % dry wt	12.92	12.21	9.23	9.52	10.14	10.57
Pyritic sulfur	0.04	0.82	1.56	1.33	1.50	1.21
Total sulfur	1.45	1.35	3.56	3.51	3.13	3.12
Waste						
Carbon	6.5	4.8	2.0	1.0	3.2	2.1
Sulfur	1.7	1.0	0.6	<0.04	1.3	1.2
SiO_2	46.8	46.9	46.1	44.6	49.6	50.7
Al_2O_3	17.7	21.9	18.1	19.3	20.5	19.1
Fe_2O_3	11.2	9.8	19.7	21.2	17.2	18.8
CaO	8.3	6.8	3.9	4.0	2.1	2.9
MgO	3.9	4.2	0.7	0.2	1.0	1.3

[a]Simulated run of mine.

Source: *Trials of American Coals in a Lurgi Gasifier at Westfield, Scotland*, FE-105, Woodall-Duckham, Ltd., Crawlye, Sussex, England, 1975.

be expected from the commercial-scale operation of that Lurgi gasifier. Engineering reports,[2] however, indicate that significant channelling of gas flow did occur in the fuel bed during these trials.

Physical and chemical properties

The Lurgi wastes can be described as heterogeneous masses of predominantly siliceous composition. Visible in these wastes is uncombusted carbonaceous material that probably resulted from the previously reported channelling. The dry density and particle density values were relatively constant. The "specific gravity >2.9" fraction differed among all three waste samples (Table 2). Sieve analysis data show that particles larger than 2 mm predominated and the wastes are well graded (Table 3). Minor variations observed in the physical characteristics of individual wastes suggest that for the samples tested, process conditions may play at least as great a role as coal type in determining the physical properties of these wastes.

Ignition weight loss data for Rosebud and Illinois 6 wastes suggest the presence of combustible carbonaceous material (Fig. 1). Relatively low particle density values further suggest the occurrence of

carbonaceous material in these wastes, and further confirm that the efficiency of the modified gasifier was not optimized during the trial runs, resulting in some incomplete conversion of the feed coals.

The pH of the Lurgi wastes ranges from acidic (pH 5.4) for Illinois 6, to about neutral (pH 6.85) for Illinois 5, and mildly basic (pH 8.8) for the western Rosebud coal (Table 4). Results of analyses of size fractions of the wastes for pH, electrical conductivity, and dissolved salts, suggest that the chemical composition of various size fractions may vary significantly. With decreasing particle size, larger surface areas are exposed, so pH is affected more, and salts dissolve more completely (Table 4). These soluble salts would be expected to be highly prone to leaching.

Significant differences in the concentrations of several minor elements were observed between the Rosebud and Illinois wastes (Table 5). Arsenic, barium, manganese, and strontium were strongly concentrated in the Rosebud waste, while cobalt, chromium, rubidium, and zinc were predominant in the Illinois wastes. Increased concentrations of a majority of these minor elements in the wastes are consistent with analytical values reported for the feed coals.[7]

Table 2. Selected physical and chemical properties of Lurgi wastes
derived from three U.S. coals

Property	Rosebud	Illinois 5	Illinois 6
Physical			
Color	Gray Black	Brown	Brown
Color after crushing	Black	Gray	Gray
Predominant size, mm	>2	>2	>2
Shape	Irregular	Irregular	Irregular
Dry density, g/cm^3	1.01	0.97	0.95
Particle density, g/cm^3	2.58	2.34	2.47
Specific gravity <2.9 fraction percent	80.7	80.1	83.1
Specific gravity >2.9 fraction percent	19.3	19.9	16.9
Chemical			
pH (1 : 1[a])	8.8	6.8	3.8
Acid or base required for neutralization, milliequivalents per gram of waste	0.024	0.148	<0.01

[a]Solid-to-solution ratio.

Table 3. Sieve analysis of Lurgi wastes as received

Size (mm)	Percent composition by size		
	Rosebud	Illinois 5	Illinois 6
>2	53.2	46.9	59.6
2−084	14.4	19.3	17.6
0.84−0.25	15.7	15.3	10.6
0.25−0.125	5.0	4.6	3.8
0.125−0.106	1.0	0.1	0.7
0.016−0.053	2.5	3.4	2.7
<0.053	8.2	10.4	5.0

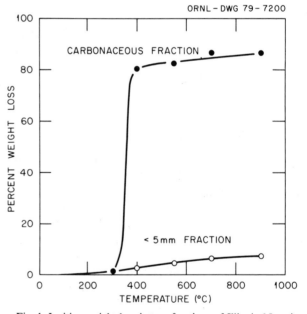

Fig. 1. Ignition weight loss in two fractions of Illinois 6 Lurgi waste.

ASH LEACHING EXPERIMENTS

The three Lurgi wastes contained an abundance and wide variety of trace elements that resulted from the composition of feed coal and from other process-related catalysts and additives. Since little is known about the leaching behavior of trace elements in these wastes, two types of leaching tests were conducted in the laboratory.

The leaching experiments were designed with potential ecological significance in mind. Batch leaching experiments are relatively simple, tend to maximize dissolution of soluble constituents, and simulate somewhat the conditions of natural flooding. Column leaching, on the other hand, was selected to simulate the movement of rainwater percolate through a landfill profile more closely. Deionized water was employed as the leaching solution to substitute for natural rainwater.

Table 4. Relationship of particle size of Lurgi waste to the pH, electrical conductivity,
and dissolved salts after 5 min of laboratory stirring

Particle size (mm)	Rosebud			Illinois 5			Illinois 6		
	pH[a]	Electrical conductivity[a,b]	Dissolved salts[c]	pH[a]	Electrical conductivity[a,b]	Dissolved salts[c]	pH[a]	Electrical conductivity[a,b]	Dissolved salts[c]
>2	6.85	0.041	ND[d]	5.1	0.074	<1	5.7	ND	1.2
0.8–2	6.8	0.105	ND	4.7	0.122	<1	4.2	0.380	1.0
0.25–0.8	7.1	0.250	<1	4.4	0.238	0.9	3.7	0.469	7.7
0.125–0.25	8.1	0.324	1.00	4.5	0.355	1.2	3.6	0.742	4.8
0.106–0.125	8.8	0.470	1.20	4.6	0.608	4.6	3.4	0.742	5.6
0.053–0.106	9.1	0.544	2.46	4.9	0.595	6.4	3.9	0.806	5.2
<0.053	9.8	0.659	5.52	9.0	0.992	9.0	4.5	1.18	10.0

[a] 1-to-25 = solid-to-liquid ratio.
[b] Electrical conductivity (mS/cm).
[c] mg/g of solid material, as determined gravimetrically.
[d] Not determined.

Table 5. Selected microelements in Lurgi conversion wastes[a]

Element	Composition, $\mu g/g$		
	Rosebud	Illinois 5[b] waste	Illinois 6[c] waste
As	19	9.6	3.4
Ba	2005	215	328
Ce	47	65	90
Co	6.8	40	36
Cr	327	846	315
Cs	23	79	8
Hf	5.9	3.5	4
La	31	37	39
Mn	891	333	194
Mo	26	67	30
Rb	33	145	147
Sb	5.85	1.85	1.55
Sc	10	17	26
Sr	726	≤300	193
Ta	1.1	1.05	0.9
Th	14	12.5	14
Ti	3705	4750	4990
U	8.2	14.3	24.6
W	5.6	4.85	3.3
Zn	≤30	1675	466
Zr	≤200	≤300	≤700

[a] As determined by NAA.
[b] Single analyses of single samples.
[c] Means of single analyses of duplicate samples.

Materials and Methods

Batch leaching

Batch mode leaching studies were conducted with deionized water (pH 5.55, electrical conductivity 2.2 μS/cm) as an extracting agent. The 25-g waste sample was placed in an Erlenmeyer flask and 250 ml of deionized water was added in a solid-to-solution ratio of 1 to 10. The suspension was shaken at room temperature on a reciprocating shaker at 120 to 128 strokes per minute. After 24 h, each sample was decanted and filtered through a 0.45-μm Millipore filter.

Solids remaining after decantation and filtration were resuspended in 250 ml of fresh, deionized water; agitated for an additional 24-h period; and filtered through a 0.45-μm Millipore filter. The two filtrate fractions were combined and analyzed for pH and electrical conductivity, and selected trace elements were analyzed by atomic absorption spectroscopy. Additional identical batch leaching tests were performed with crushed waste that was passed through a 270-mesh sieve (<0.053 mm).

Column leaching

Leaching columns were prepared by adding ~ 40 g of air-dried, unsieved Lurgi waste in glass tubes 7.5 cm in diameter and 46 cm long. The waste was tamped to maximize density. Deionized water was employed as a leaching agent. Multiple column volumes of leachate were collected under falling head conditions (flow rate 0.5–2.0 ml/min). The leachate was analyzed for pH, electrical conductivity, and selected major and trace elements.
major and trace elements.

Results and Discussion

Table 6 compares data for batch leaching runs from gasifier-run, as-received waste and waste crushed to 270-mesh-sieve size. As estimated from electrical conductivity data, large variations were evident in total dissolved salts as well as in trace elements.

Relative to the other sample, large concentrations of Cu, Cd, Pb, and Zn were observed in the leachate

Table 6. Batch mode leachate analysis of
selected trace elements from three Lurgi wastes

	pH	E.C. ms/cm[a]	Waste[a]				
			As	Cu	Cd	Pb	Zn
			(µg/g)		(ng/g)		
Rosebud, gasifier run	10.0	0.25	0.60	0.30	0.6	8.0	8.4
Rosebud, crushed	10.2	0.45	1.00	0.14	0.8	15.6	3.0
Illinois 5, gasifier run	10.4	0.38	1.00	0.09	2.4	12.6	16.6
Illinois 5, crushed	10.4	0.61	1.00	0.07	0.6	60.0	2.2
Illinois 6, gasifier run	7.4	0.54	0.6	1.06	48.0	64.0	2720.0
Illinois 6, crushed	6.9	0.33	0.6	0.54	3.6	28.0	76.0

[a]Values represent single analyses of single samples.

from Illinois 6 waste, while minimum dissolution of As, Cd, Pb, and Zn was observed in the Rosebud Lurgi leachate. Except for slightly elevated concentration of arsenic and minimum dissolution of copper, the leachate of Illinois 5 waste contained intermediate amounts of other trace elements in comparison with Illinois 6 and Rosebud Lurgi wastes under the conditions of this experiment. The differences in trace element concentration may at least in part be attributed to the equilibrium pH attained during shaking. Crushing of the ash did not, as might be expected, consistently result in higher concentrations of trace elements in the leachates, and in fact, resulted in much lower concentrations for zinc.

The concentration of selected elements present in successive column volume leachate fractions shows wide variations among the three wastes (Table 7). As with the batch leachate, concentrations of most of the elements were greatest in the Illinois 6 wastes. These data from successive leachate column volumes can be employed in identification and assessment of rates of solubility and movement of leachate components over time. Equilibrium conditions are not obtained over this short-term leaching period, but the decrease in rate of release of the elements in the successive column volumes was very sharp and appeared to approach equilibrium in several cases. Table 8 compares data for batch against data for column extractions for Illinois 6 ash. These data suggest copper and lead to be nearly completely extracted in four column volumes, while arsenic, cadmium, and zinc were much less completely extracted.

As a frame of reference from which to interpret data, the maximum observed concentrations of only five elements—Al, Fe, Mn, Se, Zn—exceeded by at least ten times the recommended levels set by federal water quality criteria, and of those five, only aluminum and iron were still more than ten times greater than that concentration by the fourth column volume. A ten-fold dilution factor was used as an estimate of landfill to groundwater dilution. Although this is the dilution factor proposed for RCRA* implementation, we emphasize that such a dilution factor is arbitrary. Actual dilutions depend heavily upon many site-specific physical, chemical, and geological characteristics not easily generalized. These data agree, however, with those of Griffin's group[5] and fall, as we might expect, on the basis of the neutral pH of the Illinois 6 leachates, between the concentrations that Griffin reported for extracts at pHs 3 and 8.

The rapid mobilization of those potentially toxic trace elements from the ash suggests that during the quenching of this material on site, many of the problem constituents might be retained in the quench water and treated with other process waters. The argument becomes circular, however, if process water

_____

*Resource Conservation and Recovery Act of 1976.

Table 7. Concentration of selected elements in column
volume leachate fractions of three Lurgi wastes

Element	Elemental concentration, $\mu g/ml$							
	Rosebud		Illinois 5		Illinois 6			
	1^a	2	1	2	1	2	3	4
Al	0.20	0.002	0.80	0.11	15.5	6.3	4.6	0.99
As	0.02	0.04	0.07	0.04	0.20	0.10	0.08	0.04
Cd	0.64^b	0.43^b	64.0^b	60.0^b	0.044	0.006	0.003	0.001
Cu	0.030	0.010	0.45	0.07	1.87	1.57	0.62	0.35
Fe	1.6	0.008	45.0	9.1	205.0	51.5	18.0	8.50
Hg	0.30^b	0.60^b	0.05^b	0.07^b				
Mn	0.50	0.017	3.0	2.1	2.18	0.90	0.38	0.18
Ni	35.0^b	24.0^b	0.96	0.52	0.91	0.26	0.06	0.05
Pb	0.09	0.15	0.21	0.11	0.27	0.10	0.03	0.002
Se	0.06	0.07	0.10	0.09	0.091	0.002	0.002	0.002
Zn	2.9^b	0.06	0.31	0.22	1.75	1.45	0.76	0.30

^{a}Sequential column volume fraction.
^{b}Concentrations = nanograms (10^{-9}g) per ml.

Table 8. Comparison of batch and column
leaching methods for efficiency of extraction
of selected trace metals from
Illinois 6 Lurgi ash

Element	Total weight fraction, $\mu g/g$, leached	
	In 4-column volumes^a	In 24-h batch test
As	0.10	0.60
Cd	0.015	0.05
Cu	1.0	1.1
Pb	0.10	0.06
Zn	1.0	2.7

^{a}Assumes density = 1; total H_2O = 4 $\times$ solid mass

evaporator bottoms are disposed of along with quenched bottom ash.

The analytical data show that in the worst case ten of eleven trace elements analyzed exceeded recommended levels for drinking water (Table 7). Further, five of the eleven elements still exceeded those levels after a 10-fold dilution factor had been applied, and aluminum and iron were still at concentrations of 10-fold or more, even after four successive column volumes. Since the extraction procedure represents perhaps the least rigorous approach one might employ, we can conclude that barring credit for soil attenuation, the levels found could require some form of waste treatment or control, such as an impermeable barrier or recycling of leachate.

ECOLOGICAL CHARACTERIZATION

Given chemical characterization data of the type described, one of the more important questions that then arises is the potential ecological significance of the movement of a leachate plume from a landfill into groundwater or surface waters.

In an attempt to analyze environmental risk, we would opt for rapidly available data that at least indicate the potential degree of hazard to the environment. The more rapidly such data can be gathered, the more efficiently environmental control technology can be employed, if it becomes necessary.

A first step in such an assessment is the use of some type of acute toxicity screening test. Development of bioassay tests for ecological purposes is an area with a great deal of current activity. Unfortunately, few, if any, of the commonly employed test methods are universally agreed on. We ourselves have employed two bioassay test systems that we have found to be relatively sensitive and reproducible to assess the biological activity of these leachate fractions initially.

Toxicity to Aquatic Biota: Short-Term Acute Toxicity Determinations

Many of the inorganic and organic components of coal and coal ash are known to be toxic to aquatic biota,[8-10] and the presence of these compounds in water leaching from coal ash may be a potential hazard to aquatic biota.[11] Use of short-term acute toxicity tests provides a starting point to answer the ultimate question, "What is the level of risk to aquatic biota resulting from this technology or process?"

In the aquatic system, we have employed a 48-h acute toxicity test using the zooplankten, *Daphnia magna,* which are relatively sensitive to a wide range of toxic pollutants. Other advantages to their use in toxicity studies include their ease of culture; their small size, which permits testing of small quantities of materials (especially important in cases where sample size is limited); and finally, the lower costs associated with using small test organisms.

These acute toxicity studies, however, provide only a gross evaluation of potential environmental hazard. Acute toxicity should immediately suggest a potential hazard, but the absence of acute toxicity cannot ensure a nonhazardous status. Various extrapolation factors have been proposed for application of acute toxicity data to chronic effects assessment. However, none have been found to be universally reliable. Our developing approach to this problem recognizes the need for a battery of tests that will include additional specific chronic toxicity tests before final hazard evaluation.

Materials and methods

A series of dilutions of the leachate were made in spring water, which has pH, 7.8; alkalinity, 119 mg/liter; and hardness, 140 gm/liter. The range of dilutions was selected to bracket 48-h LC_{50} values (the percent dilution of leachate in spring water which

caused a 50% mortality in 48h) determined from preliminary acute toxicity tests. Four *D. magna* were placed in each 100-ml breaker containing 80 ml of the leachate solution and covered with watch glasses. Control beakers with no leachate solution were included in the experiment. All tests were run in triplicate. Temperature was maintained at 22 ± 0.5°C, and lighting was maintained under a 12-h light–dark regime by placing the beakers in a controlled environment chamber. The toxicity tests were run for 48 h, after which the surviving *Daphnia* were counted. Death was defined as the immobilization of the animal. Values for 48-h LC_{50}s and 95% confidence intervals were obtained by the computerized PROBIT analytical procedures.

Results and discussion

Table 9 presents 48-h LC_{50} concentrations for these leachate materials. Several conclusions can be drawn from these data. *Daphnia magna* is very sensitive to pH, with a 48-h LC_{50} for acidity at pH 6.4 and for alkalinity at pH 10.3. This sensitivity is reflected in the toxicities of the more acidic Illinois 5 and 6 leachate fractions. Probable interactants are also the high trace metal concentrations associated with the more acidic fractions. Many of these trace metals are present in the leachates at concentrations (Table 9) that have been found to be toxic to *Daphnia magna* and other freshwater biota.[9-12] In addition, the toxicities of the individual metals may be additive or

Table 9. Acute toxicities of Lurgi ash leachates to *Daphnia magna*

Source coal	Leachate column volume	pH	Heavy metals concentration[a] (mg/liter)	48-h LC_{50}[b] (%)	95% Confidence interval	
					Upper	Lower
Rosebud	1	6.8	2.5	no mortality		
	2	8.8	0.3	no mortality		
	3	9.5		no mortality		
	4	10.0		49.1	35.1	82.9
Illinois 5	1	4.5	51.0	45.0	not determined	
	2	5.2	12.4	no mortality		
	3	5.6		no mortality		
	4	5.1		52.4	36.0	264.3
Illinois 6	1	3.2	312.2	5.5	3.7	10.1
	2	3.9		16.8	10.3	33.9
	3	4.3		60.0	26.7	1307.0
	4	5.1		no mortality		

[a]Includes Zn, Fe, Ni, Mn, Cu, Al, Hg, As, Se, Pb, Cd
[b]48-h LC_{50} = the percent dilution of the leachate in spring water which caused a 50% mortality in 48 h

otherwise interactive, which in itself strongly justifies not relying solely on chemical characterization.

The Rosebud leachate had the highest pH and the lowest toxicity of the three leachates. The first three column volumes leached from the Rosebud ash had pHs ranging from 6.8 to 9.5 and were nontoxic. The fourth column volume had a pH of 10 and had a 48-h LC_{50} at a dilution of 49%. Since the 48-h LC_{50} for alkaline pH is 10.3 and the Rosebud leachate had the lowest concentration of heavy metals of the three leachates, the toxicity of the leachates was probably caused by their high pH.

The Illinois 5 leachates had pHs ranging from 4.5 to 5.6 and had heavy metals concentrations between those of the Rosebud and the Illinois 6 leachates. These conditions were reflected in their intermediate toxicities, as compared with the other leachates.

A Phytotoxicity Screening Assay: Seed Germination Test

The genesis of bioassay systems of terrestrial biota has lagged significantly behind those for aquatic ecosystems, and only recently have efforts increased to fill this gap. Hopefully, such tests would take into consideration such factors as (1) probable pathways of exposure—above or below ground, for example; (2) the mode of phytotoxic action; and (3) the effects of the relatively greater range of physical variables—temperature, humidity, light, etc.—to which terrestrial plants are subject.

The development of various bioassay techniques for determining phytotoxic effects of substances came about primarily through weed control research.[13] Results from herbicide studies indicate that the variability in mode of phytotoxic action and in the effects of abiotic physical variables on the toxicity of many chemicals make the development of a "foolproof" phytotoxicity screening system unlikely.[14] Therefore, in our laboratory, a battery of tests is performed in addition to chemical analysis of the leachates—including seed germination tests, longer term seedling growth tests, and ultimately, field validation experiments—before the effects of a potential toxicant on the growth of terrestrial flora are finally assessed. However, if rapid and simple tests show that a substance is unquestionably phytotoxic, further testing of that substance may only be necessary to elucidate the actual toxic constituent.

The seed germination bioassay used with these materials represents the first step in a screening process still being developed for the detection of phytotoxic properties of solid waste leachates. It is simple, rapid, and inexpensive. Hopefully, such a system will prove to be sensitive enough to minimize requirements for more costly alternatives.

Materials and methods

Radish seed bioassays were completed on four sequential column volumes of each of three distilled-water, volumn-leached, solid wastes. Radish seeds were chosen for the tests because they germinate quickly and uniformly. Within 48 h after radish seeds are given adequate moisture and a suitable temperature, hypocotyl growth is well underway and measurements can be easily made on a relatively large number of seeds. Radish plants (*Raphanus sativus* L. var. Early Scarlet Globe) were used in all tests. Seeds were germinated in petri dishes containing filter paper moistened with three concentrations, respectively—100, 50, and 10%—of each leachate diluted in distilled water. Hypocotyl elongation was measured 48 h after the seeds were moistened. Thirty seeds were used for each leachate concentration. Seeds were kept at 25° C throughout the 48-h period. Distilled water, with the pH adjusted to the pH of the leachate, was used as a control (pH range 3.2–10.0).

Results and discussion

The results given in Table 10 compare the phytotoxic effects of various leachate column volumes. No effects of pH on radish hypocotyl elongation were observed between pH 3.6 and pH 9.3. However, at a pH of 3.2, growth was slightly inhibited, and a pH of 10 greatly reduced hypocotyl growth. The first column volume of Illinois 6 was the most toxic of all the leachates, causing 62% growth reduction at 10% concentration. At least two of the fractions of all leachates tested caused significant growth reduction of radish hypocotyls. However, only one of the leachate fractions, fraction 1 of Illinois 6, produced a growth reduction at the 10% concentration.

If we assume that plant materials would most likely be exposed to leachates in groundwater, either during revegetation of the landfill site or even perhaps through the use of groundwater for the irrigation of crops, the 1:10 landfill-to-groundwater dilution factor should be expected to apply. In this case, only one of the wastes appears to have potential to reduce plant growth in a worst case, and additional testing of that material in a chronic mode is advisable.

Table 10. Toxicity of Lurgi ash leachates to
Raphanus sativus L[a]

Source coal	Leachate column fraction	Concentration (%)	pH	Reduction of growth (%)
Rosebud	1	100	6.5	18
		50	6.5	18
		10	5.5	0
	2	100	9.2	52[b]
		50	9.2	14
	3	100	9.5	38[b]
		50	9.5	19
		50	9.3	9
	4	100	10.0	31[b]
		50	9.8	15
		10	9.5	9
Illinois 5	1	100	4.5	48[b]
		50	4.5	24[b]
		10	5.5	0
	2	100	5.2	11
		50	5.2	0
	3	100	5.6	12
		50	5.6	0
	4	100	5.1	86[b]
		50	5.2	72[b]
		10	5.4	0
Illinois 6	1	100	3.2	75[b]
		50	3.2	75[b]
		10	3.6	62[b]
		1	3.6	0
	2	100	3.9	45[b]
		50	4.1	55[b]
		10	4.5	5
	3	100	4.3	0
	4	100	5.1	9

[a]var. early scarlet globe
[b]Indicates statistical significance at the 0.01 probability level

Again, soil and groundwater variables at specific sites will play important roles in the actual hazard potential of such material, and additional experimentation is under development to add such variables to the assessment scheme.

CONCLUSIONS

Although physical characteristics of the three waste materials studied were relatively constant and relatively independent of the feed coal type, the elemental composition of the wastes differed significantly as a function of the composition of the source coals. Moreover, the results suggest a potentially significant subsampling problem associated with such wastes because of the wide variation in pH and soluble salts in the various size fractions. Proper subsampling techniques become extremely important to allow correct interpretation of results under these circumstances. These experiments indicate that since toxicity is related to both solid waste and leachate characteristics, better estimates of the replicability of results from different subsamples and precision of trace element analyses may be expected to be extremely important to the interpretation of future generations of tests of this type.

The ultimate significance of the acute toxicities determined here would be measured by several site-specific factors as well: (1) the dilution to surface waters—a factor of 1:100 is considered representative, but obviously is very site dependent; (2) the quality of the receiving water; and (3) the soil attenuation.

If only the dilution factor is considered, the most toxic leachate had a 48-h LC_{50} for *Daphnia* of ~5.5% or ~1:18. With the 1:100 dilution factor employed, ample dilution should occur before any trace toxicants reached surface-water bodies. Therefore, little *acute* hazard—that is, risk of morbidity or mortality—exists to sensitive aquatic and terrestrial bioassay species. The Illinois 6 material showed potential for toxicity on the basis of chemical characterization and bioassay tests at dilution levels potentially representative of groundwater concentrations. These data suggest that further study with more advanced biological systems is advisable.

Such experiments are presented for more than the data they derive: they are an illustration of methods the authors propose as a basic framework for solid waste hazard assessment. We would caution that at first, this type of evaluation must be of an iterative nature and that those who would use such data must do so with appropriate regard for the limitations of the data. On the other hand, the scientific community must begin to evaluate such approaches critically in order to initiate this iterative process, and it is in this spirit that we have presented results from characterization of unquenched bottom ash from a Lurgi gasifier.

ACKNOWLEDGMENTS

Appreciation is sincerely expressed to the central laboratories of Peabody Coal Company, Freeburg, Illinois, for supplying the ash samples used in this study. The technical assistance of A. S. Bradshaw, J. L. Forte, M. C. Reid, M. R. Todd, and H. W. Wilson is deeply appreciated.

LITERATURE CITED

1. N. F. Sather et al., *Potential Trace Element Emissions from the Gasification of Illinois Coal,* State of Illinois, Institute for Environmental Quality Document 75-08, 1975.

2. D. C. Elgin and H. R. Perks, "Results of Trials of American Coals in Lurgi Pressure Gasification Plant at Westfield, Scotland," pp. 247–68, in *Proceedings of Sixth Synthetic Pipeline Gas Symposium,* American Gas Association, Chicago, 1974.

3. *Trials of American Coals in a Lurgi Gasifier at Westfield, Scotland,* FE-105, Woodall-Duckham, Ltd., Crawlye, Sussex, England, 1975.

4. W. P. Van Meter and R. E. Erickson, *Environmental Effects from Leaching of Coal Conversion By-Products. Interim Reports October 1975–March 1977,* FF-2019, University of Montana, Missoula, Mont., 1977.

5. R. A. Griffin et al., "Solubility Attenuation and Toxicity of Accessory Elements in Coal Wastes and Leachates," unpublished report to the Committee on Accessory Elements in Coal Board of Mineral Resources, National Research Council, National Academy of Sciences, Washington, D.C., March 1977.

6. H. S. Arora, J. B. Dixon, and L. R. Hossner, "Pyrite Morphology in Lignitic Coal and Associated Strata of East Texas," *Soil Sci.* **125,** 151–59 (March 1978).

7. H. J. Gloskoter et al., *Trace Elements in Coal: Occurrence and Distribution,* Circular 499, Illinois State Geological Survey, Urbana, Ill., 1977.

8. D. J. Wilkes, "Bioenvironmental Effects," pp. 9-1–9-144 in *Environmental, Health, and Control Aspects of Coal Conversion: An Information Overview,* vol. 2, ed. H. M. Braunstein, E. D. Copenhaver, and H. A. Pfuderer, ORNL/EIS-95, Oak Ridge National Laboratory, Oak Ridge, Tenn., April 1977.

9. J. E. McKee and H. W. Wolf, *Water Quality Criteria,* 2nd ed., Publ. 3-A, State of California, Water Resources Control Board, 1963.

10. C. D. Becker and T. O. Thatcher, *Toxicity of Power Plant Chemicals to Aquatic Life,* WASH-1249, United State Atomic Energy Commission, June 1973.

11. M. H. Somerville, J. L. Edler, and R. G. Todd, *Trace Elements: Analysis of Their Potential Impact from a Coal Gasification Facility,* Engineering Experiment Station Bulletin 77-05-EES-01, University of North Dakota, Grand Forks, N.D., 1977.

12. S. G. Hildebrand, R. M. Cushman, and J. A. Carter, "The Potential Toxicity and Bioaccumulation in Aquatic Systems of Trace Elements Present in Aqueous Coal Conversion Effluents, pp. 305–13 in *Trace Substances in Environmental Health-X. A Symposium,* ed. D. D. Hemphill, University of Missouri, Columbia, Mo., 1976.

13. B. A. Krathy and G. F. Warren, "The Use of Three Simple, Rapid Bioassays on Fourth-Two Herbicides," *Weed Res.* **11,** 257–62 (1971).

14. R. P. Upchurch and D. D. Mason, "The Influence of Soil Organic Matter on Phytotoxicity of Herbicides," *Weeds* **10,** 9–14 (1962).

DISCUSSION

Phil Singer, University of North Carolina: In your *Daphnia* studies, when you looked at the various fractions, were they conducted with a test carried out at the same pH as that at which the wastes were generated, or were they neutralized as part of the testing?

Dave Shriner, Oak Ridge National Laboratory: These were neutralized to eliminate the pH effect.

Phil Singer: Did you find that the metals precipitated out upon neutralization?

Dave Shriner: We have experienced that problem with some samples, but as I recall that was not a particular problem with these samples.

Phil Singer: Okay, I guess I'm thinking that the pHs are very important to the parameter in terms of controlling metals distribution.

Dave Shriner: Yes, they are.

Phil Singer: Have you made any kind of equilibrium calculations on the concentrations of metals coming out of the leaching columns at the given pH values to see if there were any equilibrium controls on their solubility?

Dave Shriner: No, we have not done that up to this time.

Ron Neufeld, University of Pittsburgh: I'd like to restate Phil Singer's precipitation question. Have you noticed any subsequent changes of samples of the leachates, particularly precipitation, as you store the samples over a period of time?

Dave Shriner: That is something that we haven't really looked at from the standpoint that we try to use these samples for the biological tests as rapidly as possible after they are generated.

FIELD LEVEL CHARACTERIZATION OF SOLID RESIDUES FROM OIL SHALE RETORTING*

R. K. Skogerboe†
Department of Chemistry

W. A. Berg†
Department of Agronomy

D. B. McWhorter†
Department of Agricultural and Chemical Engineering

INTRODUCTION

Interest in developing western oil shale reserves has been encouraged by the limited domestic oil and natural gas reserves and the increased price of imported oil. The oil shale reserves located mainly in Colorado, Utah, and Wyoming could supply U.S. needs for 100 years at 1977 consumption rates. The development of an oil shale industry will require that many environmental and technical problems be solved.

Oil can be recovered from shale in three general ways: (1) open pit or room and pillar mining followed by aboveground retorting, (2) fracturing of the underground shale beds followed by in-place (in situ) retorting, and (3) modified in situ recovery in which a fraction of the shale is mined conventionally followed by caving to create a permeable system for in situ retorting.[1] The modified in situ technology is the only approach now under development on a commercial scale.

Environmental concerns associated with oil shale development include control of product and by-product discharges, effects on water resources, aquatic and terrestrial ecological effects, health and safety effects, and socioeconomic impacts.[2] The most significant environmental and technological problems relate to water [i.e., the impact(s) on water resources as well as the control and use of the coproduced (process) waters]. The water problem is foremost because the development is in a region where water is a valuable commodity in short supply and because of the number of uncertainties yet to be resolved. Several studies[1,3] have been made of the process (coproduced) water,[1] but will not be discussed here.

Although a mature oil shale industry could produce a million barrels/day,[4] it would also produce approximately 200 ha·m of retorted oil shale (ROS) wastes, which would be either left underground or disposed of in surface depositories. The environmental problem linked to ROS is water pollution by leaching, because it will be impossible to prevent ROS contact with surface and ground waters totally. The primary technological problem is the management of ROS deposits to minimize subsurface leaching.[5,6] In either case, a knowledge is essential of (1) the chemical characteristics of ROS leachates, (2) the factors which control leaching, (3) the environmental lifetimes of the leachate constituents and their degradation products, and (4) the possible modes of impact on the environment. The present report summarizes the results of investigations designed to provide such information.

---

*Research supported by the Environmental Protection Agency under Grant Numbers R 803788-03 (Environmental Research Laboratory, Cincinnati, Ohio) and R 803950 (Environmental Research Laboratory, Duluth, Minnesota).

†Colorado State University, Fort Collins, Colorado 80523

INVESTIGATIVE STRATEGY

To study these four subjects, the most common investigative approach is to do leaching studies on a micro or semimicro scale under laboratory conditions. This is partly because soil chemists, geochemists, and other chemists (as well as scientists from other disciplines) are laboratory oriented. Moreover, the experiments can be better controlled in the laboratory. Laboratory-scale leaching studies provide reliable answers if carried out under conditions properly simulative of the environment in question. This requirement is often not met, so laboratory studies serve only as approximations. Principles and approaches drawn from the field of hydrologic mining show that characterization of leaching in field experiments is more accurate.

The scientist dealing with hydrologic mining starts with the a priori recognition that he is dealing with a complex physical and chemical system that is difficult to model. Factors affecting interactions of water-mineral systems include:

1. identities, concentrations, and relative solubilities of the minerals present;

2. the size distributions and physical forms of the respective minerals;

3. the prevalent (limiting) solution equilibria including those associated with dissolved oxygen and carbon dioxide;

4. the degree of compaction of the mineral deposit;

5. seasonal precipitation patterns;

6. the degree of surface runoff and/or run-in;

7. the aspect, slope, and elevation of the dump surface;

8. nature and extent of plant cover;

9. the extent of cementation; and

10. the frequencies and durations of freeze-thaw and wet-dry cycles.

Although this list is not all-inclusive, it emphasizes the importance of the physical and environmental, as well as chemical, characteristics of the mineral deposit—characteristics difficult to simulate in a laboratory.

Leaching of a mineral deposit (e.g., ROS dump) is affected by physical dimensions at four different orders of magnitude, those of (1) the deposit itself; (2) the particle array (interparticle porosity), which limits solution flow, liquid retention, and accessible surface area; (3) the individual particle (intraparticle porosity), which influences the rate of extraction; and (4) the reaction layer of the individual particle (Fig. 1).

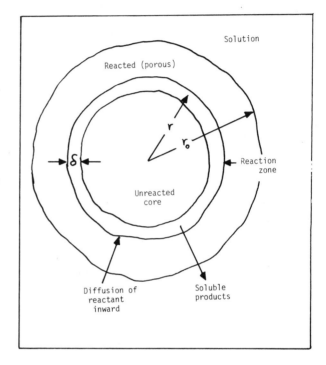

Fig. 1. Mixed mineral particle showing diffusion path and reaction zone.

As particles consisting of a mixture (aggregate) of soluble and insoluble minerals dissolve, the water must diffuse through a porous layer (from which the soluble minerals have been dissolved) to the reaction layer and then back out to carry the dissolved products into the bulk solution.[7] As a result, the rate of leaching is limited by the interparticle porosity and its influence on the mass transport of solution through the mineral dump. Since water will move through at least a portion of the dump, the rate and extent of leaching will ultimately be controlled by the intraparticle porosity of the system in question. The leaching kinetics can be described by:

$$\text{Rate} = c \cdot f(\alpha, d, r)h^n A k_a , \qquad (1)$$

where c is the concentration of the soluble mineral(s) available, α is the fraction of mineral dissolved, d is the size dimension of kinetic importance (δ in Fig. 1), r is the dimension of the porous zone that must be penetrated to leach the reaction layer (e.g., $r_0 - r$ in Fig. 1), h^n is a depth dimension function, A is the effective cross-sectional area of the dump (as influenced by the interparticle porosity), and k_a is the apparent rate constant. Of particular importance is the geometric factor function, $f(\alpha, d, r)$, which is greatly affected by limitations on the rate at which water moves through the porous layer to the reaction layer and carries soluble products back to the mobile bulk solution. Unless the particle is composed entirely of readily soluble minerals, this function will be highly nonlinear and could be so sharply affected by retention of water in capillary pores that the leaching process would be entirely limited to ionic diffusion.[7]

These factors suggest the complexity of mineral leaching and the difficulty of modeling it. For these reasons leaching studies representative of the actual ROS should be based on a large-scale system. While such investigations are more expensive than laboratory-scale studies, the increased costs may be offset by combining several investigative efforts on the same system. This was done in the present case.

GENERAL OBJECTIVES

The present program has five objectives:

1. to investigate the movement of water and the salts leached from the ROS through compacted and uncompacted zones;

2. to model leaching from the ROS in relation to the pertinent chemical and physical parameters including coverage of the deposits with varying depths of local soils;

3. to define and demonstrate means for, and the effectiveness of, vegetative stabilization of ROS disposal sites;

4. to identify leachate constituents potentially toxic to aquatic life; and

5. to characterize the transformation and transport behaviors (environmental lifetimes) of toxic constituents present.

EXPERIMENTAL SYSTEM

Each unit of the large-scale lysimeters[5,6] is 13.1 m wide by 40 m long and 2.4 m deep and has a 25% slope representing the dam face of a possible disposal configuration and a 2% slope representing the fill area (Fig. 2). Retorted shale was placed in each lysimeter to give a 90-cm-deep compacted zone (1.5–1.6 g/ml) under a 150-cm uncompacted zone (1.2–1.4 g/ml); other treatments included coverage with 0, 20, 40, 60, and 80 cm of soil. The units have drainage systems which allow measurement and collection of surface runoff, subsurface flow at the compacted-uncompacted interface, and subsurface flow from below the compacted layer (Fig. 2). After a measured amount of leaching water had been applied by sprinkler irrigation, the hydrologic flow through the system was then measured with

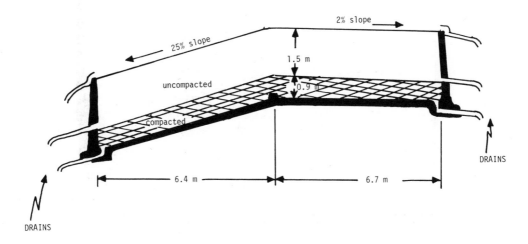

Fig. 2. Cross section of the ROS lysimeters.

tensiometers, piezometers, and neutron probe access tubes in each lysimeter. Moreover, in situ salinity sensors have been used as temporal indicators of saline leaching processes.[5,6] One irrigation leaching method uses intensive irrigation at the start to leach out excess salts in preparation for planting of vegetation followed by continued irrigation as required to establish plant growth. Thus, the lysimeter systems allow parallel investigations under conditions quite representative of a probable ROS disposal site. The relatively large size of the units allows the collection of appreciable quantities of leachate waters for further characterization studies (e.g., biological testing).

RESULTS

The chemical and physical properties of ROS will vary considerably with the retorting process used. In this case the ROS used in the lysimeters was produced by a direct gas combustion process.

More than 90% of the water recovered was collected from the drainage below the compacted

layer. Appreciable amounts of water were retained by the ROS, including the compacted zone. Although covering the ROS to increasing depths of soil caused increased delays in penetration by the water, the general water quality of the leachates collected was uniform.

A general overview of the leachate composition and the changes in it may be deduced from test data presented in Fig. 3. Because the water used for irrigation had a pH of ~7.8 (hardness, ~200 ppm as $CaCO_3$), the lower pH observed early in the cycle implies the occurrence of acid production (e.g., perhaps by dissolution of sulfide minerals or dissolution of acidic salts). This trend was rapidly offset by the dissolution of basic materials which raised the pH to between 10 and 11 for long periods. The pH ultimately returned to the baseline level of the water used (~7.8). The sulfate concentration of the leachate paralleled these pH trends, achieving very high levels and dropping off sharply as the most soluble sulfate salts were removed by leaching.

The trends of the major and minor leachate constituents (Table 1) provide further general

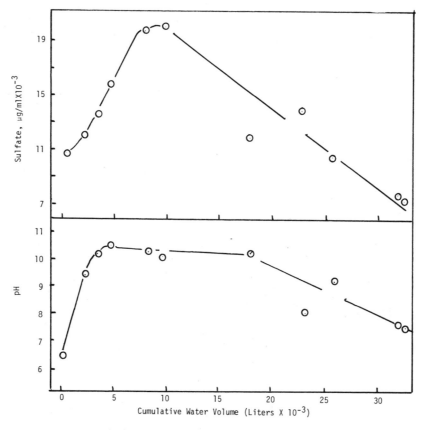

Fig. 3. Changes in sulfate concentration and pH.

insights. Those constituents showing concentration decreases early in the leaching cycle must be associated with highly soluble and accessible minerals (e.g., NaCl). Those showing a slower rise with a later decrease are associated with less soluble and/or less accessible minerals (e.g., a soluble salt in an aggregated particle). Constituents appearing in the leachates at relatively constant concentrations are largely those present in lower solubility forms (e.g., $MgSO_4$), in particle forms that lower their accessibility to water, or in soluble forms which dissolve and lead to reprecipitation of the cation by the anionic composition of the leachate waters. An example of the latter is the dissolution of soluble forms of Ca, Mg, or Sr followed by their reprecipitation as sulfates due to the excess of sulfate present. The constituent trends (Table 1) can be explained in terms of the mineral forms expected for ROS. For example, a correlation analysis between the Na plus K and the SO_4 concentrations suggests that more than 95% of the sulfate present could be accounted for by the presence of Na_2SO_4 and K_2SO_4.

The trends of trace constituents (Table 2) give qualitative information regarding the solubilities and/or accessibilities of the substances (relative to local surface waters) present. The presence of high and constant concentrations of As, B, F, Mo, Se, and V suggests that relatively large quantities of these

Table 1. Summary of typical leachate concentrations
and trends for major constituents

Constituent	Leaching trend	Typical concentrations (mg/l)
C (organic)	Decreasing	$250 \rightarrow 5$
Ca	Relatively constant	300–500
Cl	Decreasing	$2,000 \rightarrow 50$
K	Initial rising, later decreasing	300–1,500
Mg	Relatively constant	1–30
Na	Initial rising, later decreasing	3,000–15,000
SiO_2	Relatively constant	10–30
SO_4^{2-}	High, decreasing	$20,000 \rightarrow 5,000$
$S_2O_3^{2-}$	High, decreasing	$40,000 \rightarrow 5,000$
Sr	Relatively constant	5–15
Hardness	Initial rising, later decreasing	100–500
pH	Initial rising, later decreasing	$10.5 \rightarrow 7.5$

Table 2. Summary of typical leachate concentrations
and trends for trace constituents

Constituent	Leaching trend	Typical concentration ($\mu g/l$)
Al	Inconsistent	$150 \rightarrow <10$
As	Relatively constant	40–100
B	Relatively constant	1,000–2,000
Ba	Decreasing	$150 \rightarrow 40$
Be		<10
Cr	Decreasing	$50 \rightarrow 1$
Cu	Decreasing	$200 \rightarrow 2$
F	Relatively constant	1,000–2,000
Fe	Decreasing	$300 \rightarrow <10$
Mn	Decreasing	$100 \rightarrow <10$
Mo	Relatively constant	1,000–10,000
Ni	Decreasing	$50 \rightarrow <10$
NO_3^-		<100
Pb	Decreasing	$50 \rightarrow <5$
Se	Relatively constant	10–30
V	Relatively constant	100–300
Zn	Decreasing	$1,000 \rightarrow 50$

constituents are retained in the ROS in moderately soluble forms. Because these elements have harmful effects, they are likely to be of great concern.

Large concentrations of thiosulfate are present; charge- and mass-balance calculations for these thiosulfate leachates show a large deficiency in cations which suggests that the thiosulfate (or another anionic species) may originate from an organic rather than an inorganic compound. This requires further experimental validation.

The presence of thiosulfate has distinct biological consequences because it imparts a high chemical oxygen demand to the water. The addition of as little as 0.1–0.5 v/v% of these leachates to a typical western surface water completely removes the dissolved oxygen and quickly results in the death of oxygen-dependent organisms including fish. Similar problems have been found with the oil shale retort process (coproduced) waters.[1] Present investigations have relied on extensive aeration to offset the destruction of dissolved oxygen by thiosulfate. Under such circumstances the effects on aquatic life are sharply reduced.

The present program is moving ahead with modeling the physical and chemical parameters which control leaching in large-scale systems, in identifying the long-term leaching effects, in characterizing the ways in which environmental factors such as freeze-thaw and wet-dry cycles affect the system, in evaluating vegetative stabilization, and in identifying the potential impacts of ROS disposal on the aquatic system. The investigations planned are comprehensive; their completion appears essential to the development of an environmentally acceptable oil shale industry.

REFERENCES

1. D. S. Farrier et al., "Environmental Research for In Situ Oil Shale Processing," in *Proceedings of 11th Annual Oil Shale Symposium,* Colorado State University Press, Golden, Colo., 1978.

2. J. M. Powderly, J. Reilly, H. Thomas, and C. Wilson, "Environmental Planning for ERDA's Oil Shale Technology Program: Statement of the Problem," in *Proceedings of 2nd Pacific Chemical Engineering Congress,* U.S. Energy Research and Development Administration, San Francisco, Calif., 1977.

3. H. L. Bergman et al., "Effects of Complex Effluents from In Situ Fossil Fuel Processing of Aquatic Biota," in these proceedings.

4. U.S. Department of Interior, *Final Environmental Statement for the Prototype Oil Shale Leasing Program,* Vol. I, Washington, D.C., 1973.

5. H. P. Harbert III, W. A. Berg, and D. McWhorter, "Vegetative Stabilization of Paraho Spent Oil Shales," Annual Report on U.S.E.P.A. Grant No. R 803788-03, 1978.

6. H. P. Harbert III, "Lysimeter Study on the Disposal of Paraho Retorted Shale," M.S. Thesis, Colorado State University, 1978.

7. M. Wadsworth, "Hydrologic Extraction of Minerals," presented at Denver University Conference on Mining and the Environment, Denver, Colo., 1975.

DISCUSSION

Phil Singer, University of North Carolina: Is there much pyritic sulfur in these samples?

Rod Skogerboe, Colorado State University: There is a reasonable amount of pyritic sulfur, in several forms, although most of the sulfur is organic.

Don Gardiner: I would like to compliment the speaker on his graphs. When I see graphs with data points on them, no matter how crude the curve or how naive the models, somehow I have more confidence in those curves when I see the points through which they pass.

Rod Skogerboe: Thank you, I appreciate that.

ECOLOGICAL EFFECTS OF ATMOSPHERIC RELEASES
FROM SYNTHETIC FUELS PROCESSES*

J. H. Shinn†
J. R. Kercher†

ABSTRACT

The primary ecological effects of exposure to synthetic fuels gas-stream emissions are discussed in this paper. The primary impacts are on vegetation caused by a combination of phytotoxic gases. Little is known about the effects of potential trace gas emissions such as COS, CS_2, hydrocarbons, and volatile metals, but the most important phytoactive gases are CO_2 and H_2S. With the aid of a model, photosynthesis and growth effects are investigated on the basis of experimental, bioassay dose-response data with mixtures of gases. The threshold concentration for growth injury (dry matter production) is an order of magnitude lower than for photosynthesis. CO_2 ameliorates H_2S effects; other environmental influences on dose response were demonstrated by the model. Urban smog (ozone) increases the toxicity of H_2S. However, under worst-case conditions (chronic exposure, no CO_2 present), the threshold for leaf injury is about 0.2 ppm H_2S.

BACKGROUND

This paper presents data on the primary ecological effects of atmospheric emissions from synthetic fuels (synfuel) processes. In particular, the important inorganic gases that may have an impact on plant growth are examined.

The synfuel effluent gas streams are process specific, especially in the minor constituents. However, we may characterize the inorganic constituents as high in N_2, CO_2, and H_2S with lesser amounts of CO, SO_2, COS, CS_2, and NO_x, and traces of NH_3, volatile metals, hydrogen and others. Each particular synfuel process may have its own fingerprint in terms of the ratios of these constituents and the types and amount of hydrocarbon emissions. For purposes of ecological impact assessment, it is important to determine the ratios of the minor gases to CO_2.

We have been particularly interested in the gas streams from modified in situ (MIS) oil shale retorts that, for example, produced the concentrations of gases shown in Table 1 during preliminary tests.[1] In this case, the ratios of the important sulfur gas H_2S to CO_2 is about 1:100.

Table 1. Constituents in the gas stream
of modified in situ oil shale retorts
during preliminary tests at
Lawrence Livermore Laboratory

Constituent	Concentration
N_2	55–80%
CO_2	15–35%
CO	1–7%
H_2	1–4%
Hydrocarbons	1–3%
H_2S	500–4000 ppm
SO_2	3–5 ppm
NH_3	50–300 ppm
COS	65–125 ppm
As, Se	Trace
NO_x	Unknown[a]

[a]Concentration of NO_x is highly dependent on the input of air and N_2.

*Research sponsored by the U.S. Department of Energy under contract W-7405-eng-48 with the University of California, Lawrence Livermore Laboratory.

†Lawrence Livermore Laboratory.

Target Organisms

Because of the nature of the synfuel emissions, it is important to examine the impact on vegetation. Experience has shown that low levels of SO_2 injure plants, but until recently little was known about H_2S. Furthermore, certain trace metals may accumulate in vegetation and be injurious to consumer animals further up the food chain.

Influences of gaseous emissions on animal populations may be largely indirect rather than direct. For example, loss of habitat may be caused by injury to vegetation. On the other hand, subtle, direct influences may occur. It has been long known that CO_2 anesthesizes pollinator insects, and, in combination with H_2S, it has been shown to increase mortality and decrease the life span of honeybees.[2] The latter effect may decrease seed production in plant communities.

We will present data only on the primary effects, which are expected to be impacts on vegetation.

Expected Impacts on Vegetation

Let us examine the possible constituents of a synfuel gas stream that may impact vegetation. Nitrogen is a harmless constituent of air. Carbon dioxide is the essential ingredient of photosynthesis; many green plants are deficient in CO_2 (during high light and optimum nutrition) up to ambient concentrations near 1000 ppm. Carbon dioxide is therefore usually a benefit to plants. Carbon monoxide is harmful to plants only at extremely high levels; the control strategy based on the occupational health problem will greatly supersede concern for vegetation. Hydrogen is harmless to plants at concentrations below the flammability limit. Ammonia produces visible injury to sensitive plants after a 4-h exposure to 16 ppm or a 1-h exposure to 40 ppm.[3] Thus NH_3 is not normally a problem if emissions are in the "trace" category.

Our major concerns must be with the phytotoxic effects of the gases H_2S, SO_2, COS, CS_2, NO_x, trace metals, and hydrocarbons. Sulfur dioxide and NO_x effects have been studied extensively, as for example in a review edited by Mudd and Kozlowski,[4] and air quality standards exist to provide adequate protection for vegetation. On the other hand, effects of COS and CS_2 are largely unknown. The hydrocarbons are of concern also. Coal conversion processes release benzene, xylene, and toluene.[5] It has been shown that toluene, acetone, tertiary butanol, and others are

volatile components of oil shale retort water.[6] In addition to direct emissions of the gas stream, it is possible that slow volatilization of organic constituents may occur from the wastewater. Pellizzari[6] detected a number of volatile organics lost from in situ oil shale retort water after 2 months; these included 2-heptanone, cyanobenzene, 2-octanone, and benzene as the major constituents volatilized. At low levels, the emitted hydrocarbons will be harmlessly deposited on and metabolized by vegetation, but the thresholds of their phytotoxic effects are unknown. The volatile trace metals, once their speciation becomes determined, should be studied further in terms of accumulation in vegetation.

It is the combination of gases that will produce a net effect, and the most important of the phytoactive gases in synfuels are CO_2 and H_2S.

STUDIES OF CO_2-H_2S EFFECTS ON VEGETATION

Recent studies of H_2S alone and in gaseous mixtures have shown that it is necessary to understand the mechanism of plant response to air pollutants to provide accurate predictions of injury. We have done this with the aid of a model, GROW1, which explains some contradictions we observed in experiments.[7] To save going into the details of the model here, it is necessary to point out that single exposures of a few hours at sublethal levels can result in either a gain or a loss of photosynthetic production. This may be a slight effect unless the exposure is repeated frequently and thus would produce a significant change in the growth and energy allocation strategy of the plant. The model permits consideration of complex interaction of these effects.

For example, Shinn et al.[8] determined that photosynthesis of lettuce was stimulated when exposed to a 15:1 mixture of CO_2 and H_2S at concentrations lower than a threshold concentration of about 2 ppm H_2S, as shown in Fig. 1. The same response was found for sugar beets. Coyne and Bingham[9] found the same response and approximate threshold for photosynthesis of snap beans exposed to H_2S alone. On the other hand, studies of long-term, chronic exposure to H_2S by Thompson and Kats[10] have shown that growth (i.e. production of leaf dry matter) is much more sensitive than photosynthesis. Table 2 shows their data for growth where the threshold concentration for injury is closer to 0.2 ppm H_2S (an order of magnitude lower than the

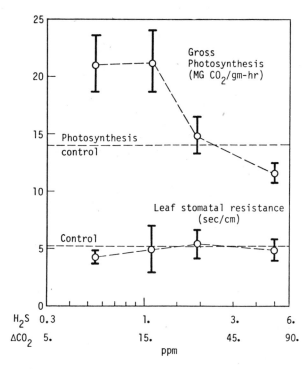

Fig. 1. Photosynthetic and stomatal response of field-grown lettuce to H₂S and CO₂ during 3-h exposures at 30°C and high light.

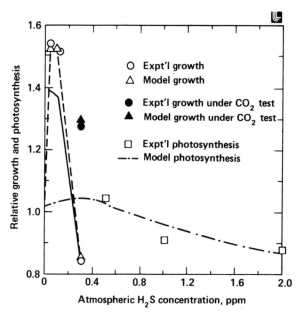

Fig. 2. Dose-response of sugar beets to H₂S. A comparison between growth response (open circles) and photosynthetic response (open squares) with model generated curves. Solid points represent addition of 50 ppm CO_2.

Table 2. Effects of H₂S on growth of leaves

H₂S level	Dry matter production		
	Sugar beet (%)	Lettuce (%)	Alfalfa (%)
0 ppm	100	100	100
0.03 ppm	147	112	108
0.10 ppm	137	109	93
0.30 ppm	97	82	79

Source: C. R. Thompson and G. Kats, "Effects of Continuous H₂S Fumigation on Crop and Forest Plants," *Environ. Sci. Technol.* **12**(5) 550–53 (1978).

threshold for photosynthesis). Now we understand with the aid of GROW1 that these two different sensitivities are compatible and are the result of both the plant metabolic processes and the integration of stress over time. Our model uses the dose-response observations shown in Fig. 2 and, by incorporation of basic physiological mechanisms and integrating from the leaf level over the whole plant, predicts growth and many complex interactions of response.

The presence of CO_2 ameliorates H₂S effects. Later work in collaboration with Lawrence Livermore Laboratory by Thompson and Kats[11] showed that addition of 50 ppm CO_2 overcame the effect of H₂S as shown in Table 3. This protection by CO_2 is demonstrated by our model as shown in Fig. 3.

There are other environmental influences on H₂S stress, which we will not quantify here. In the short term, temperature, light levels, and leaf stomatal diffusion resistance determine how much injury H₂S causes. Over the growing season these environmental

Table 3. Reduction of H₂S injury to leaves by added CO₂

Treatment	Dry matter production			
	Sugar beet (%)	Lettuce (%)	Alfalfa (%)	Cotton (%)
Control	100	100	100	100
0.3 ppm H₂S	110	82	82	90
0.3 ppm H₂S + 50 ppm CO₂	134	85	104	108

Source: C. R. Thompson and G. Kats, *Supplemental Studies of Air Pollution Effects for Imperial Valley Environmental Program, Effects of H₂S, H₂S + CO₂, and SO₂ on Lettuce, Sugar Beets, Alfalfa, and Cotton,* Contractor's Report UCRL-13782, Lawrence Livermore Laboratory, Livermore, Calif. (1977).

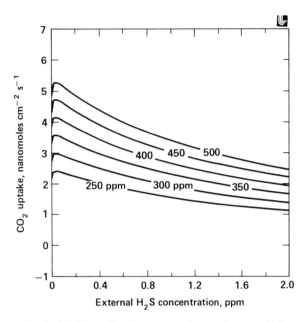

Fig. 3. Model-predicted response of sugar beets to H_2S at various levels of atmospheric CO_2. With a mixing ratio for CO_2/H_2S at 100:1 typical of *in situ* oil shale retorts, the level of atmospheric CO_2 will be increased over the normal background of 325 ppm CO_2.

influences average out. For the sake of brevity, we do not present here how, for example, soil moisture plays a role in causing stress avoidance.

Hydrogen sulfide injury is accentuated by ozone and probably by other air pollutants. Coyne and Bingham[9] showed that the presence of 0.07 ppm ozone (typical of urban smog) increases the toxicity of H_2S to photosynthesis as shown in Table 4.

In summary, under the worst-case conditions, the threshold for injury (10% loss of leaf dry matter) is about 0.2 ppm H_2S, assuming chronic exposure and no CO_2 present to ameliorate the effect. Realistically,

CO_2 and other pollutants such as ozone should be included in the assessment. Finally, a model such as GROW1 is essential to realistically interpret experimental bioassay dose-response data and to predict the growth effect, which is really a multivariate function of environmental parameters in addition to the air pollutant concentration.

REFERENCES

1. J. H. Raley, private communication, Lawrence Livermore Laboratory, Livermore, Calif. (1978).

2. E. L. Atkins, "Apiary Response to Air Pollution," p. 51 in *Potential Effects of Geothermal Energy Conversion on Imperial Valley Ecosystems*, ed. J. H. Shinn, Tech. Rep. UCRL-52196, Lawrence Livermore Laboratory, Livermore, Calif., Dec. 17, 1976.

3. J. S. Jacobson and A. C. Hill, eds., *Recognition of Air Pollution Injury to Vegetation: A Pictorial Atlas*, p. F6, Air Pollution Control Association, Pittsburgh, Pa., 1970.

4. J. B. Mudd and T. T. Kozlowski, eds., *Responses of Plants to Air Pollution*, Academic Press, New York, 1975.

5. D. S. Shriner, "Research Needs Related to Air Pollution from Coal Conversion Processes," in *AIChE Proc. 82nd National Meeting*, Atlantic City, N.J., Aug. 29–Sept. 1, 1976.

6. E. D. Pellizzari, *Identification of Components of Energy-Related Wastes and Effluents*, Research Triangle Institute Report, EPA-600/7-78-004, Research Triangle Park, N.C., 1977.

7. J. R. Kercher, *GROW1: A Crop Growth Model for Assessing Impacts of Gaseous Pollutants from Geothermal Technologies*. Tech. Rep. UCRL-52247, Lawrence Livermore Laboratory, Livermore, Calif., Mar. 17, 1977.

8. J. H. Shinn, B. R. Clegg, M. L. Stuart, and S. E. Thompson, "Exposures of Field-Grown Lettuce to Geothermal Air Pollution—Photosynthetic and Stomatal Responses," *J. Environ. Sci. Health*, **A11**(10 & 11), 603–12 (1976).

9. P. I. Coyne and G. E. Bingham, "Photosynthesis and Stomatal Light Responses in Snap Beans Exposed to Hydrogen Sulfide and Ozone," *J. Air Pollut. Control Assoc.* **28**(11), 1119–23 (1978).

10. C. R. Thompson and G. Kats, "Effects of Continuous H_2S Fumigation on Crop and Forest Plants," *Environ. Sci. Technol.* **12**(5), 550–53 (1978).

Table 4. Increases in toxicity of H_2S to photosynthesis of snap beans with exposure to ozone

H₂S alone		H₂S + ozone	
Concentration (ppm)	Maximum photosynthesis (mg CO_2 dm^{-2} h^{-1})	Concentration (ppm)	Maximum photosynthesis (mg CO_2 dm^{-2} h^{-1})
0	29	0 + 0.07	24
0.74	32	0.74 + 0.07	27
3.3	19	3.3 + 0.07	15
5.0	16	5.0 + 0.07	15

Source: P. I. Coyne and G. E. Bingham, "Photosynthesis and Stomatal Light Responses in Snap Beans Exposed to Hydrogen Sulfide and Ozone," *J. Air Pollut. Control Assoc.* **28**,(11), 1119-23 (1978).

11. C. R. Thompson and G. Kats, *Supplemental Studies of Air Pollution Effects for Imperial Valley Environmental Program, Effects of H_2S, $H_2S + CO_2$, and SO_2 on Lettuce, Sugar Beets, Alfalfa, and Cotton,* Contractor's Report UCRL-13782, Lawrence Livermore Laboratory, Livermore, Calif., 1977.

DISCUSSION

Ron Filby, Washington State University: What control do you have over oxidation of H_2S to SO_2 in your experiment? I'm particularly concerned about the mixture with ozone. I think, from a mechanical point of view, you would simply oxidize H_2S into SO_2.

Joe Shinn: In high concentrations, of course, ozone does oxidize H_2S. In the normal atmosphere, without a high presence of ozone, that is, less than 0.07 parts per million, we believe that the oxidation rate to SO_2 is very slow, on the order of a day. None of these experiments I discussed are conducted in a static situation. In the open-top chambers H_2S is added at a tremendous rate with perhaps two turnovers per minute in the chamber. The residence time of the H_2S in the chamber is, therefore, the order of seconds, certainly little more than 30 s. Hence, there is no real important decrease of H_2S in terms of SO_2 formation. There is also essentially no SO_2 present in the chambers because the incoming air is scrubbed by charcoal filters, and when the H_2S is added, it has insufficient time to produce SO_2. Therefore, there cannot be a cotoxic effect of SO_2 present. I think that in the Thompson and Katz study where H_2S is added continuously, they also had a fairly high turnover rate of H_2S in the chambers.

EFFECTS OF COMPLEX EFFLUENTS FROM IN SITU FOSSIL FUEL PROCESSING ON AQUATIC BIOTA*

H. L. Bergman[†]   G. M. DeGraeve[†]

A. D. Anderson[‡]

D. S. Farrier[§]

ABSTRACT

In situ processes for advanced fossil fuel recovery elicit major concerns about contamination of surface and groundwater with potential effects on water uses and aquatic biota. Acute and limited-chronic toxicity bioassays were conducted to determine the effects of process waters produced from two in situ technologies being developed by DOE's Laramie Energy Technology Center: (1) Omega-9 retort water from the Rock Springs Site 9 in situ oil shale processing experiment; and (2) Hanna-3 condenser water from the Hanna underground coal gasification experiment. The acute TL_{50} dilution (50% toxicity level) for oil shale retort water was about 0.5% for rainbow trout and fathead minnows (96-h) and for *Daphnia pulex* (48-h), and the threshold effect dilutions for limited-chronic bioassays with rainbow trout were 0.3% for egg hatchability and 0.1% for fry growth; the principal solutes affecting the acute toxicity were probably inorganic. The acute TL_{50} dilution for underground coal gasification condenser water was about 0.1% for rainbow trout and fathead minnows (96-h) and 0.2% for *Daphnia pulex* (48-h), while the threshold effect dilutions for limited-chronic bioassays with fathead minnows were 0.04% for egg hatchability and 0.02% for fry growth; the principal solutes influencing acute toxicity were probably phenolics and ammonia. Possible groundwater contamination is a critical concern with in situ technologies, and studies are continuing on chemical characterization and aquatic toxicity of these process waters.

INTRODUCTION

A number of processes are being developed to extract synthetic fuels from oil shale, coal, and tar sands.[1-4] Among these the in situ technologies offer the potential for certain environmental and economic advantages over alternate recovery schemes relying on mining and surface processing.[5] Nevertheless, in-situ processes elicit major concern about contamination of surface and groundwaters potentially affecting water uses and aquatic biota.

With surface waters the problems involve containment, treatment, use, and effects of waters which, co-produced with the oil or gas, are brought to the surface heavily laden with organic and inorganic solutes. This problem, however, is essentially identical to the water quality problem faced with aboveground processing technologies. As pointed out by Bostwick in Session I of this symposium,

the current development plans for aboveground coal conversion, at least, envision no discharge of contaminated water. Rather, these waters are expected to be treated for reuse within the plant. Whether this is ultimately done or not, these waters can be contained and either treated before discharge or evaporated from shallow holding ponds. This

---

*The work upon which this publication is based was performed pursuant to an Interagency Agreement between the U.S. Department of Energy and the U.S. Environmental Protection Agency under Contract No. ET-77-S-03-1761 to the Rocky Mountain Institute of Energy and Environment, University of Wyoming.

[†] Department of Zoology and Physiology, The University of Wyoming.

[‡] School of Pharmacy, The University of Wyoming.

[§] Laramie Energy Technology Center, U.S. DOE, Laramie, Wyoming 82071.

processing would presumably eliminate or at least restrict potential aquatic ecosystem contamination problems at aboveground conversion facilities. With true or modified in situ technologies, the co-produced water could be handled similarly.

The principal concerns about water contamination with in situ technologies, then, are associated with groundwater. It is conceivable that groundwaters in the region could become contaminated either by passing through the underground reaction cavity or by movement of product gases through fracture zones. The groundwater might then migrate and eventually contaminate wells or surface waters, ultimately affecting water use and aquatic ecosystems. In Session II of this symposium, Pellizzari described his work related to this problem at Lawrence Livermore Laboratory's Hoe Creek underground coal gasification project. He concluded from chemical analyses of groundwater that volatile, water-soluble organic compounds migrate quickly through fractures in the coal and overburden to contaminate groundwater at least 45 ft from the reaction zone. But preliminary results at Hoe Creek reported by Lawrence Livermore Laboratory[6] also suggest that phenolics and other contaminants may be adsorbed onto coal as the groundwater moves through the coal seam downflow from the gasification zone, which tends to confine the contaminants.

Our study characterizes the potential aquatic ecosystem effects of process waters from the three in situ technologies being developed by DOE's Laramie Energy Technology Center (LETC). These technologies are: (1) true in situ oil shale retorting being developed experimentally near Rock Springs, Wyoming; (2) in situ or underground coal gasification being conducted in sub-bituminous coal seams near Hanna, Wyoming; and (3) in situ tar sands extraction at LETC's site near Vernal, Utah. Thus far all of our work has been on LETC's in situ oil shale retorting and underground coal gasification technologies. Summaries have been published previously on these processes[7-10] and on associated environmental research.[5,11,12]

In the aquatic studies now underway, we are using toxicity, degradation, and bioaccumulation tests to evaluate: (1) whole process waters, including co-produced water and groundwater; (2) process water fractions; and (3) important single constituents. Since other tests are still underway, only summary data are presented at this time for tests with co-produced process waters. More detailed coverage on the

toxicity of these process waters will be forthcoming.[13,14]

The results reported here are not necessarily representative of the chemical and biological characteristics of process waters from in situ oil shale retorting and underground coal gasification. These characteristics will vary somewhat with both the process and the site, and insufficient comparative research has been completed thus far to allow generalizations about the process waters produced eventually by the industry.

MATERIALS AND METHOD

Materials

Process waters

Oil shale retort water from the Laramie Energy Technology Center (LETC) Site-9 in situ retort experiment near Rock Springs, Wyoming, was provided by LETC from their stock reserve. This oil shale retort water, designated "Omega-9 water" by LETC, had been settled and pressure-filtered to remove suspended oils and particles larger than 0.4 μm and held until use at 4°C. Handling, filtration, and storage conditions are described in greater detail elsewhere.[15] General chemical characteristics of the Omega-9 retort water[16] (Tables 1 and 2) and organic constituents of Omega-9 water[17-19] (Fig. 1) have been described in recent reports.

Condenser water from LETC's UCG site near Hanna, Wyoming, was obtained throughout the Hanna-3 gasification experiment. After oily residue

Table 1. General chemical characteristics of Omega-9 oil shale retort water

Water quality parameter	Concentration or value
pH	8.7
Alkalinity (mg/liter as CaCO₃)	16,200
Hardness (mg/liter as CaCO₃)	110
Conductivity (μS/cm at 25°C)	20,400
Total dissolved solids (mg/liter)	14,210
Total organic carbon (mg/liter)	1,003
Ammonia (mg/liter as NH₄⁺)	3,470
Cyanide (mg/liter)	0.42

Source: J. P. Fox, D. S. Farrier, and R. E. Poulson, *Chemical Characterization and Analytical Considerations for an In Situ Oil Shale Process Water*, Laramie Energy Technology Center Report No. LETC/RI-78/7, 1978.

206   H. L. Bergman et al.

Table 2. Principal inorganic constituents
in Omega-9 oil shale retort water

Constituent	Concentration (mg/liter)
HCO_3^-	15,940
Na^+	4,333
NH_4^+	3,470
$S_2O_3^{2-}$	2,740
SO_4^{3-}	1,990
Cl^-	824
CO_3^{2-}	500
$S_4O_6^{2-}$	280
SCN^-	123
F^-	60
K^+	47
Mg^{2+}	20
Ca^{2+}	12

Source: J. P. Fox, D. S. Farrier, and
R. E. Poulson, *Chemical Characterization
and Analytical Considerations for an In
Situ Oil Shale Process Water,* Laramie
Energy Technology Center Report No.
LETC/RI-78/7, 1978.

and tars were removed using separatory funnels, all the remainder of the condenser water sample (about 75 liters) was mixed thoroughly and stored at 4°C. Although extensive chemical analyses of Hanna-3 condenser water are not available, elemental and volatile organic analyses of condenser water from the Hanna-2 underground coal gasification (UCG) experiment are available[17] and should be qualitatively similar. General chemical characteristics of Hanna-3 UCG condenser water are summarized in Table 3, and a high performance liquid chromatograph (HPLC) gradient elution of organics from Hanna-4 UCG condenser water is shown in Fig. 2.

Table 3. General chemical characteristics of
Hanna-3 UCG condenser water

Water quality parameter	Concentration or value
pH	9.0
Alkalinity (mg/liter as $CaCO_3$)	11,200
Hardness (mg/liter as $CaCO_3$)	22
Total organic carbon (mg/liter)	8,900
Ammonia (mg/liter as NH_4^+)	21,000
Cyanide (mg/liter)	0.05

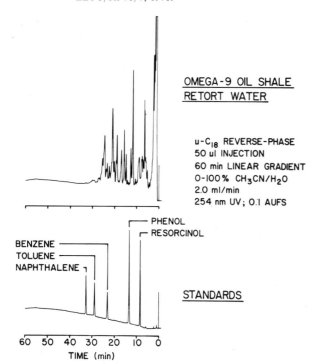

Fig. 1. HPLC chromatogram of Omega-9 oil shale retort water. Concentrations of standards are all 100 mg/liter except naphthalene, which is 10 mg/liter.

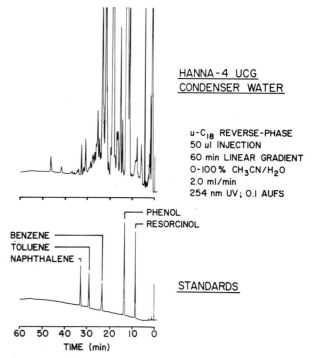

Fig. 2. HPLC chromatogram of Hanna-4 UCG condenser water. Concentrations of standards are all 100 mg/liter except naphthalene, which is 10 mg/liter. (Hanna-3 UCG water was qualitatively similar.)

Test waters

Dilution water used in the toxicity tests was dechlorinated tap water in some of the oil shale retort water tests and artesian well water for all other tests (see Table 4 for characteristics of dilution water).

Experimental animals

Juvenile rainbow trout (*Salmo gairdneri*) used in acute toxicity tests were obtained from the Wyoming Game and Fish Department or, in the case of some of the oil shale retort water tests, from a commercial hatchery and held in the laboratory water supply for a minimum of 10 d before testing. Eyed rainbow trout eggs used in limited-chronic tests with oil shale retort water were obtained from a U.S. Fish and Wildlife Service hatchery. Adult fathead minnows (*Pimephales promelas*) were obtained from a commercial hatchery and, again, held for a minimum 10-d acclimation period. Newly fertilized fathead minnow eggs for limited-chronic tests of UCG condenser water were obtained from our laboratory brood stock. *Daphnia pulex* used in all acute toxicity tests were also obtained from laboratory brood cultures. The original fathead minnow and *Daphnia pulex* stocks were both provided by the U.S. Environmental Protection Agency's Duluth Environmental Research Laboratory.

Method

Toxicity bioassays

For Omega-9 and Hanna-3 96-h bioassays with fish and 48-h bioassays with *Daphnia pulex*, we used intermittent-flow diluters modified from those originally described by Mount and Brungs.[20] With well water (Table 4) as dilution and control water, rainbow trout were tested in 28-liter tanks and fathead minnows in 14-liter tanks, each with a maximum of 10 fish per tank; *Daphnia pulex* were tested in small chambers suspended in the tanks.[21] In one series of 96-h tests with rainbow trout exposed to Omega-9 water, dechlorinated tap water (Table 4) was used for the control and dilution water in a continuous-flow diluter with 50 juvenile trout per 188-liter tank. In all cases dead fish were removed at least daily. The TL_{50} (50% toxicity level) values were determined either graphically or by computation.

In the Omega-9 limited-chronic tests 20-d, eyed rainbow trout eggs were placed 50 per tank in 5 dilutions and a dechlorinated tap water control. Dead eggs were removed daily. After surviving eggs hatched, the experiment was continued until the 69th day when the fry were measured and weighed.

Newly fertilized fathead minnow eggs were used in all of the Hanna-3 limited-chronic bioassays. In these tests well water served as dilution and control water and the tests were terminated at 30 d; other test conditions were similar to those described above for trout. After all bioassays, analysis of variance or paired Student's t tests were run to determine threshold effect concentrations for hatchability and fry growth.

RESULTS AND DISCUSSION

The results reported here for oil shale retort water and underground coal gasification condenser water

Table 4. Water quality characteristics of dilution water used in toxicity bioassays

Water quality parameter	Concentration or value	
	Dechlorinated tap water	Well water
Temperature (°C)	10	14 or 25[a]
Dissolved oxygen (mg/liter)	8.8	6.7
Dissolved oxygen (% saturation)	100	94
Free CO_2 (mg/liter)	15.1	3.0
pH	7.6	8.1
Alkalinity (mg $CaCO_3$/liter)	167	148
Hardness (mg $CaCO_3$/liter)	230	703
Conductivity (μS/cm at 25°C)	362	1195

[a]14°C for 96-h acute tests and 25°C for fathead minnow embryo-larval tests.

are preliminary, and additional work is required before definitive conclusions about potential aquatic ecosystem effects of these in situ technologies can be drawn. However, the general impressions gained from the aquatic toxicity results at this stage are instructive.

In Situ Oil Shale Retort Water

Acute toxicity of the retort water from LETC's Omega-9 oil shale experiment was similar for all aquatic organisms tested. The 96-h TL_{50} for rainbow trout and fathead minnows and the 48-h TL_{50} for *Daphnia pulex* were all at a retort water dilution of about 0.5% (Fig. 3). In limited-chronic bioassays the threshold effect level for rainbow trout egg hatchability was at about 0.3% retort water, and growth of rainbow trout fry was reduced at approximately 0.1% retort water (Fig. 3).

At the TL_{50} dilution of 0.5% retort water, the total organic carbon (TOC) concentration would only have been about 5 mg/liter, but the ammonia and sulfur species concentrations would have been quite high (Tables 1 and 2). Thus, it seems possible that inorganic, rather than organic, constituents might be responsible for much of the acute toxicity of Omega-9 retort water. In fact, preliminary tests of the acute toxicity of a synthetic mixture of inorganic constituents at levels found in Omega-9 water (Table 2) support this possibility. The 96-h TL_{50} dilution of this synthetic mixture was roughly 0.5 to 0.6% for rainbow trout, approximately the same as the TL_{50} of whole Omega-9 water with organic compounds present.[13] The relative contribution of inorganic and organic compounds in the Omega-9 water toxicity to rainbow trout egg hatchability and fry growth is not known, but inorganic compounds were no doubt important.

Underground Coal Gasification Condenser Water

Acute toxicity bioassays with condenser water from LETC's Hanna-3 UCG experiment resulted in 96-h TL_{50} dilutions of about 0.1% for rainbow trout and fathead minnows and a 48-h TL_{50} of 0.2% for *Daphnia pulex* (Fig. 4). The threshold level effects

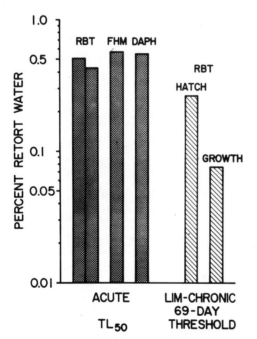

Fig. 3. **Acute and limited-chronic toxicity of Omega-9 oil shale retort water to aquatic biota.** The two bars for RBT acute toxicity show results for bioassays in dechlorinated tap water (left bar) and well water (right bar). Acute bioassays with rainbow trout (RBT) and fathead minnows (FHM) were for 96 h and *Daphnia pulex* (DAPH) acute bioassays were for 48 h. Source: Adapted from A. D. Anderson et al., "Toxicity of an In Situ Oil Shale Process Water to Aquatic Species," School of Pharmacy, University of Wyoming, Laramie, Wyoming (in press).

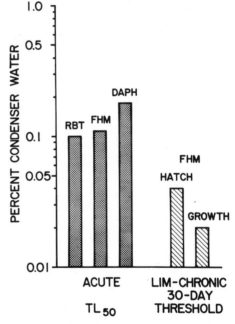

Fig. 4. **Acute and limited-chronic toxicity of Hanna-3 UCG condenser water to aquatic biota.** Acute bioassays with rainbow trout (RBT) and fathead minnows (FHM) were for 96 h, and acute bioassays with *Daphnia pulex* (DAPH) were for 48 h. Source: Adapted from G. M. DeGrave et al., "Toxicity of Underground Coal Gasification Condenser Water and Selected Organic Constituents to Aquatic Biota," Department of Zoology and Physiology, University of Wyoming, Laramie, Wyoming (in press).

for fathead minnow egg hatchability and fry growth were observed at dilutions of about 0.04 and 0.02% respectively (Fig. 4).

At a dilution of 0.1% Hanna-3 condenser water, the 96-h TL_{50} for rainbow and fatheads, the TOC concentration would have been about 9 mg/liter (Table 3). In this UCG condenser water one-half or more of the TOC consisted of phenolic compounds (Fig. 2), with phenol itself present at 2.3 g/liter. Thus, at the 0.1% TL_{50} dilution the phenol concentration would have been about 2.3 mg/liter and the total phenolic concentration would have been roughly 5 mg/liter, assuming that 50% of the TOC was phenolics. In separate toxicity tests of phenol under the same laboratory conditions, the 96-h TL_{50} was measured at 9 and 68 mg/liter for rainbow trout and fathead minnows respectively.[14] Therefore, although phenolics no doubt contributed to the toxicity of Hanna-3 condenser water, other constituents were important as well.

Although we do not have extensive inorganic analyses of the Hanna-3 UCG condenser water comparable to those for the Omega-9 water (Table 2), the total ammonia concentration of Hanna-3 UCG water was about 21 g/liter (Table 3). Therefore, at the 0.1% dilution of condenser water, the TL_{50} for rainbow trout and fathead minnows, the total ammonia might then have been about 21 mg/liter. At the test water pH of 8.1 and temperature of 14°C (Table 4), un-ionized ammonia, the toxic species, would have been 3.5% of the total ammonia concentration[22] or about 0.74 mg/liter. The TL_{50} concentration of un-ionized ammonia has been reported at about 0.4 mg/liter for rainbow trout in 48-h and 60-h acute tests.[23,24] The un-ionized ammonia concentrations at the Hanna water TL_{50} dilution were almost twice this reported value. Thus the ammonia undoubtedly contributed the major portion of the observed condenser water toxicity with some contribution by phenolics. Confirmation that ammonia and phenolics were principal toxic constituents will have to come from additional chemical characterization in conjunction with toxicity tests of UCG condenser water fractions.

Process Water Comparison

In acute and limited-chronic bioassays with aquatic organisms, Hanna-3 UCG condenser water was slightly more toxic than Omega-9 oil shale retort water. For Hanna UCG water, ammonia and phenolics were important constituents contributing to the observed acute toxicity. With Omega-9 water,

which has a TOC concentration only 12% of Hanna-3 water (Tables 1 and 3) and much lower concentrations of phenolics (Figs. 1 and 20), the principal toxic constituents in the acute tests were inorganic. Ammonia and inorganic sulfur compounds were no doubt important.

Thus far, there are few published reports which describe aquatic toxicity of complex process waters from advanced fossil fuel technologies, and they deal with aboveground technologies. Schultz et al. evaluated the toxicity of condenser water from a laboratory-scale synthane process for coal gasification in acute bioassays with fathead minnows where the test solutions were replaced daily.[25] They determined the 96-h TL_{50} dilution as 0.056%, about one-half our reported 96-h toxicity level for fathead minnows (0.1% dilution) in flow-through tests of Hanna-3 UCG water. However, considering that both the coal gasification technologies and the bioassay procedures were different, the TL_{50} dilutions from the present study and that by Shultz et al.[25] are remarkably similar.

The first study[25] concluded that the principal toxicity was probably due to phenolics which were present in the original Synthane condenser water at a concentration of 5.22 g/liter. Schultz's group discounted ammonia toxicity as significant even though the ammonia concentration was about 9 g/liter in the original Synthane water. This contrasts with our conclusion that ammonia was an important contributor to Hanna UCG condenser water toxicity, in addition to the phenolics.

A 48-h TL_{50} dilution of 15.7% (54.8 mg dissolved material/liter) of a solvent refined coal (SRC) aqueous effluent in static bioassays with *Daphnia magna* by Parkhurst et al. revealed that the most toxic fractions were: (1) the neutral fraction, containing naphthalene, methylnaphthalene, phenanthrene, anthracene, pyrene, and other neutral organics, and (2) the acid fraction which contained the phenolics.[26] Together the neutral and acid fractions accounted for most of the toxicity of the whole SRC effluent. In contrast to our conclusions about the importance of inorganic constituents in Omega-9 retort water and Hanna-3 UCG condenser water, Parkhurst et al. found that the inorganic fraction of SRC water contributed little to observed whole-water toxicity. Although their fractionation procedure could have lost volatile or highly reactive inorganics such as ammonia, thiosulfate, or thiocyanate, if present, this does not appear to have been a problem because the recombined fractions gave a toxicity value essentially equivalent to that of the original non-fractionated SRC effluent. It is also possible that any volatile or reactive inorganic or organic compounds could have been lost during pretest storage of the SRC effluent, but the storage procedures are not described. On the other hand, another study by Gehrs et al. reports that for a hydrocarbonization process water the inorganic fraction, with a high ammonia concentration, was the most toxic component in 48-h static bioassays with *Daphnia magna.*[27]

Next year additional studies will be completed on the chemical characteristics and aquatic toxicity of other complex process waters from advanced fossil fuel technologies. It may be possible then to compare the potential aquatic ecosystem hazards of different technologies. On the basis of the information that is available now, we expect that constituents of principal concern for many process waters will be phenolic and neutral organic compounds or inorganic chemicals such as ammonia, cyanide, and sulfur compounds. If this is so, the complex problems of hazard evaluation and waste treatment design to avoid those hazards may be simplified.

Future Research Needs

For in situ technologies the potential contamination of groundwater remains the most serious aquatic ecosystem and water use problem. A full evaluation of groundwater contamination requires information about the chemical characteristics and aquatic toxicity of the groundwater and toxicity, degradation, bioaccumulation, and fate of important constituents in the groundwater. These data are needed for different technologies and for different geological and groundwater environments, and they are needed over time and distance from the underground reaction zones. Long-term hazard potential must be evaluated carefully for both contaminated groundwater and co-produced water. This, of course, requires degradation and bioaccumulation studies along with additional chronic toxicity bioassays. Such studies now underway for the LETC in situ technologies will provide the necessary information to meet the overall objectives of helping to assess potential aquatic ecosystem impacts from in situ processing technologies and to set goals for water treatment system design.

ACKNOWLEDGMENTS

For their assistance in this work we wish to thank M. L. Lebsack, D. Woods, J. Meyer, R. Overcast, D. Geiger, and M. Marcus.

REFERENCES

1. T. H. Maugh, "Oil Shale Prospects on the Upswing Again," *Science* 198, 1023–27 (1977).

2. T. H. Maugh, "Gasification: A Rediscovered Source of Clean Fuel," *Science* 178, 44–45 (1972).

3. T. H. Maugh, "Underground Gasification: An Alternate Way to Exploit Coal," *Science* 198, 1132–34 (1977).

4. T. H. Maugh, "Tar Sands: A New Fuels Industry Takes Shape," *Science* 199, 756–60 (1978).

5. D. S. Farrier et al., "Environmental Research for In Situ Oil Shale Processing," in *Eleventh Oil Shale Symposium Proceedings,* Colorado School of Mines Press, Golden, Colorado, 1978.

6. J. H. Campbell, E. Pellizzari, and S. Santor, "Results of a Groundwater Coal Gasification Experiment (Hoe Creek I)," UCRL-52405, Lawrence Livermore Laboratory, University of California, Livermore, California, 1978.

7. E. L. Burwell, T. E. Sterner, and H. C. Carpenter, *In Situ Retorting of Oil Shale: Results of Two Field Experiments,* U.S. Bureau of Mines RI7783, 1973.

8. A. Long, Jr., N. W. Merriam, and C. G. Mones, "Evaluation of Rock Springs Site 9 In Situ Oil Shale Retorting Experiment," in *Tenth Oil Shale Symposium Proceedings,* Colorado School of Mines Press, Golden, Colorado, 1977.

9. C. F. Brandenburg et al., "The Underground Gasification of a Subbituminous Coal," *Am. Chem. Soc., Div. Fuel Chem.* 20, 3–10 (1975).

10. T. C. Bartke et al., "Status Report on the Hanna III and Hanna IV Underground Coal Gasification Experiments," in *Fourth Underground Coal Conversion Symposium Proceedings,* Sandia Laboratories, Albuquerque, New Mexico, 1978.

11. J. E. Virgona, "Environmental Research for Underground Coal Gasification," *Environmental Assessment of Solid Fuel Processes Symposium Proceedings,* 71st AICHE Meetings, Miami Beach, Florida, 1978.

12. D. S. Farrier, L. W. Harrington, and R. E. Poulson, "Integrated Compliance and Control Technology Research Activities for In Situ Fossil Fuel Processing Experiments," presented at U.S. DOE Environmental Control Symposium, Washington, D.C., November 28–30, 1978.

13. A. D. Anderson et al., "Toxicity of an In-Situ Oil Shale Process Water to Aquatic Species," School of Pharmacy, University of Wyoming, Laramie, Wyoming, in press.

14. G. M. DeGraeve et al., "Toxicity of Underground Coal Gasification Condenser Water and Selected Organic Constituents to Aquatic Biota," Department of Zoology and Physiology, University of Wyoming, Laramie, Wyoming, in press.

15. D. S. Farrier et al., "Acquisition, Processing and Storage for Environmental Research of Aqueous Effluents Derived from In Situ Oil Shale Processing," *Second Pacific Chemical Engineering Congress Proceedings* 2, 1031–35 (1977).

16. J. P. Fox, D. S. Farrier, and R. E. Poulson, *Chemical Characterization and Analytical Considerations for an In Situ Oil Shale Process Water,* Laramie Energy Technology Center Report No. LETC/RI-78/7, 1978.

17. E. D. Pellizzari, *Identification of Energy-Related Wastes and Effluents,* Interagency Energy-Environment Research and Development Program Report EPA-600/7-78-004, 1978.

18. W. D. Felix, D. S. Farrier, and R. E. Poulson, "High Performance Liquid Chromatographic Characterization of Oil Shale Retort Waters," *Second Pacific Chemical Engineering Congress Proceedings* 1, 480–85 (1977).

19. H. A. Stuber and J. A. Leenheer, "Fractionation of Organic Solutes in Oil Shale Retort Waters for Sorption Studies on Processed Shale," *Am. Chem. Soc., Div. Fuel Chem. Prepr.* 23(2), 165–74 (1978).

20. D. E. Mount and W. Brungs, "A Simplified Dosing Apparatus for Fish Toxicology Studies," *Water Res.* 1, 21–29 (1967).

21. G. M. DeGraeve, T. Cruzan, and R. W. Ward, "Chamber for Holding Aquatic Macroinvertebrates During Toxicity Tests in a Flow-Through Diluter System," *Prog. Fish Cult.* 39, 100–01 (1977).

22. K. Emerson et al., "Aqueous Ammonia Equilibrium Calculations: Effect of pH and Temperature," *J. Fish. Res. Board Can.* 32, 2379–83 (1975).

23. I. R. Ball, "The Relative Susceptibilities of Some Species of Freshwater Fish to Poisons—I. Ammonia," *Water Res.* 1, 767–75 (1967).

24. R. Lloyd and L. D. Orr, "The Diuretic Response by Rainbow Trout to Sublethal Concentrations of Ammonia," *Water Res.* 3, 335–44 (1969).

25. T. W. Schultz, S. Davis, and J. N. Dumont, "Toxicity of Coal-Conversion Gasifier Condensate to the Fathead Minnow," *Bull. Environ. Cont. Tox.* 19, 237–43 (1978).

26. B. R. Parkhurst, C. W. Gehrs, and I. W. Rubin, "The Value of Chemical Fractionation for Identifying the Toxic Components of Complex Aqueous Effluents," *ASTM Second Annual Symposium on Aquatic Toxicology,* Cleveland, Ohio, 1977.

27. C. W. Gehrs, B. R. Parkhurst, and D. S. Shriner, "Environmental Testing," *Symposium on Application of Short-Term Bioassays in the Fractionation and Analysis of Complex Environmental Mixtures,* 1978.

RESPONSE OF AQUATIC BIOTA TO EFFLUENTS FROM THE FORT LEWIS, WASHINGTON, SOLVENT REFINED COAL PILOT PLANT*

J. B. States†   M. J. Schneider†
C. D. Becker†   J. A. Strand†

ABSTRACT

Solvent refined coal is briefly described as one of the most promising coal liquefaction technologies being developed to provide substitutes for foreign oil and fuels which burn more cleanly than center coals. Results of static toxicity tests run on pilot plant effluents reveal that (1) untreated liquid effluents originating within the pilot plant contain chemical features, such as the heavy loading of organic materials, which make them highly toxic to aquatic life and (2) treated effluent is relatively low in toxicity when treatment facilities are operating properly. A strategy is presented to screen all potentially damaging source materials initially, for both acute and chronic ecological effects at several levels of biological organization, before turning to more detailed chemical and biological investigations into exact causes of the most serious effects revealed.

INTRODUCTION

In his environmental message of May 1977, the President of the United States declared a national commitment to reduce reliance on foreign oil through increasing utilization of our own vast coal reserves. He admonished, however, that we were to do this in a way that protects the quality of our environment. To meet this challenge, government and industry have entered into an unprecedented series of cooperative efforts to develop new energy technologies. Their purpose is to convert coal to oil substitutes and alternate fuels that burn more cleanly than their parent coals. Among these coal conversion technologies are those in which slurried coal is subjected to high heat and pressure, causing the coal to liquefy.

The technology development time lines for three of the most promising coal liquefaction technologies are shown in Figure 1. Note that in all three cases, commercial plants are expected to be built and in operation by 1987. This means some major decisions on commercialization may have to be made fairly quickly, in 1980 to 1981, and that the time frame for conducting the necessary research to assure that those decisions are environmentally sound is uncomfortably short.

Of the three most promising coal liquefaction technologies shown in Figure 1, Solvent Refined Coal, or SRC, is the first expected to go to the demonstration level of development. Design of the demonstration facility will derive largely from experience gained at a 50-ton-per-day SRC pilot plant at Fort Lewis, Washington. The purpose of this paper is twofold—to describe research results on the aquatic effects of wastewaters from the pilot plant and to describe how this research is being modified and expanded to provide more information of use in designing a safe demonstration facility.

The work described is Battelle Pacific Northwest Laboratories' part of a larger research program on coal synfuels being conducted by the Department of Energy under the supervision of Dr. Robert A. Lewis.

THE SOLVENT REFINED COAL PROCESS

The research and its results are best understood if one knows something about the SRC process. Two

---

*Prepared for the U.S. Department of Energy under Contract No. EY-76-C-06-1830.

†Battelle, Pacific Northwest Laboratories, Richland, Washington 99357.

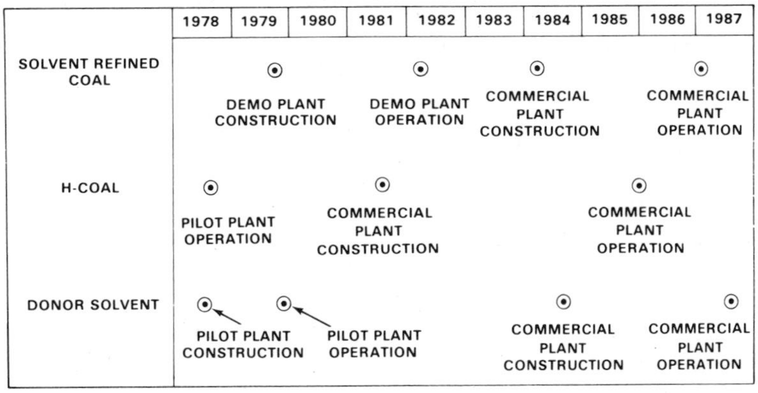

	1978	1979	1980	1981	1982	1983	1984	1985	1986	1987

SOLVENT REFINED COAL: ⊙ DEMO PLANT CONSTRUCTION · ⊙ DEMO PLANT OPERATION · ⊙ COMMERCIAL PLANT CONSTRUCTION · ⊙ COMMERCIAL PLANT OPERATION

H-COAL: ⊙ PILOT PLANT OPERATION · ⊙ COMMERCIAL PLANT CONSTRUCTION · ⊙ COMMERCIAL PLANT OPERATION

DONOR SOLVENT: ⊙ PILOT PLANT CONSTRUCTION · ⊙ PILOT PLANT OPERATION · ⊙ COMMERCIAL PLANT CONSTRUCTION · ⊙ COMMERCIAL PLANT OPERATION

Fig. 1. Coal liquefaction processes: expected development time lines. Source: Department of Energy Environmental Readiness Document for Coal Liquefaction, 1977.

operating modes, SRC I and II, are now undergoing pilot-plant-phase testing as diagrammed in Figure 2. The two processes are essentially the same in the initial steps. The feed coal is first pulverized and enters a preheater unit as a slurry of coal and solvent. Hydrogen is injected into the preheater and the liquefaction process initiated. From the preheater, the coal-solvent slurry moves into the dissolver unit where liquefaction is taken to completion under high-pressure and -temperature conditions. From here the liquefied coal goes through a series of separation and fractionation steps, which differ between the two processes.

In SRC I, the slurry passes to a filter which separates out a mineral residue that represents the large bulk of solid waste produced by the process.

The remaining filtrate is delivered to a vacuum flashing unit where a distillate fraction is separated from a residual fraction. The residual fraction is diverted to a Sandvik belt where it is cooled to form the *solid* product, one which burns much more cleanly than its parent coal.

In the SRC II process, there is no filtration step. The slurry leaving the gas vapor separator is split, with the main stream being recycled to the preheater while a small portion is delivered directly to vacuum flashing. The distillate from this step proceeds to the fractionation column where the major products of the SRC II process, a heavy and a middle distillate, are separated. A solid waste by-product, vacuum bottoms, is recovered from the vacuum flashing step and cooled on the Sandvik belt. These vacuum

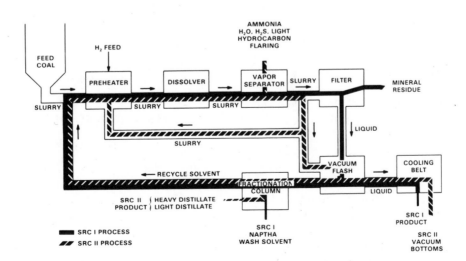

Fig. 2. Flow diagram of SRC I and II processes, Fort Lewis Pilot Plant.

bottoms are higher in organic matter than the SRC I mineral residue and will probably be gasified for hydrogen recovery.

A review of the major differences shows briefly that the SRC I process produces a solid product, and its solid waste is a low-carbon residue removed through a filtration process. The SRC II process produces a liquid, oil-like, product, and its solid wastes are the high-carbon bottoms from the vacuum flash unit.

A major concern with coal conversion processes is potential for release of toxic materials, particularly complex organics, to surface waters.

The basic technique for reducing toxicity of plant waste effluent in both the SRC I and SRC II processes is similar. A biodigester containing phenol-attaching microorganisms is employed to break up phenols and other hydrocarbons after neutralization of process waste streams in a surge reservoir. Following treatment in the biodigester, the wastewater passes through a sand filter, to reduce particulates, and an activated carbon filter, to lower amounts of remaining contaminants. The treated process effluent is expected to be relatively low in direct toxicity to aquatic organisms, particularly when greatly diluted with either tap or cooling system water before discharge to a receiving pond.

MATERIAL AND METHODS

One-hundred-gallon samples of waste effluent were obtained from the pilot plant and transported to Richland for toxicity tests which used mostly fish as test organisms. Samples were transported in large glass carboys that were sealed with rubber stoppers wrapped in teflon tape to avoid contamination and reduce the rate of potential volatilization. The carboys were thoroughly cleaned before filling. Effluent was passed into the carboys by means of a galvanized metal bucket and funnel or, in the case of the 1978 samples, by filling directly from the biodigester discharge line.

All toxicity tests were static, of 96-h duration, and conducted in 54-liter glass aquaria. Although flow-through toxicity tests were desirable, they would require a greater volume of effluent than could be transported to our laboratory. The exposure aquaria were arrayed randomly in a water bath for temperature control at 17–19°C. Effluent dilutions of 100, 80, 60, 40, and 0% were obtained by adding Columbia River water. Temperature, pH, dissolved oxygen, fish mortality, and observations on behavior were recorded frequently during the initial hours of each test. Subsequent recordings were made twice daily on the 2nd, 3rd, and 4th days.

We avoided aeration of aquaria in most tests to restrict possible loss of volatile compounds from solution. But use of fingerling fish in a few early tests exceeded aquarium "loading capacity," and low dissolved oxygen caused complications. According to accepted procedures for toxicity testing, aeration is required in static tests when a specific organism weight is exceeded for a given solution volume and test duration in each aquarium. Accordingly, only young fish or "fry" were used.

Rainbow trout, *Salmo gairdneri,* was the primary test organism. The species is commonly used in toxicological studies, so the responses we observed could be evaluated in relation to their response to other chemicals. Rainbow trout stocks were reared in the Pacific Northwest Laboratories hatchery–wet laboratory, and their thermal history, growth, metabolism, disease state, etc., were controlled and documented. Fingerling largemouth bass (*Micropeterus salmoides*) and crayfish (*Pacifasticus leniusculus*) collected from the Columbia River were also used in some early tests. The need for comparative testing between successive samples led eventually to exclusion of these latter species from further testing.

Methods used for chemical analysis of inorganic and organic compounds in the SRC effluent discharges varied. Development of new analytical methods for dealing with highly complex effluents in a reliable, efficient, and economic fashion remains a major research objective. In general, basic water quality analyses (i.e., pH, total alkalinity, EDTA hardness) were performed onsite. Inorganic determinations were done by X-ray fluorescence. Organic analyses were done by a combination of one or more procedures, including gas chromatography and high-pressure liquid chromatography. Separation of sample extracts into acid, neutral, and basic fractions has proved an essential step in arriving at more detailed chemical analyses.

RESULTS AND DISCUSSION

Toxicity Tests

Static, short-term toxicity tests are useful in the preliminary evaluation of impacts from a discharge arising from a new technological process and usually are among the initial evaluation techniques employed.

In several static bioassays, Becker et al.[1] used various dilutions of liquid wastes from the SRC pilot plant to observe mortality of test organisms ranging from 0 to 100% (Table 1).

Table 1. Summary of results from 96-h static laboratory toxicity tests with treated SRC process effluent at 17–19°C[a]

Date assay conducted	Test organisms	Assessment[b]
	SRC I	
Aug. 3–7, 1976	Rainbow trout fingerlings	No relationship between mortality and effluent concentration. Losses due to low dissolved oxygen.
Aug. 24–28, 1976	Largemouth bass fingerlings	No mortality attributable to effluent concentration. Some losses due to disease. Some stress displayed by fish.
Oct. 12–16, 1976	Rainbow trout fingerlings	Significant mortality (>50%) at effluent concentrations of 80 and 100%.
Nov. 16–20, 1976	Rainbow trout and crayfish	Significant mortality (>50%) at effluent concentrations of 40, 60, 80, and 100% for trout, >60% for crayfish; narcosis evident among crayfish.
Jan. 27–31, 1977	Rainbow trout fry	No mortality in 96 h but some fish stressed at 80 and 100% effluent concentrations.
	SRC II	
June 21–25, 1977	Rainbow trout fingerlings	Some mortality at higher effluent concentration.
Dec. 14–18, 1977	Rainbow trout fry	No mortality attributable to effluent concentration.
Feb. 17, 1978	Rainbow trout fry	Total mortality at all effluent dilutions. Test terminated after 30 min and repeated with higher dilutions.
Feb. 17–21, 1978	Rainbow trout fry	Over 80% mortality at low effluent concentrations of 6, 8, and 10% (repeat).
Mar. 24–28, 1978	Rainbow trout fry	Over 90% mortality at effluent concentrations of 60, 80, and 100%.

[a]Effluent was collected at the Pittsburg & Midway Coal Co. SRC pilot plant, Fort Lewis, Washington, one day before start of each assay.

[b]See text for further explanation of factors influencing relative toxicity.

The inconsistency in relative toxicity results from factors that modify the organic and/or inorganic composition of the effluent at the time and point of sampling. These factors include (1) the condition of the charcoal filter below the biodigester (effectiveness in sorbing low molecular weight, soluble organics), (2) the pH level in the biodigester that controls the abundance and effectiveness of microorganisms breaking down hydrocarbons, (3) the stage of the plant operational run when the effluent sample is obtained, (4) the extent of effluent dilution with cooling water or tap water, and (5) the specific process—SRC I or SRC II—that is under way.

During the SRC I mode of plant operation, we used treated process effluent obtained from the discharge pipe leading to the discharge pond. The tests were done with treated process effluent as it issued from the pipe, undiluted or "unconcentrated," whereas normally, the treated process effluent was routinely diluted manyfold with tap water before discharge to the pond.

The results indicated that, under normal conditions, the treated SRC I effluents were low in acute toxicity, as measured by short-term static tests. The lack of toxicity was apparently due to the effectiveness of the biodigester action followed by sand and activated carbon filtration. However, a slow narcosis of test organisms or an impairment of their functioning was noted in some of these tests, implying that longer exposure might have proved lethal. These results, however, should be kept in perspective by remembering that the pilot plant routinely dilutes its treated process effluent with tap water before discharge to the pond.

Tests run in October and November 1976 under the SRC I process were influenced by abnormal plant operations producing effluent that did not pass through the biodigester and filter to reduce hydrocarbon loading. In the October case, the effluent collected for toxicity tests was taken when the carbon filters were being backflushed. In the November case, effluent from the backwash tank was added to increase the available flow. These two toxicity tests yielded greater than 50% mortality at effluent concentrations ≥40%, demonstrating that incompletely treated or untreated SRC I effluent was toxic to aquatic life.

Changeover to the SRC II process involved a fundamental change in mode of plant operation. Dilution water for the process effluent was taken from the plant cooling system so that effective dilution was higher and more water was passed from the discharge pipe to the pond. Since the cooling water discharge could not be regulated, undiluted treated process effluent was no longer available at the discharge pipe outlet but instead was taken within the plant below the charcoal filter.

Results of the first two acute toxicity tests conducted during SRC II operation indicated low toxicity but a narcotic effect similar to tests conducted during normal SRC I operation. However, considerable toxicity of untreated process effluent was subsequently indicated when abnormal plant operating conditions prevailed.

Abnormal conditions under the SRC II mode of plant operation were reflected in the February 1978 acute toxicity tests. Before the effluent samples were obtained, the surge reservoir was cleaned, and highly acidic wastes inadvertently entered the biodigester, killing most of the microorganisms. Thus, the effluent used in the acute toxicity test was undigested and, as might be expected, highly toxic. In the first test, all fish soon died, and the test was terminated in 30 min. A subsequent test employed greater dilutions of the same effluent sample (6, 8, and 10%) and again, high toxicity was indicated with over 80% mortality. The March sampling was undertaken at a time when the biodigester was

functioning and the plant was believed by plant engineers to be operating in a stable mode. Engineers later expressed doubts as to the representativeness of this run, and the observed (90%) mortality at 0–40% dilutions lends credence to these doubts.

Short-term toxicity tests with SRC plant effluent indicate, in summary:

1. Treated process effluent passing through the biodigester, sand filter, and activated charcoal filter is relatively low in toxicity to aquatic life.

2. Untreated effluent originating within the plant contains chemical features that make it highly toxic to aquatic life.

Chemical Characterization

For causes of the observed toxicity, one looks to the chemical composition of the effluent. Chemical analysis of such materials, particularly of their organic content, is a complex procedure requiring development of new sampling and analytical methodologies. With the development of such methods, one hopes to trace an observed effect to its chemical agent and to the source of that agent in the process. Armed with this information, a process engineer is able to design appropriate control technology.

Without going into long lists of chemical contents and their concentrations, analyses to date have yielded some interesting results. For example, much of the toxicity observed in untreated effluent samples may be attributed to high concentrations of phenols. When the biodigestor was "up" and toxicity was low, analyses showed that phenols had been very effectively removed. A look at Fig. 3 illustrates not only the complexity of the effluent but the effectiveness of the biodigester, especially when one considers that the bottom sample is concentrated 15 times. One observation worthy of note is the apparent similarity between SRC effluents and petroleum-contaminated water. These differences include the smaller-than-expected quantities of alkylated compounds relative to the parent hydrocarbons and the presence of two major oxygenates, phenyl ether and dibenzofuran.

To seek an explanation, we went to the plant engineers for a better understanding of the process. What we learned is illustrated in Fig. 4. In a gross way, this diagram depicts a water balance for the

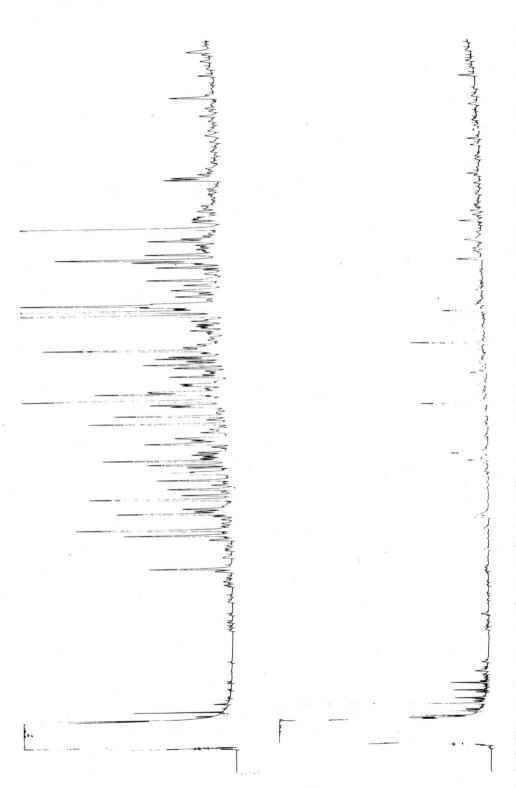

Fig. 3. Capillary gas chromatograms of solvent refined coal effluent, aromatic hydrocarbons. Above, untreated effluent; below, treated effluent. Treated effluent concentration 15 times that of the untreated effluent.

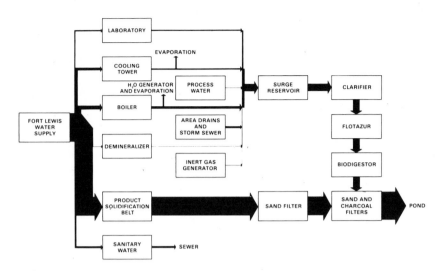

Fig. 4. Flow diagram of water and wastewater at Fort Lewis SRC Pilot Plant.

Fort Lewis plant. The likely explanation for the similarity to petroleum-contaminated water may be traced to the box labeled "area drains and storm sewer." Literally everything that is dumped on the pilot plant "pad," possibly including petroleum solvents, crankcase oil, etc., is washed into the wastewater stream which feeds into the biodigester.

PROSPECTUS

Although this information is highly useful with regard to understanding the toxicities observed at the only water treatment system at a liquefaction plant in the country, it also poses a problem. During the design of the initial research, the dual objectives were to evaluate the effects of the pilot plant and develop a data base for predicting effects at the demonstration stage of development. It soon became clear that predicting effects at the next stage of development should become the primary goal. Thus, selection of source streams that are representative of or can be scaled up to the demonstration scale becomes of paramount importance. From this perspective, sampling a stream that may be contaminated with water from the pilot plant is not very useful. However, sampling further upstream in the process, such as from the process water which comes directly from the process itself and likely contains most of the complex hydrocarbons developed from the process (Fig. 4), would be very useful.

As a result of these and other considerations, we have revised our research strategy in a way that may be of general interest. Keep in mind that the objective of this strategy is to provide feedback on ecological effects to process engineers in a way that does not delay but rather hopefully facilitates technology development in an environmentally safe manner. As a basis for the new research strategy, it is assumed that a developing technology can be described in two ways—in terms of physical and chemical properties of the liquid, solid, and gaseous materials it produces (source materials) and in terms of the environmental effects or perturbations that these materials may cause. The wide variety of source terms which may require characterization for assessment purposes is illustrated in Fig. 5. The summarized note at the bottom left of the figure shows a variety of factors that may cause significant changes in each source term and the several levels at which each variant could be chemically described.

In general, source materials in Fig. 5 are listed from top to bottom in decreasing order of environmental concern (amount of material to be released to environment, lack of knowledge about and harshness of material, etc.). Solid waste and product spill problems are given highest priority, and liquid waste and product spills are given second priority. Environmental effects of gaseous emissions cannot usefully be studied at the pilot plant.

Each source term in Fig. 5 may have a wide variety of effects at several levels of biological organization. These may be ecologically significant

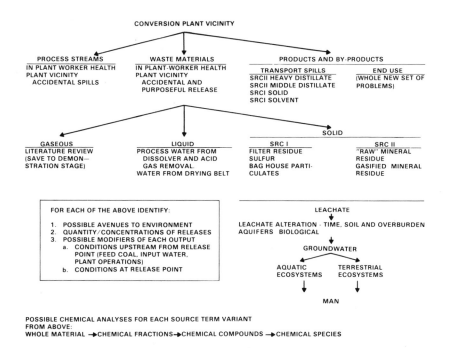

Fig. 5. Source terms of concern in the solvent refining of coal.

if they cause either the death of individual animals or a functional impairment that effectively decreases reproductive success. The wide variety of possible source terms, coupled with a multitude of possible biological effects from each, makes it nearly impossible to characterize a technology completely, both chemically and biologically. Instead it is necessary to employ a "top-down strategy," by which we first ask "is there an effect?" This question is answered through appropriate biological screening methods, which may be categorized in order of increasing realism but also increasing experimental complexity as shown in Fig. 6.

Figure 6 lists ecological screening procedures from top to bottom in terms of experiment complexity (static to flow-through mode of testing, acute to chronic toxicity, test species to simple microcosms, "controlled" laboratory to "real world" field conditions). In general, planned research focuses on the higher priority source materials and covers the largest number of plant operating conditions in the simpler experiments. Fewer operating conditions can be researched where there is a lower degree of concern or increased experimental complexity because of time and resource limitations.

The environmental experiments to be run are further reduced during program planning by

A. UNDER "CONTROLLED" LABORATORY CONDITIONS

    1. ORGANISM AND POPULATION LEVEL OF ORGANIZATION FOR SELECTED TEST SPECIES

        a. STATIC (AQUATIC) OR DIRECT CONTACT (TERRESTRIAL) EXPERIMENTS

            (1) ACUTE TOXICITY
            (2) CHRONIC TOXICITY

        b. FLOW-THROUGH EXPERIMENTS
            (1) ACUTE TOXICITY
            (2) CHRONIC TOXICITY

    2. POPULATION AND COMMUNITY LEVEL OF ORGANIZATION- SIMPLE MICROCOSMS IN FLOW THROUGH MODE (AQUATIC) OR GROWTH CHAMBER (TERRESTRIAL)

        a. ACUTE TOXICITY
        b. CHRONIC TOXICITY
            (1) BIOCONCENTRATION
            (2) FOOD CHAIN TRANSFERS

B. CONFIRMATION UNDER "FIELD" CONDITIONS

    1. ENDEMIC SPECIES
        a. ACUTE TOXICITY
        b. CHRONIC TOXICITY

    2. ECOSYSTEMS
        a. ACUTE TOXICITY
        b. CHRONIC TOXICITY
            (1) BIOCONCENTRATION
            (2) FOOD CHAIN TRANSFERS

Fig. 6. Ecological screening procedures for characterizing SRC source materials.

committing to utilize results from the simpler tests on whole effluents or materials. Decisions are then made on whether to proceed with more detailed analysis of chemical fractions or species, or to move

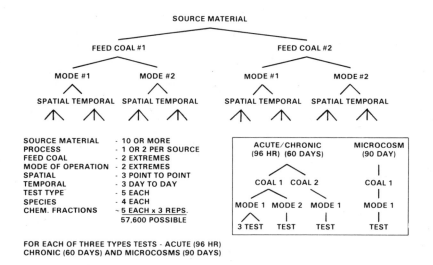

SOURCE MATERIAL

FEED COAL #1   FEED COAL #2

MODE #1   MODE #2   MODE #1   MODE #2

SPATIAL TEMPORAL SPATIAL TEMPORAL   SPATIAL TEMPORAL SPATIAL TEMPORAL

SOURCE MATERIAL   - 10 OR MORE
PROCESS   - 1 OR 2 PER SOURCE
FEED COAL   - 2 EXTREMES
MODE OF OPERATION - 2 EXTREMES
SPATIAL   - 3 POINT TO POINT
TEMPORAL   - 3 DAY TO DAY
TEST TYPE   - 5 EACH
SPECIES   - 4 EACH
CHEM. FRACTIONS   - 5 EACH x 3 REPS.
                    57,600 POSSIBLE

ACUTE/CHRONIC   MICROCOSM
(96 HR) (60 DAYS)   (90 DAY)

COAL 1 COAL 2   COAL 1

MODE 1 MODE 2 MODE 1   MODE 1

3 TEST TEST   TEST   TEST

FOR EACH OF THREE TYPES TESTS - ACUTE (96 HR)
CHRONIC (60 DAYS) AND MICROCOSMS (90 DAYS)

Fig. 7. An illustration of the large numbers of tests which could be run along with one strategy for reducing these to manageable proportions (see inset).

toward assessment of more complex ecological interactions.

Chemical analysis costs will be reduced by archiving samples of source materials under nondegradative conditions, to be tested later if initial ecological screening suggests potential problems. Even with this reduction, there still may be a large number of tests to be run on each source material. My own research staff went through this exercise for SRC and came up with a possible 57,600 experiments that might be run, as shown in Fig.7. We are now reducing the number of possible tests to manageable numbers by strategies like the one in the inset. This approach allows us to evaluate at each step the need to go farther in relation to the apparent information value of the other investigations we have under way.

Thus, as we are faced with the impossibility of looking at everything, we must play the odds carefully by attacking problems in this stepwise fashion. At each step, we focus on the most likely avenues for the most serious potential environmental effects (to the exclusion of others). Viewed positively, this gives us the greatest chance of flagging the most serious effects. Viewed negatively, by leaving many avenues unexplored, we risk missing something significant.

Nonetheless, if such a strategy of selective risk-taking is not adopted, we will become so hopelessly bogged down as to end up accepting the greatest risk of all: technology development may proceed but without adequate controls protecting the environment.

REFERENCE

1. C. Dale Becker, W. G. Woodfield, and J. A. Strand, *Solvent Refined Coal Studies: Effects and Characterization of Treated Solvent Refined Coal Effluent,* Progress Report FY 1977, PNL-2608, Battelle, Pacific Northwest Laboratories, Richland, Wash., July 1978.

FIELD-SITE EVALUATION OF AQUATIC TRANSPORT OF POLYCYCLIC AROMATIC HYDROCARBONS*

S. E. Herbes†   W. H. Griest†
G. R. Southworth†

ABSTRACT

An ongoing project is monitoring the transport behavior of polycyclic aromatic hydrocarbons (PAH) discharged in coking plant effluent. Water and sediments at 11 sites extending 11.4 km downstream from the outfall were sampled in April and July 1978. Following solvent extraction and extract purification, PAH concentrations in samples were measured by gas-liquid chromatography (GLC). Concentrations of 10 representative PAH effluent samples ranged from 0.67 to 18.3 μg/liter in April 1978. The estimated flux of benzo(a)pyrene in the effluent was 36 g/day. PAH concentrations in water from an upstream control site were higher than anticipated and suggest the presence of a second source upstream from the effluent outfall.

INTRODUCTION

Objectives of the Study

The transport and fate of trace organic contaminants released into natural waters in effluent streams will influence ecological and human health hazards associated with commercial coal conversion plants. Laboratory studies, which permit determination of transport rate constants for representative compounds, facilitate, to an extent, the prediction of transport behavior (both through semiquantitative comparisons within compound classes and through formulation of transport models). Field measurements are necessary, however, to confirm laboratory studies, validate models, and to permit identification of potentially critical transport pathways and parameters which might otherwise be overlooked. The goal of this study was to provide directly applicable information for predicting the impact of commercial-scale coal conversion facilities by characterizing the transport behavior of potentially hazardous trace organic contaminants in a fluvial environment from a point source.

The class of organic contaminants upon which initial studies have been focused are polycyclic aromatic hydrocarbons (PAH). These are fused-ring compounds, most commonly ranging from three to six rings in size, which are produced during high-temperature pyrolysis processes.[1] Several PAHs have been identified in product water samples for small-scale coal conversion processes.[2,3] Some PAHs are highly carcinogenic,[4] and thus their potential uptake by humans (either directly from contaminated water or through ingestion of contaminated fish or shellfish) may present an immediate and long-term health hazard.

This study of the disposition of PAH in receiving waters was designed with the following objectives:

1. evaluation of the persistence of PAH in a fluvial ecosystem,
2. determination of the extent of accumulation of PAH in sediments, and
3. validation of predictions of PAH transport behavior based on laboratory studies.

Persistence of waterborne PAH was determined by the comparison of PAH concentration decreases within a discrete effluent "slug" as it moved downstream, with those decreases anticipated on the basis of effluent dilution and dispersion alone (determined by use of a conservative tracer injected into the effluent "slug"). Sediment PAH accumulation was assessed by comparison of sediment PAH

*Research sponsored by the Office of Environmental Research, U.S. Department of Energy, under contract W-7405-eng-26 with Union Carbide Corporation. Environmental Sciences Division Publication No. 1267.

†Oak Ridge National Laboratory, Oak Ridge, Tennessee 37830.

levels with those of the overlying water. Finally, laboratory-derived predictions were to be compared with onsite process rate measurements (photolysis and microbial transformation) and with mathematical modeling. Because the study is not yet complete, only the first two objectives are discussed in this presentation.

Site Selection

Several criteria involving both the nature of the point source and the receiving riverine environment were established to aid selection of a field study site.

Nature of point source

1. The plant must provide a single industrial discharge of PAH with effluent composition similar to that anticipated from commercial coal conversion plants.
2. The effluent must contain sufficiently high PAH levels to produce detectable PAH concentrations in receiving waters.
3. Wastewater treatment procedures must be similar to those anticipated at commercial coal conversion plants.
4. The plant must have been operating under steady-state conditions for a number of years, so that transport processes in receiving waters would represent those affecting PAH transport during most of the 40- to 50-year lifetime of a commercial coal conversion plant.

Environment

5. The outfall must discharge into a relatively small riverine system with simple hydrological characteristics (to simplify data interpretation).
6. The receiving river must receive no, or minimal, discharges of similar industrial effluents.
7. The river water should be of high quality.
8. Historical hydrologic and water quality information should be available from state and Federal agencies.

Because no commercial coal conversion facilities are presently operational, alternatives for selection of a field study site were (1) a small-scale pilot coal conversion facility or (2) a large-scale plant of an industry similar to coal conversion. Although a pilot or process development plant would produce wastewater from a bona fide coal conversion process, most of the criteria outlined for an acceptable field site are not met by any small-scale unit in operation

or presently planned. Most important, most wastewaters produced are treated quite differently than will likely be the case in commercial plants (criterion 3), where wastewater volumes will necessitate physicochemical and biological treatment procedures prior to partial recycle or discharge. Second, most small-scale facilities in advanced stages of planning will be sited in heavily industrialized areas bordering rivers that are already highly degraded (criteria 6 and 7). The possibility of monitoring transport and transformations of contaminants released in a low-flow effluent stream under such conditions is remote. Third, pilot and process development units operate intermittently and with frequent changes in operating conditions, producing wastewater which may vary considerably in quantity and quality; little opportunity would exist to achieve an approximation of a steady-state discharge or effluent composition under such conditions (criterion 4).

As an alternative to a small-scale coal conversion plant, a coal coking plant is the most similar existing industry, both in process conditions employed (high-temperature, oxygen-limited pyrolysis of coal) and in wastewater constituents produced[5] (Table 1). A large number of by-product coke oven plants exist in the United States, most of which have coal-coking capacities in excess of 1000 tons/day. Many have operated continuously for many years using effective physicochemical and biological wastewater treatment procedures to meet state and Federal standards.

Table 1. Composition of wastewaters representative of coal conversion

Pollutant	Coke plant waste ammonia liquor (mg/liter)	Synthane process by-product water (mg/liter)
pH	8.3–9.1	7.9–9.3
COD	2,500–10,000	1,700–43,000
Ammonia	1,800–4,300	2,500–11,000
Cyanide	10–37	0.1–0.6
Thiocyanate	100–1,500	21–200
Phenols	410–2,400	200–6,600
Sulfide	0–50	N/D[a]
Alkalinity (as CaCO₃)	1,200–2,700	N/D
Specific conductance (as μmho/cm)	11,000–32,000	N/D

[a]Not determined.

Source: Reprinted by permission from E. S. Rubin and F. C. McMichael, "Impact of Regulations on Coal Conversion Plants," *Environ. Sci. Tech.* **9**, 112–117 (1975). Copyright by the American Chemical Society.

Although most of these plants are located in highly industrialized cities, some are situated in smaller towns in which the coking wastewater discharge may represent a major PAH source to the receiving river. From an initial listing[6] of 75 by-product coke oven plants in the United States, the plant which best satisfied the selection criteria was the Bethlehem Steel Corporation coke works in Bethlehem, Pennsylvania.

Site Description

The Bethlehem Steel Corporation coke works in Pennsylvania is one of the five largest in the United States. Approximately 9×10^5 liters/d of highly contaminated wastewater are treated by an activated sludge unit which removes phenols and cyanide prior to discharge. This unit was one of the first of its kind, and has been operating continuously with high efficiency since the early 1960s.[7]

Until late July 1978, the treated coke effluent was diluted and discharged into Saucon Creek, a shallow, rapidly flowing stream about 15 m wide and 30 cm deep (average flow approximately 2.5 m³/sec) which originates as groundwater 15 km upstream. An outfall from the Bethlehem Municipal Wastewater Treatment Plant is located 1.2 km below the coke outfall, and the confluence with the Lehigh River is 0.3 km farther downstream (Figs. 1 and 2). Sediments are generally gravel, with pockets of sand and silt. The river at the point of confluence is 60 m wide and 2 m deep (average flow approximately 50 m³/sec). The water quality of the river is highly degraded, primarily from municipal wastes from a dense upstream population and multiple outfalls of the Bethlehem Steel plant which lines the south shore of the river for several km upstream. Surface oil sheens are often visible. No other pollutant inputs occur downstream of the Saucon-Lehigh confluence for a distance of 10 km to a low dam (Glendon Dam), which forms the downstream boundary of the study area; through this reach the Lehigh is bordered primarily by fields and forests. The bottom is composed primarily of boulders and rubble with occasional pockets of silt and sand; silt and organic muck occur directly above the dam.

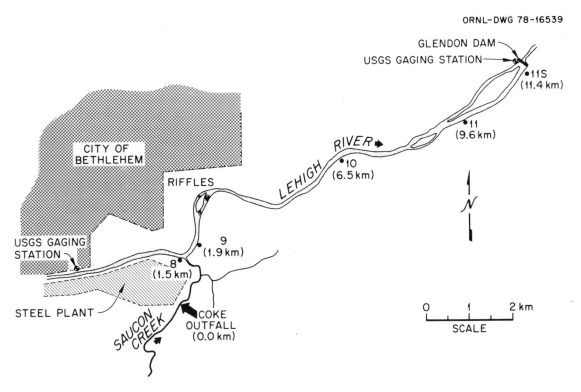

ORNL-DWG 78-16539

Fig. 1. **Coking plant study area, Bethlehem, Pennsylvania.** Black circles indicate sampling sites on the Lehigh River; distances from outfall (km) in parentheses.

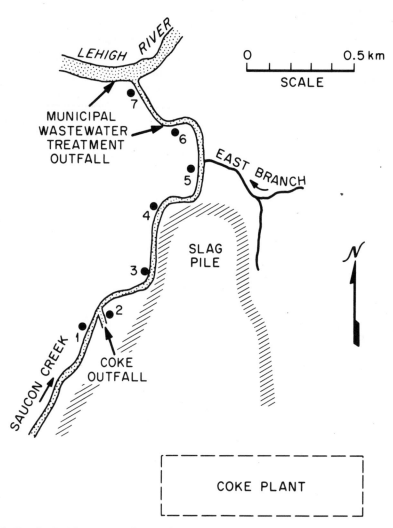

ORNL-DWG 78-16538

Fig. 2. Detail of coking plant outfall and Saucon Creek showing sampling locations 1 through 7.

METHODS AND MATERIALS

Sample Collection

Seven sampling sites were selected on Saucon Creek and four on the Lehigh River; upstream control sites were included on each (Figs. 1 and 2). Water samples were taken on April 27 and July 18, 1978. Water samples at both times were collected in 3.8-liter amber glass bottles. During the April study, samples were passed within 3 hr of collection through resin sorption columns (Rohm and Haas XAD-2) by gravity flow (6–12 ml/min) to concentrate PAH; columns were frozen and shipped on dry ice back to Oak Ridge National Laboratory (ORNL), where they were stored frozen until extracted and analyzed.

When later results demonstrated that column retention of three-ring PAH was incomplete, samples from the July study were sterilized by addition of 40 ml of chloroform to each bottle and shipped back to ORNL for analysis.

In each study period, replicate 7.6-liter samples were collected at sites 1 and 3–7 on Saucon Creek. Six 3.8-liter samples were collected in the effluent discharge channel. Larger samples (11.4 liters) were collected from each of 5 transect positions across the Lehigh River at sites 8–11. All samples were obtained at a depth of 20 cm.

Sediment samples were obtained in Saucon Creek by scooping the upper 2 cm of streambed at several locations, mixing in a stainless steel pan, and removing a 100-g sample for PAH analysis and a 50-g sample for determination of particle size distribution. Replicates were obtained from each side of Saucon Creek at each site. Lehigh River samples were obtained at each transect position by the use of a small Ponar dredge. All sediment samples were sieved through 6.4-mm mesh prior to mixing and sample removal. Sediment samples for PAH analysis were frozen within 2 h, shipped to ORNL on dry ice, and stored under acetone at $-20°C$ until analysis; samples for particle size analysis were maintained at room temperature.

Use of LiCl Tracer

To provide an internal standard for effects of dilution and dispersion of the effluent plume, a constant concentration (160 g/liter) of LiCl dissolved in 180 liters of Saucon Creek water was metered by siphon into the effluent channel at a rate of 2 liters/min from a 210-liter drum for 90 min. The travel time of the Li "slug" was estimated by prior injection of a Rhodamine B tracer. Sampling times at each site were selected such that the midpoint of the Li "slug" would pass the site when PAH sampling occurred. Additional 100-ml samples in polyethylene bottles were obtained at each site at 10-min intervals for 50 min before and after PAH sampling and were preserved with 0.1 ml of concentrated HNO_3. Samples were analyzed at ORNL for Li content by atomic absorption spectrophotometry (detection limit: 0.01 mg/liter).

PAH Analytical Procedures

The analytical procedure consisted of elution of material sorbed by resin columns, preparation of PAH isolate, and determination of the identities and concentrations of PAH by gas-liquid chromatography (GLC).[8] After thawing, resin packings were reverse-transferred to larger elution columns, and sorbed materials were eluted (in the reverse direction to water collection flow) sequentially with methanol (50 ml), tetrahydrofurane (100 ml), diethyl ether (150 ml), and methylene chloride (30 ml). The aqueous phase eluted from each column was extracted four times with redistilled cyclohexane. The latter was combined with the organic phase and the combined organic extracts were reduced in volume to approximately 10 ml with dry flowing nitrogen gas under reduced pressure and temperature. This extract was added to a column containing 10 g of Florisil which was eluted with 150 ml of 6:1 hexane-benzene (v/v).

The eluate was again reduced in volume to 10 ml and was passed through 20 g of neutral alumina (4% moisture) with a hexane-benzene step gradient of 6:1 (100 ml) and then 2:1 (300 ml). The polyaromatic fraction defined by radiotracer elution experiments was collected and concentrated to 0.1 ml after addition of a hexadecane internal standard. GLC analysis was carried out on a Perkin-Elmer model 3920 gas chromatograph equipped with a flame ionization detector and a 3-mm-OD glass column, 3 m long, packed with 3% Dexsil 400 on 80/100 mesh HP Chromosorb G. The column temperature was raised from 110°C to 320°C at 2°C/min. PAH identifications were made by comparison of retention times with those of standards. PAH recoveries were estimated in laboratory experiments using water samples containing PAH standards and radiolabeled tracers.

RESULTS

Substantial amounts of PAH are discharged into Saucon Creek via the coke plant effluent. Of 20 compounds identified, concentrations of 10 are shown in Table 2. At least three known carcinogens were tentatively identified by co-chromatography (pending mass spectral confirmation): benzo(c)phenanthrene, benzo(a)pyrene, and dibenzanthracene (possibly the a,c and/or a,h isomers). The observed concentrations of PAH in the effluent were slightly less than the aqueous solubility limit for several of the PAHs, including benzo(a)pyrene and perylene.[9] Those PAHs having higher aqueous solubility, such

Table 2. Concentrations of PAH in Bethlehem, Pa., coke outfall water samples

	April 1978 (mean of 5 samples) (μg/liter)		April 1977 (one sample) (μg/liter)
Fluoranthene	18.3	(4.3)[a]	2.8
Pyrene	14.4	(3.2)	4.0
Benzo(a)fluorene	1.4	(0.3)	0.80
Benzo(b)fluorene	1.3	(0.3)	0.81
Benzo(c)phenanthrene	0.67	(0.13)	N/D[b]
Benzo(b)fluoranthene	5.4	(1.6)	N/D
Benzo(a)pyrene	2.0	(0.5)	1.8
Perylene	0.72	(0.24)	N/D
Dibenzanthracenes	1.4	(0.3)	0.95
Benzo(ghi)perylene	1.2	(0.1)	N/D

[a]Standard error.
[b]Not detected.

as fluoranthene and pyrene, were found at concentrations considerably greater than other PAHs in the effluent.

The replicability of the analyses was reasonable, yielding a standard error of about 20% of the mean for the population of five effluent samples. Preliminary analyses of the effluent composition made one year earlier (April 1977) agreed well with results from April 1978 for most compounds (Table 2).

Discharge of the effluent into Saucon Creek provided an input sufficient to raise total PAH concentrations in the creek to at least 5 μg/liter at the point of discharge (assuming instantaneous mixing). On the basis of effluent flow rate, the effluent contributed approximately 36 g/d of benzo(a)pyrene to the creek.

Water samples from only three Saucon Creek sites (in addition to the effluent) have been analyzed at this time; concentrations of several representative PAHs determined in those samples are presented in Table 3. Results of the lithium tracer analysis indicated that the effluent is diluted about 13 times between the effluent channel (site 2) and site 4 and then diluted another 25% before reaching the confluence with the Lehigh River (site 7). After complete mixing in the Lehigh River (achieved at station 10), lithium concentrations are below the limit of detection (10 μg/liter).

PAH concentrations at site 1 (Saucon Creek control, upstream from the effluent) were considerably higher than anticipated from preliminary studies. Levels observed at site 1 are characteristic of highly contaminated systems.[10] At site 4, downstream from the effluent, observed concentrations did not reflect the input of additional PAH from the effluent, with concentrations generally lower than were found at site 1. At site 7, the confluence with the Lehigh River, PAH concentrations were higher than at site 4 but still were roughly equivalent to levels observed at the control site. With the exception of benzo(a)pyrene, the fluxes of the representative PAH entering the Lehigh River via Saucon Creek at site 7 were nearly twice those discharged by the coking plant (site 2). Analytical replicability in Saucon Creek samples was not so good as noted in effluent samples [about 10 to 20% for fluoranthene and pyrene and 50% for benzo(a)pyrene].

In the Lehigh River water samples, a combination of high background PAH, high dilution, and wide variation among samples from the same site made quantification of the PAH contribution from Saucon Creek unreliable.

Although sediment samples obtained in April 1978 have not yet been analyzed, sediment PAH concentrations in samples taken in April 1977 (Table 4) show a pattern similar to that observed for aqueous PAH in April 1978. Levels in the sediments at the upstream control site (site 1) were approximately equal to levels found at site 4. At station 7, PAH concentrations were about 20 times higher than at stations 1 and 4, possibly due largely to the highly organic nature of the sediment at that site, which might be expected to increase PAH sorption. Concentrations observed within the effluent channel sediment were similar to those at site 7, although the effluent concentrations were 5 to 25 times higher than

Table 4. Concentrations of representative PAH in sediment samples from Saucon Creek, April 1977

	Site 1	Site 2	Site 4	Site 7
Fluoranthene	1.3[a]	30.8 (9.1)[b]	1.1	27.4
Pyrene	1.2	23.5 (8.6)	0.93	2.8
Benzo(a)pyrene	0.57	19.4 (5.4)	0.53	13.2
Dibenzanthracenes	0.04	8.6 (4.5)	0.05	1.2

[a]Concentrations in mg/kg dry weight.
[b]Standard error of two replicate samples.

Table 3. Concentrations of LiCl tracer and selected PAH in water samples from Saucon Creek, April 1978

	Site 1, mean of 2 samples	Site 2, mean of 5 samples	Site 4, mean of 4 samples	Site 7, mean of 2 samples
Li, mg/liter	<0.01	3.32 (0.14)[a]	0.26 (0.01)	0.19 (0.01)
Fluoranthene, μg/liter	2.3 (0.2)	18.3 (4.3)	1.0 (0.1)	1.8 (0.1)
Pyrene, μg/liter	1.9 (0.1)	14.4 (3.2)	0.67 (0.05)	2.0 (0.1)
Benzo(a)pyrene, μg/liter	0.12 (0.05)	2.0 (0.5)	0.14 (0.03)	0.08 (0.11)
Dibenzanthracenes, μg/liter	0.10 (0.06)	1.4 (0.3)	0.11 (0.01)	0.23 (0.08)

[a]Standard error.

water at site 7. The packed clay sediment in the channel would, however, be expected to have a lower affinity for PAH than the organic muck at site 7. The higher ratio of concentration of benzo(*a*)pyrene to pyrene observed in sediments when compared with water is illustrative of the higher degree of partitioning of the larger, more hydrophobic PAH into sediments.

DISCUSSION

The high concentrations of PAH observed in control samples from both Saucon Creek and Lehigh River make interpretation of PAH transport uncertain until analyses from additional sites are completed. The high background PAH and apparent nonhomogeneity of water samples were not entirely unanticipated in the Lehigh River, where surface oil and grease contamination is evident. The conclusion (at least at the time of sampling) that the Bethlehem Steel coke works effluent was not the sole source of PAH contamination in Saucon Creek was certainly unexpected. From the Li data, upstream diffusive transport from the outfall could not contribute to the PAH levels observed at site 1. No obvious, large potential source was observed upstream of site 2 on Saucon Creek, although groundwater percolating through the large mountain of slag bordering the creek is a possibility. Stream flow measurements of the Lehigh River (Fig. 3) indicate that the month

preceding the sampling was a period of substantial precipitation and snow melt. This observation supports the likelihood of substantial percolation through the slag pile. Other possible sources of PAH include runoff from the watershed or some unknown effluent from a point source farther upstream. Airborne particle fallout may also be significant: Saucon Creek is located 1 km directly east of the Bethlehem Steel plant, 2 km to the southeast of Bethlehem (a city of approximately 75,000), and 1 km northeast of the coke ovens. Airborne emissions of coke plants have been demonstrated to contain high levels of PAH.[11]

The increase in PAH noted between sites 4 and 7 suggest either additional input from the east branch of Saucon Creek or redissolution/resuspension of PAH from the highly contaminated sediments at site 7 (which may have accumulated at an earlier date). Future studies will clarify whether contaminated sediments may sometimes serve as a source (rather than a sink) for PAH, thus maintaining waterborne PAH at levels higher than anticipated.

The initial PAH analyses indicate the need for further information to clarify the role of PAH sources on Saucon Creek in addition to the effluent outfall. Completion of April sample analyses will permit evaluation of trends in Saucon Creek PAH concentrations and will identify any anomalous data points. Analyses of samples collected on July 18,

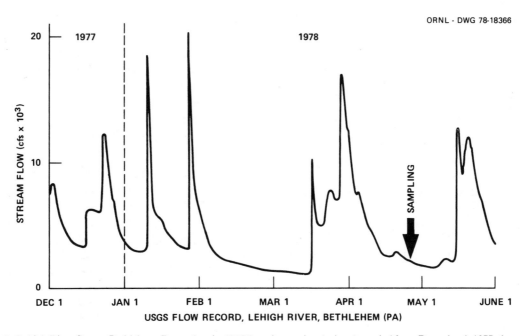

Fig. 3. **Lehigh River flow at Bethlehem, Pennsylvania.** USGS gaging station during the period from December 1, 1977, through June 1, 1978.

1978, during a period of low river flow, should provide additional information on the possible roles of groundwater influx, runoff, and snow-melt percolation through the slag pile. In addition, the use of direct solvent extraction of July water samples, rather than use of XAD-2 sorption columns, will permit evaluation of the relative precision and possible interferences inherent in each method.

On July 28, 1978, construction of a pipeline running parallel to Saucon Creek was completed, and the coking effluent discharge was diverted to an outfall on the Lehigh River approximately 20 m east of the Saucon-Lehigh confluence. The diversion presents an opportunity to evaluate the rate of removal of PAH from contaminated sediments in Saucon Creek. Sampling of water and sediments in both Saucon Creek and in the Lehigh River is planned for the fall of 1978 and the spring and summer of 1979 to assess both the rate of recovery of the creek and PAH transport in the river following direct effluent discharge. Data accumulated during the next year will be used to validate mathematical models of PAH transport behavior being formulated and will assist in evaluating potential hazards of wastewater discharges from commercial coal conversion plants in the future.

ACKNOWLEDGMENTS

The authors gratefully acknowledge the cooperation of J. W. Flecksteiner, Director, Department of Environmental Control, Bethlehem Steel Corporation. Messrs. J. Smurda, L. Oberdick, E. Coppenhaver, and N. George of the Pennsylvania Department of Environmental Resources (Bureau of Water Quality) have continuously provided advice and invaluable assistance in the planning and conduction of the sampling program. We thank W. Grim, Superintendent of the Bethlehem Municipal Wastewater Treatment Plant, who has generously furnished building space for sample treatment and equipment storage. G. R. Jenkins, Director, Institute of Research at Lehigh University, provided advice on sampling strategy and mediated contacts with five Lehigh University students who assisted in the samplings.

S. Watson and R. Reagan (Analytical Chemistry Division, ORNL) provided excellent technical assistance in sample preparation and PAH analysis.

REFERENCES

1. J. B. Andelman and M. J. Suess, "Polynuclear Aromatic Hydrocarbons in the Water Environment," *Bull. W. H. O.* **43**, 479 (1970).

2. M. R. Petersen, J. S. Fruchter, and J. C. Laul, *Characterization of Substances in Products, Effluents, and Wastes from Synthetic Fuel Production Tests,* BNWL-2131, Battelle Pacific Northwest Laboratories, Richland, Wash., September 1976.

3. W. D. Shults (ed.), *Preliminary Results: Chemical and Biological Examination of Coal-Derived Materials,* ORNL/NSF/EATC-18, Oak Ridge, Tenn., March 1976.

4. H. E. Christensen and T. T. Luginbyhl, eds., *Registry of Toxic Effects of Chemical Substances: 1975 Edition,* U.S. Department of Health, Education, and Welfare, Rockville, Md., June 1975.

5. E. S. Rubin and F. C. McMichael, "Impact of Regulations on Coal Conversion Plants," *Environ. Sci. Tech.* **9**(2), 112 (1975).

6. G. F. Nielsen (ed.), *1978 Keystone Coal Industry Manual,* McGraw-Hill, New York, 1978, p. 428.

7. P. D. Kostenbader and J. W. Flecksteiner, "Biological Oxidation of Coke Plant Weak Ammonia Liquor," *J. Water Pollut. Control Fed.* **41**(2), 199 (1969).

8. W. H. Griest, "Multicomponent Polycyclic Aromatic Hydrocarbon Analysis of Inland Water and Sediment," presented at the International Symposium on the Analysis of Hydrocarbons and Halogenated Hydrocarbons in the Aquatic Environment, Hamilton, Ontario, Canada, May 23–25, 1978.

9. D. Mackay and W. Y. Shiu, "Aqueous Solubility of Polynuclear Aromatic Hydrocarbons," *J. Chem. Eng. Data* **22**(4), 399–402 (1977).

10. D. K. Basu and J. Saxena, "Polynuclear Aromatic Hydrocarbons in Selected U.S. Drinking Waters and Their Raw Water Sources," *Environ. Sci. Tech.* **12**(7), 795–98 (1978).

11. R. C. Lao, R. S. Thomas, and J. L. Monkman, "Computerized Gas Chromatographic–Mass Spectrometric Analysis of Polynuclear Aromatic Hydrocarbons in Environmental Samples," *J. Chromatogr.* **112**, 681–700 (1975).

ADDENDUM

Although most of the information presented at the Symposium remains valid, the acquisition of additional data since that time requires revision of some of our original conclusions:

1. Gas chromatographic–mass spectroscopic examination of the site 1 sample revealed no detectable PAH. The site 1 gas chromatographic peaks originally assigned as PAH on the basis of chromatographic retention time actually were not PAH. Therefore, we have no basis to assume the presence of an additional PAH source upstream of the coking plant.

2. Gas chromatographic–mass spectroscopic examination of the site 2 samples confirmed most preliminary PAH identifications, including the carcinogen BaP and the cocarcinogens fluoranthene and pyrene. Additional unsubstituted or monomethylated PAH were identified also. The coking plant effluent channel thus appears to be the major PAH input to Saucon Creek down to this point.

3. Extraction and analysis of samples from two PAH-spiked control XAD-2 columns handled in the same matter as the exposed XAD-2 columns (except that no water sample was passed through the control XAD-2 columns) yielded essentially quantitative recoveries of 4 to 6-ring PAH and gave no indication of a contamination or background problem. This result suggests that the XAD-2 column preparation, handling, storage, and extraction, PAH isolation, and analytical procedures were performed properly.

4. Extraction and analysis of two nonspiked blank XAD-2 columns handled differently from the exposed XAD-2 columns (inadvertently, blanks were not frozen or shipped with the exposed columns) yielded a significant and complex gas chromatographic background which would interfere with PAH identification and measurement. Although the analytical results from such XAD-2 columns normally would be disregarded due to different handling, they may be relevant in view of the similarity of their background to the appearance of the PAH isolates from several of the exposed XAD-2 column samples. This similarity suggests contamination from some unidentified, uncontrolled, and possibly nonconstant factor in the handling or extraction of the XAD-2 columns. The discrepancy in the results between the blank and control columns would seem best attributed to differences in their shipping and storage, but the exposed XAD-2 columns were shipped in the same manner as the controls and thus should not have been contaminated. We cannot resolve this apparent conflict at the present.

The remaining Saucon Creek samples have not yet been fully examined by gas chromatography–mass spectroscopy for interferences in the PAH analyses, and corrected PAH concentrations are not available. However, the evidence described above suggests that the PAH data for sites 4 and 7 in Table 3 must be treated as maximum possible PAH concentrations. Decreases in PAH concentration downstream of the coking plant effluent channel may thus be more rapid than indicated in Table 3.

DISCUSSION

M. A. Shapiro, University of Pittsburgh: I think one conclusion you can draw is that your so-called simple stream turns out to be very complex.

Steve Herbes: Yes, that's true.

Rod Skogerboe, Colorado State University: First of all I'd like to compliment you on these studies. It's a type of study that needs to be done, and needs to be done extensively all over this country, because PAH transport through natural water systems is going to be more and more important as time goes on. Secondly, however, I would like to comment that there is one mechanism that you have overlooked in your exploration of possibilities. That is that you are in a downwind configuration within the meteorological plume of the major city that produces a lot of PAH compounds. Based on some of our experiences, those experiences of Dave Natusch, and experiences of the water chemistry laboratory at the University of Wisconsin, you will find major amounts of these materials deposited on the ground and flushed into the system during precipitation events, particularly those that are associated with snowfall. So you are going to be in a configuration when you are going to have to worry about correcting for that.

Harold Bergman, University of Wyoming: Those are chlorinated outfalls, I presume?

Steve Herbes: Yes. They practice spot chlorination; I believe it is on an 8-hour cycle.

Harold Bergman: You people, of course, are familiar with chlorination problems associated with these chemicals.

Steve Herbes: During our sampling period the environmental control department of Bethlehem Steel assured us that they would not chlorinate the effluent.

Harold Bergman: How about the municipal outfalls?

Steve Herbes: Yes, they do practice chlorination.

Harold Bergman: Probably one milligram per liter and you have not found chlorinated PAHs at all?

Steve Herbes: We haven't looked for them. I am not sure whether they would be removed during our clean-up procedures or not. Dr. Griest?

Wayne Griest: They may be there, but we have not looked for them.

Steve Herbes: Before I close, I should have acknowledged, and I neglected to, the cooperation of John Flecksteiner, Director of Environmental Control at the Bethlehem Steel plant. I would also like to acknowledge the help and cooperation of the Department of Environmental Resources of the State of Pennsylvania, and finally, superintendent Bill Grim and the people at the Bethlehem Municipal Wastewater Treatment Plant, which served as a base of operation during this study.

SESSION V: OCCUPATIONAL HEALTH CONTROL TECHNOLOGY

Chairman: James M. Evans
Enviro Control, Inc.
Cochairman: R. D. Gammage
Oak Ridge National Laboratory

SESSION V: OCCUPATIONAL HEALTH CONTROL TECHNOLOGY

SUMMARY

James M. Evans*

The preceding four sessions dealt with technology, chemical characterization, biological effects, and environmental and ecological effects. This session deals with something closer to home—effects on humans, and in particular, on those people who may potentially be affected by the toxic materials that they produce. Of all the ecological, environmental, and community problems that may result from coal conversion, occupational health effects may be the most difficult to cope with.

The papers presented in this session have been written by a concerned group of activists who have widely differing points of view, who come from different parts of the country and of the world; yet these papers are a cohesive whole.

"Synthetic" fuels from tar sands, shale oil, and coal all contain more of the carcinogenic PNA materials than do the fuels derived from natural gas and crude oil that we are used to. Epidemiological studies on coal tars have shown them to be carcinogenic. However, actual epidemiological studies on coal conversion processes—liquefaction or gasification —have been restricted to the single study reported by Sexton et al. in 1960. To date, there have also been only three somewhat restricted toxicological studies by Weil and Condra on materials from the Institute plant, by Huper on materials from the Louisiana, Missouri, plant, and by Bingham on materials from the Cresap plant.

Epidemiological studies on coke oven workers in the United States show definitively that the workers in these plants, especially in certain categories, are without question at high risk from pyrolysis products of coal. Many of these pyrolysis products are found in products from coal liquefaction. However, both Dr. Jackson and Dr. Archibald point out that the processing methodology for synfuels is vastly

different from that utilized for the coking of coal and thus the occupational exposure will be different. Yet Dr. Jackson shows a real similarity between the operations that will be or are being utilized and those that produce "synthetic" fuels from a variety of materials such as tar sands, shale oil, and coal.

Dr. Palmer's and Dr. Archibald's papers indicate the difficulty in making meaningful epidemiological studies, especially in extrapolating the result of these studies into the future. Dr. Palmer's paper adds to our existing knowledge by presenting information even though it might be open to question. Without this publication these data would be hidden. Dr. Palmer's paper does us the extreme service of being useful, especially when there is such a need as now for this type of information.

Dr. Yodaiken points out the need for medical surveillance programs that are both adequate and simple. He would also like to use a first-things-first approach. Doctors Yodaiken and Archibald both express the need for adequate monitoring of the workplace and say that sophisticated tools are not necessarily required, but rather that simple indicators may be used to estimate the exposure to other toxicants in the workplace. Dr. Archibald cautions, however, that worker exposure is not necessarily defined by the concentration of hazardous materials in the total workplace area.

Doctors Yodaiken and Archibald both express the need for adequate engineering control technology, and Dr. Jackson indicates that the development of control technology for one unit operation in one process may well be applied to that of another process. Despite good engineering control technology, Dr. Archibald indicates that there are problems,

---

*Enviro Control, Inc.

233

even with improved engineering control technology, that worker exposure may continue to be high due to inadequate or deteriorating work habits.

The papers given in this session imply that the equipment available to the industrial hygienist for monitoring and to the work force for respiratory protection is inadequate, bulky, uncomfortable, or all of these. The papers pointed out that until now, those of us who are in the field have had to use minimally performing equipment. Doctors Archibald and Hawthorne tell us that this is rapidly coming to an end; that new concepts in respiratory protection are being developed, and that new monitoring instruments, particularly those that are real-time, multicomponent, easily operated, and inexpensive are being defined and developed in the laboratory.

I would suggest that the papers given in the session on occupational health define both the enormity and the heart of the problem with which we are dealing. They also offer us the hope that with a practical, straightforward, first-things-first approach we can more effectively protect the workers. They indicate that this can be done both as quickly as the rapid advance of technology demands, and that it will be cost effective for the employee, the employer, and the public.

MORTALITY EXPERIENCE OF 50 WORKERS WITH OCCUPATIONAL EXPOSURE TO THE PRODUCTS OF COAL HYDROGENATION PROCESSES

Alan Palmer*

ABSTRACT

A study was undertaken to determine the mortality experience of 50 workers diagnosed with either skin cancer or precancerous lesions. These workers were first diagnosed between 1955 and 1959 as part of a screening program for workers employed in a coal hydrogenation plant. Workers were traced and death certificates obtained for those who were deceased. Findings revealed that five of the workers had died from noncancerous causes, 16 had retired, 28 were still working, and one subject was lost to follow-up. Because of the limited scope of the study [(it included only those workers with skin lesions) and the small sample size (50)], no firm conclusions can be made, although it would appear that there is no increase of death caused by systemic cancers.

INTRODUCTION

The coal hydrogenation process was first patented in Germany in 1917 by Bergius.[1] He demonstrated that bituminous coal treated with hydrogen at high pressures and high temperatures yielded 85% of its weight in soluble or liquid products. These products consist of phenolic substances and hydrocarbons that are similar to the petroleum hydrocarbons. This process was first tried experimentally in the United States in 1930 by Union Carbide Corporation and commercially in 1952.[2]

The chief health hazards associated with the many chemical compounds produced by this process are essentially produced by exposure to the oil products that boil at temperatures greater than 260°C, i.e., polycyclic aromatic hydrocarbons (3,4-benzopyrene), aromatic amines, toxic metals, and organometallic compounds.[2,3]

Relationships have already been established between many of these compounds and occupational cancers of the skin, of the upper and lower respiratory tract, and of the urinary tract. For example, machine finishing workers exposed to machine cutting oils containing benzopyrene showed an excess of dermal cancers (spinocellular epithelioma) of the scrotum, forearms, and hands in addition to precancerous lesions such as papillomas and keratoacanthanomas.[4] A mortality study of coal gas workers demonstrated that those workers exposed to tars and tar-containing vapors had a slight increase in urinary tract carcinoma over those not exposed.[5] In another group exposed to coal and tar gas, deaths from lung cancer were found to be double the normal rate.[6]

In this mortality study, we have determined the vital status of a subgroup of 50 out of 359 Union Carbide employees. All were exposed to high-temperature boiling oils containing polycyclic hydrocarbons, coal tar, and pitch—the by-products of the coal hydrogenation process initiated at the Institute, West Virginia, plant in May 1952. Recognizing the potential health hazard of such exposures, the Union Carbide Corporation instituted a cancer-control program in 1952 whereby workers exposed to this process were given an annual physical examination and, beginning in 1955, quarterly skin inspections.

A paper written by Dr. Richard Sexton,[2] the Medical Director of the Institute plant, describes the findings of the surveillance program, which was continued until 1960. Dr. Sexton found one worker with two cutaneous cancers, nine workers with one cutaneous cancer each, and forty workers with one cutaneous precancerous lesion each.

*SRI International, 333 Ravenswood Ave., Menlo Park, California 94025.

To the degree that the workers manifesting the cutaneous cancers and precancerous lesions may have experienced much heavier exposures than other workers in the plant, the question was posed "Do these workers represent a special group of high risk in regard to developing cancers of other organ systems?" Therefore, an investigation of the mortality experience of these 50 workers appeared warranted as a preliminary step in the search for possible long-term effects through a definitive epidemiological study.

The specific objectives of this limited study have been to (1) identify and trace the original 50 workers to ascertain their vital status, (2) determine the cause of death as precisely as possible for those verified to be dead as of July 1977, and (3) collect available information on malignancies present at the time of death.

This investigation did not attempt to ascertain the vital status of the remainder of the work force who did not have skin lesions, nor were there attempts to document occupational exposures for these 50 workers.

METHOD

Records of the original cohort of 50 workers were reviewed at the medical department of the Union Carbide plant in Institute, West Virginia, and compared with those reported in Sexton's paper (Tables 4, 5, and 6).[2] Original pathology reports were also reviewed for each member of the cohort to verify the diagnosis. A close examination of the data revealed that Sexton's Table 4 included two people with cutaneous cancer in the 42 precancerous cases. These two had already been accounted for in Sexton's Table 5. After correcting this redundancy, the distribution of cases was:

- precancerous lesions on 40 people,
- 11 cancerous lesions on 10 people, or
- 51 lesions on 50 people out of an exposed work force of 359.

Each subject was listed, including the person's full name, date of birth, Social Security number, continuous service date, payroll number, job description at the time of diagnosis, and the skin lesion diagnosed.

A July 1977 listing of workers was obtained from the plant payroll department, from which it was determined that 28 of the 50 were still working. To check on the status of the remaining 22 people, company retirement and claims records were examined. The record showed that 11 of the 22 were currently receiving retirement benefits, which was taken as preliminary evidence that these 11 were still alive. Furthermore, the records showed that four would be applying for a pension as soon as their vestment periods had matured, although they were not currently receiving checks.

The plant retirement office confirmed that these four were still alive, based on contacts that had been made over the course of each year. The last known addresses of the four were recorded for subsequent follow-up. The disposition of two workers remained unsolved, although their last-known addresses, termination dates, and Social Security numbers were available. Five workers were known to have died. The copies of their death certificates, which were on file in the claims office, were reviewed, and causes of death were noted. No autopsies were performed on any of the deceased.

To establish the vital status of 15 of 17 workers thought to be alive (two of whom were confirmed to be alive through a recent company contact), the tracing facilities of Equifax Corporation were employed. Using its national network of field investigators, Equifax confirmed the vital status of 14 of the 15 subjects. Routine techniques (i.e., Post Office, letter carrier interviews, circuit court clerks, tax assessment office, neighbors, dwelling house managers, and city directories) were used to confirm that the 14 subjects were still alive. One subject, thought to be living in Florida, was sought through a review of all Equifax branch claim files for that state and through the Florida Department of Motor Vehicles record department. No trace of this individual was found.

RESULTS

Table 1 shows the distribution of the 40 subjects originally identified in Sexton's original study as having precursors of skin cancer, with the redundancy error corrected.

Table 1. Precursors of skin cancer cases:
white males only[a]

| Number of cases | Mean age[b] | Length of exposure [years (months)] | | Histological diagnoses |
		Minimum	Maximum	
3	30	(10)	6 (2)	Pitch acne[c]
3	33	(3.5)	3 (6)	Chondrodermatitis helicis[c]
17	38	(10)	9 (8)	Keratoses[d]
8	44	1 (5)	8	Keratoses
9	40	(5)	9	Acanthoses and hyperkeratoses
Total 40				

[a]From 359 workers examined over 5 years.
[b]Mean age at time of diagnosis.
[c]Clinical diagnosis only.
[d]Diagnoses by a single pathologist only.

Table 2 shows the distribution of cases of confirmed cutaneous cancers found on ten subjects. Of these subjects, two were confirmed to be dead, and one could not be traced.

• Case 4 died 8/12/66 at the age of 47, 11 years after the initial diagnosis of skin cancer. This person was exposed for 4 years before being

diagnosed in 1955. The death certificate cited the cause of death as acute pulmonary infection, emphysema, arteriosclerotic heart disease, and cor-pulmonale.

• Case 9 died 12/13/73 at the age of 67, 18 years after the initial diagnosis of skin cancer. This person was exposed for 5 years before diagnosis

Table 2. Coal hydrogenation cutaneous cancer cases: white males only[a]

Case	Age[b]	Length of exposure [years (months)]	Job assignment	Body site	Local pathologist	Out-of-town pathologist
1	30	3 (5)	Operations	Forearm, right	Basal-cell carcinoma	Malherbe's calcifying epithelioma
2	33	5 (2)	Maintenance	Cheek, left	Basal-cell carcinoma	Basal-cell epithelioma
3	29	1	Maintenance	Cheek, left	Squamous-cell carcinoma	No pathology done, clinical diagnosis only
4	37	4	Maintenance	Buttock, left	Squamous-cell carcinoma	Inverted follicular keratoma type of seborrheic keratosis
5	43	(9)	Maintenance	Hand, left	Squamous-cell carcinoma	Squamous-cell carcinoma
5	43	(11)	Maintenance	Ear, right	Mixed basal- and squamous-cell carcinoma	Metatypical carcinoma
6	34	6 (11)	Operations	Hand, left	Intraepithelial squamous-cell carcinoma	Bowenoid keratosis
7	46	6 (11)	Operations	Neck, posteriolateral, right	Squamous-cell carcinoma	Prickle-cell epithelioma
8	40	9 (8)	Operations	Ear, right	Squamous-cell carcinoma	Keratoacanthoma
9	33	5	Maintenance	Leg, left	Keratosis	Intraepithelial squamous-cell carcinoma
10	38	6	Maintenance	Leg, right	Keratosis	Squamous-cell carcinoma

[a]From 359 workers examined over 5 years.
[b]Age at time of diagnosis.

in 1955. The death certificate listed the cause of death as myocardial infarction.

From the group with precancerous lesions, 3 persons had died:

- Subject 1 died 5/11/76 at the age of 68, 18 years following the initial diagnosis. This person had been exposed for 6 years before diagnosis in 1958. The cause of death was coronary disease, arteriosclerotic heart disease.

- Subject 2 died 4/28/71 at the age of 67, 16 years following the initial diagnosis. This person had been exposed for 6 years before diagnosis in 1955. The cause of death was pulmonary embolus, myocardial failure.

- Subject 3 died 8/31/73, at the age of 67, 17 years following the initial diagnosis. This person had been exposed for 6 years before diagnosis in 1956. The cause of death was acute myocardial infarction.

Although a precursory review of worker medical records was made on these 50 people, no attempt was made to exhaustively pursue the health history of each worker. However, this review did reveal that one retired worker with precancerous lesions was currently ill with Parkinson's disease and cancer of the prostate and that one retired worker with cancerous skin lesions was ill with lung cancer.

Table 3 shows that, of the original 10 workers with diagnosed skin cancer, two have died from non-cancerous causes, two have retired, five are still working, and one was lost to follow-up. The 40 subjects diagnosed as having precancerous skin lesions are distributed as follows: three have died from noncancer causes, 14 have retired, and 23 are still working. Because of the small number of workers in the cohort and the paucity of deaths that occurred, these data were not analyzed statistically.

DISCUSSION

Review of the findings of this limited study does not appear to support the initial assumption that prompted the research—that coal hydrogenation process workers with evidence of cancerous skin lesions may be at increased risk of developing other organ system cancers. This observation is based on the marked lack of cancer-related deaths in both the confirmed skin cancer group and the precursor group, after a latency period of 18 to 20 years.

These findings are similar to those reported by others. A study of 462 workers from 25 refineries who were exposed for at least 5 years to saturated cyclic compounds in bitumins and aromatic compounds in tar did not show any increase in malignancy rate when compared to unexposed controls.[7] Likewise, a study of 1077 matched pairs of refinery workers exposed to high-boiling aromatic petroleum fractions did not show any increase of skin lesions when compared to controls.[8]

A further review of the original group of records used as a basis for the original study, however, revealed that one case of prostate cancer was diagnosed in 1956 but was not included in Sexton's paper. Also not included in the original cohort were five workers diagnosed as having mouth (velum) lesions. These were not included because Sexton believed this finding was probably not associated with the exposures in question. The five deaths reported revealed that all five causes of death, as reported on the death certificate, were cardiac related (i.e., coronary disease, arteriosclerotic disease, cor pulmonale, and myocardial disease). Two of these subjects also demonstrated pulmonary involvement.

Nevertheless, the possibility of systemic cancers still exists because known exposures have been shown to be carcinogenic and should be investigated further. Perhaps the most glaring epidemiological

Table 3. Distribution of original cohort of 50 subjects

	Deaths		Retired	Still working	Lost to follow-up	Total
	Noncancer	Cancer				
Confirmed skin cancers	2	0	2	5	1	10
Confirmed precursor of skin cancer	3	0	14	23	0	40
Total	5	0	16	28	1	50

shortcoming in this study was the highly selected group that comprised the cohort under study (i.e., those that developed skin lesions), whereas the disposition of those workers who did not develop lesions is still unknown. This problem could be overcome by identifying the remainder of the work force (309) who were also exposed to the by-products of the coal hydrogenation process but were not included in the study cohort because they did not develop skin lesions. Their vital status, morbidity, mortality experience, and work history could be determined and these findings compared with a matched group of workers without chemical exposures. The possibility also exists that some of the workers who had not developed skin lesions at that time may have developed more serious organ cancers after the surveillance program was discontinued in 1960.

SUMMARY

A review of an original cohort of 10 subjects with diagnosed skin cancers revealed that death certificates on the two that died indicated that neither had died of organ cancers. However, no autopsies were performed that might have found underlying pathology. The remainder consisted of three retirees, one of which was lost to follow-up, and five still working at the plant. Of the 40 confirmed precancerous cases, three had died from noncancerous causes (no autopsies performed), 14 had retired, and 23 were still working in the plant. These data, therefore, do not demonstrate an increase of chronic disease and death, particularly from systemic cancers, from heavy exposure of high boiling oils containing polycyclic hydrocarbons, coal tar, and pitch.

ACKNOWLEDGMENT

The author thanks Dr. Richard Sexton, Medical Director, Union Carbide Chemicals Company, Division of Union Carbide Corporation, for his energetic assistance during this project.

REFERENCES

1. H. Lowry, *Chemistry of Coal Utilization*, John Wiley & Sons Inc., New York, 1945, pp. 377–78.
2. R. J. Sexton, "The Hazards to Health in the Hydrogenation of Coal," *Arch. Environ. Health* 1, 181–233 (1960).
3. D. W. Koppenaal and S. E. Manahan, "Hazardous Chemicals from Coal Conversion Processes?," *Environ. Sci. Technol.* 10, 1104–07 (1976).
4. C. Thony, J. Thony, M. Lafontaine, and J. C. Limarret, "Carcinogenic Polycyclic Aromatic Hydrocarbons in Petroleum Products. Possible Prevention of Mineral Oil Cancer," *Institut National de la Santé et de la Recherche Médicale Symposia Series* 52, 165–70 (1976).
5. A. Manz, "Urinary Tract Carcinoma in Gas Industry Employees," *Muench. Med. Wochenschr.* 118, 65–68 (1976).
6. R. L. Carter and F. J. Roe, "Chemical Carcinogens in Industry," *J. Soc. Occup. Med.* 25, 86–94 (1975).
7. G. Siou, "Is There a Risk of Carcinogenesis by Bitumens?," *Rev. Pathol. Comp. Med. Exp.* 72, 65–70 (1972).
8. L. Wade, "Observations on Skin Cancer among Refinery Workers. Limited to Men Exposed to High Boiling Fractions," *Arch. Environ. Health* 6, 730–35 (1963).

DISCUSSION

J. M. Evans read A. Palmer's paper.

Ralph Yodaiken, National Institute for Occupational Safety and Health: I would just like to make two comments. First of all the skin lesions were looked at by two pathologists. A brief look at that chart that you flashed up there shows that even under these optimum circumstances, there was a considerable difference of opinion as to whether or not these were cancerous lesions or precancerous lesions. It underlines what I've said earlier about the difficulty of making a decision on what might be inadequate pathological grounds. Secondly, I don't know how the ages of these people line up with the life expectancies in the United States but the striking feature, as I saw it, was that in most of the cases, people seemed to die of coronary artery diseases. On our trip, I was struck by the number of cases of people who had died of coronary artery disease or related blood circulatory problems. We tend to be so overawed by the problem of cancer that sometimes we can't see the wood for the trees, and I think that this is an aspect that we have to look at very much more closely.

Mike Holland, Oak Ridge National Laboratory: It is unfortunate that Dr. Palmer couldn't be here. I'd prefer addressing these comments to him. If I remember correctly, Dr. Palmer concluded that there was no evidence that the presence of skin abnormalities (non-neoplastic and neoplastic) in workers, following exposure to coal liquification materials, predisposed these workers to cancer at other (noncutaneous) sites. I question whether the data available to Dr. Palmer were adequate to assess the question, let alone draw conclusions either way. Since death certificates were apparently the basis for this study, there is a possibility that clinically silent neoplasms were undetected. Systematic autopsies of each individual would have been necessary to approximate the true frequency of neoplasms. One value of this study is that is suggests that whether there are systemic late effects associated with occupational exposure to coal liquids or not, these effects are too subtle to be detected in a study of this type.

Jim Evans, Enviro Control: I believe that the comments are well made. As you are well aware, the amount of data that is available in this country, in Europe, and in South Africa on the epidemiology of diseases induced by coal conversion in almost zero. It takes a great deal of courage for an epidemiologist to put out a report that really does not have all of the support that a truly scientific paper might have. I know from my own personal knowledge that Morgantown has been somewhat questioning about the desirability of letting this paper be published because of the fact that it is not a complete study and there is no way that it can be. Yet it is another piece of information that we otherwise wouldn't have. We will just have to go piecing information as quickly and as best we can.

WORKER PROTECTION IN THE COAL CARBONISATION INDUSTRY IN THE UNITED KINGDOM

R. M. Archibald*   J. L. M. Launder*
J. D. Watt*

INTRODUCTION

In 1976 the European Community's demand for coke stood at 76 million tonnes, equivalent to 99 million tonnes of coal, of which the U.K. provided a major share. In the United Kingdom nearly all the coke used is produced either by the British Steel Corporation, to supply their own blast furnaces, or by National Smokeless Fuels (NSF), a subsidiary of the National Coal Board, to supply domestic, foundry, or blast furnace cokes to the market at large. The world-wide recession in the steel industry has reduced the demand for coke in the United Kingdom, but it remains a large industry employing many thousands of men and having very large capital investments.

In common with coking plants elsewhere in the world, those in the United Kingdom face formidable problems of pollution arising from the clouds of dust, vapours, and gases which are generated by the coking process, creating nuisance and possible health hazards to the workers in the plant and to residents outside. In the past it has been suggested several times that the prolonged exposure to high concentrations of these atmospheric pollutants might lead to an increased incidence of lung cancer among coke oven workers, as it had been shown to do for workers in the gasmaking industry where conditions are somewhat similar.[1-6] However, a careful epidemiological study of mortality among coke oven workers employed by NSF[7] produced no evidence of increased lung cancer, though a later study[8] did show increased incidence of bronchitis and other non-malignant respiratory diseases among these men, particularly if they smoked.

In 1970 these conclusions, reassuring so far as lung cancer was concerned at least, were disturbed by the results of an investigation that Lloyd and his co-workers had been making into the health of U.S. steelworkers.[9] These showed, apparently convincingly, that the incidence of lung cancer among men who had worked five years or more on the tops of the ovens (i.e., exposed to the heaviest pollution) was ten times as high as that for men in the rest of the steel plant; men who had worked for a shorter time on the oven tops and men who had worked on the ovens at less polluted locations also showed a significant but less marked increase in lung cancer.

These conclusions clearly called for action within NSF to reduce pollution at the plant and thereby reduce this potential health risk. This was given greater force by the enactment in the United Kingdom of the Health and Safety at Work Act 1974, which corresponds to the U.S. Occupational Safety and Health Act and enjoins similar strict obligations on the employer to provide a safe and healthy working environment. The consequent establishment of the Health and Safety Executive, incorporating the previous Factory Inspectorate and Alkali Inspectorate, has led to an ongoing dialogue between the coke industry and the government on the questions of monitoring and controlling the working environment and protecting the worker. The present paper gives a brief account of recent developments in these fields within the U.K. coking industry.

*National Coal Board, Hobart House, Grosvenor Place, London SW1X7AE.

THE COKING PROCESS: A BRIEF ACCOUNT OF THE PLANT AND ITS OPERATION

To form a clearer idea of the nature of pollution in the coking industry, it is necessary first to give a brief account of the coking process (Fig. 1). Basically, this consists in heating a charge of prepared coal (15–25 tons) to a temperature of 1200–1350°C for 16–30 hours, the time and temperature depending on the type of coke being produced. This operation is carried out in a 'coke oven,' a narrow, vertical, refractory-lined, slot-like chamber, 12–15 m long, 3–7 m high, and 40–50 cm wide, heated by gas burnt in flues passing through the brickwork walls. In a coking plant, up to 100 or even more of these ovens may be placed side by side, forming a structure up to 100 m long and 12 m high.

Among the essential features of this plant (Fig. 2) are (a) the openings in the roofs of the ovens, the 'charge-holes' through which the crushed coal charge is introduced, the holes being closed at other times by heavy lids; (b) the 'larry car' which travels on rails along the top of the oven from the bunkers, where it is filled with coal, to the oven being charged, where it discharges this load; (c) the doors at each end of the oven which fit tightly to contain the gases evolved during heating but which can be removed by machinery at the end of the coking cycle to allow the coke to be discharged; (d) the 'ram car' which actuates the ram which pushes the slab of coke horizontally through the oven into the 'coke car' placed to receive it on the other side; and (e) the 'guide car' which directs the slab of coke during this operation. The coke in the coke car, red hot and broken up during the discharge from the oven, is carried along rails to the 'quencher' at the end of the oven where it is cooled rapidly by sluicing it with a controlled amount of water. It is then brought back and discharged on to the 'wharf,' a sloping surface at the side of the oven. From the wharf it is transported by conveyor belt to other parts of the plant for screening, stocking and eventual dispatch.

SOURCES AND FORMS OF POLLUTION ON COKE OVENS

The coal from the larry car is charged directly into the hot oven, and formerly this operation was accompanied by the emission of a very heavy cloud of coal dust and by gases and vapours, inorganic and organic, representing all the products that can be produced by the decomposition or combustion of coal (Fig. 3). Pollution from this source has now been almost entirely avoided in NSF plants by the use of a procedure known as 'sequential charging' developed within NSF; with proper operation, the pollution during this operation is now negligible (Fig. 4). Less easily controllable is the cloud emitted during the discharge of the coke into the coke car (Fig. 5); this probably contributes less to the operator's personal hazards, but because it is a principal source of the neighbourhood pollution, methods for controlling it are now being developed.

Besides these sources of pollution, there are many others which, although individually smaller, together contribute very significantly to a man's overall exposure. These include the leaks from the oven doors that cannot in practice be made quite gas tight (Fig. 6), leaks from the heavy cast-iron 'oven lids' that cover the charge holes after charging is complete (Fig. 7), leaks from the 'ascension pipes' that carry off the streams of gases and vapours evolved during carbonisation (Fig. 8), and leaks from a number of other vents and access ports that cannot readily be maintained in a gas-tight condition (Fig. 9), particularly on an aging plant.

ORGANISATION OF ENVIRONMENTAL CONTROL WITHIN NSF

As the first step in the more rigorous control of pollution, National Smokeless Fuels established in late 1970 an Environmental Control Committee with a commission to concern itself with the whole problem of pollution in plants operated by the Company and, in particular, to identify, characterise, and monitor all such pollution, to interpret the findings in terms of potential health hazard, and to encourage and where necessary initiate experimental and development work to reduce pollution.

Monitoring Men's Personal Exposures to Dusts and Gases

The first stage in implementing this commission was to establish a programme of routine monitoring by which each of the 13 plants in the organisation was visited several times a year by a specially trained team: at each visit, lasting a week, four shift samples were taken by personal samplers from each of the principal workers, usually ten, on the ovens. This sampling gave information on the levels of dust each man was exposed to and on the amount of benzene-soluble material (BSM) in the dust, for comparison with the American Conference of Government Industrial Hygienists (ACGIH) standard: the Health

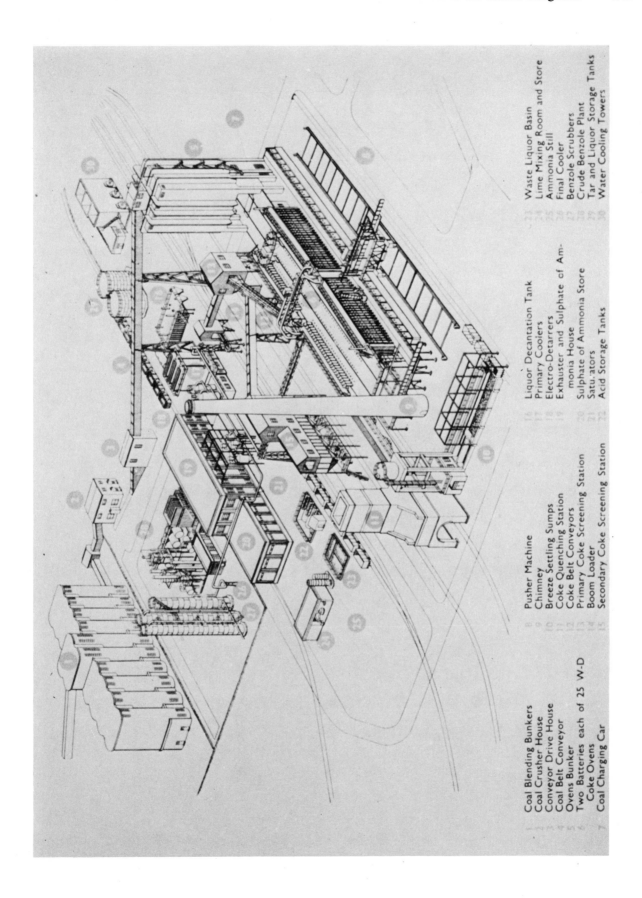

Coal Blending Bunkers
Coal Crusher House
Conveyor Drive House
Coal Belt Conveyor
Ovens Bunker
Two Batteries each of 25 W-D Coke Ovens
Coal Charging Car

Pusher Machine
Chimney
Breeze Settling Sumps
Coke Quenching Station
Coke Belt Conveyors
Primary Coke Screening Station
Boom Loader
Secondary Coke Screening Station

Liquor Decantation Tank
Primary Coolers
Electro-Detarrers
Exhauster and Sulphate of Ammonia House
Sulphate of Ammonia Store
Saturators
Acid Storage Tanks

Waste Liquor Basin
Lime Mixing Room and Store
Ammonia Still
Final Cooler
Benzole Scrubbers
Crude Benzole Plant
Tar and Liquor Storage Tanks
Water Cooling Towers

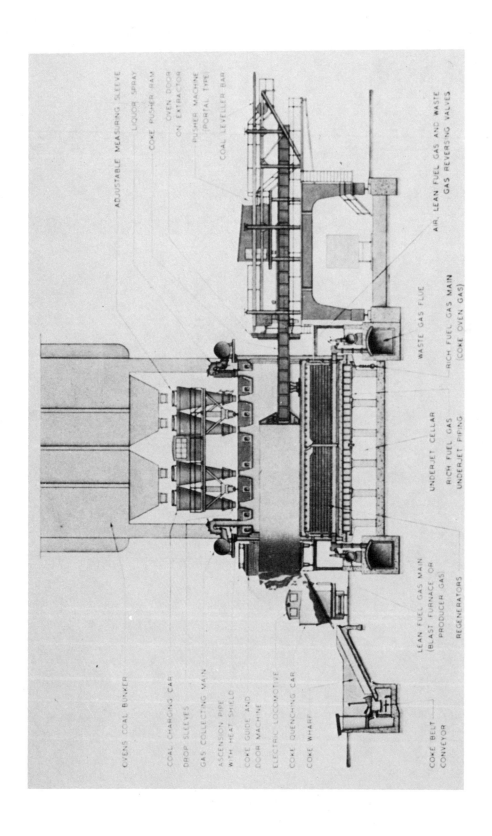

ADJUSTABLE MEASURING SLEEVE

LIQUOR SPRAY

COKE PUSHER RAM

OVEN DOOR
ON EXTRACTOR

PUSHER MACHINE
(PORTAL TYPE)

COAL LEVELLER BAR

AIR, LEAN FUEL GAS AND WASTE
GAS REVERSING VALVES

WASTE GAS FLUE

RICH FUEL GAS MAIN
(COKE OVEN GAS)

UNDERJET CELLAR

RICH FUEL GAS

UNDERJET PIPING

LEAN FUEL GAS MAIN
(BLAST FURNACE OR
PRODUCER GAS)

REGENERATORS

OVENS COAL BUNKER

COAL CHARGING CAR

DROP SLEEVES

GAS COLLECTING MAIN

ASCENSION PIPE
WITH HEAT SHIELD

COKE GUIDE AND
DOOR MACHINE

ELECTRIC LOCOMOTIVE

COKE QUENCHING CAR

COKE WHARF

COKE BELT
CONVEYOR

Fig. 3. Cloud of gas and particles rising from coke oven during charging.

Fig. 4. Charging procedure for coke oven.

Fig. 5. Coke being discharged from oven.

Fig. 6. Leaks from coke oven doors during operation.

and Safety Executive had decided to follow U.S. practice in this respect although there were many strong arguments against the use of BSM as a measure of the carcinogenic potential of the atmosphere and against the 0.2 mg/m^3 value quoted by the ACGIH. To provide supplementary information likely to be useful in interpreting the personal sampler measurements, static samplers were also used (Hexhlets and Mining Research Establishment samplers) to measure local concentrations of dust and to estimate the proportion of respirable components in typical samples. Later in the programme, samples were collected also by personal samplers incorporating a cyclone to give a direct measure of the respirable material inhaled.

Although the ACGIH threshold limit value (TLV) had been accepted in the United Kingdom, for various reasons it was thought desirable not to use the United States methods of sampling and analysis but to

develop our own method. This was done in collaboration with the British Steel Corporation and the Factory Inspectorate. The principal difference from the United States method of sampling was in the use of a larger sampling filter (47 mm) to prevent blocking during the 8-hour sampling exposure. In analysis, the U.K. method differed mainly in the determination of BSM both by measuring the weight loss of the extracted material and by weighing the amount of material recovered to give an inbuilt check on the accuracy of the determination. Results by these two methods that differed by more than a certain amount were discarded. Some other minor changes were made to improve the analytical accuracy but at best the determination was a difficult one; BSM is not a precisely definable material, it can be determined only gravimetrically and the amounts present are very small—0.2 mg in an average 8-hour sample if the plant operations conform to the TLV.

Fig. 7. Vapours rising from oven lids.

Fig. 8. Ascension pipes with leaking vapours.

Fig. 9. Top of coke ovens in operation.

As already mentioned, it was recognised that the fumes on the oven top contained not only dust but significant proportions of many toxic gases and vapours, both inorganic and organic; since these might have a connection with the increased incidence of non-malignant respiratory diseases[3] they were monitored also. No convenient personal samplers were available for this purpose at that time, and so an eight-port static sampler, which allowed samples of each of any eight gases to be sampled repeatedly in turn and analysed by wet chemical methods, was adapted for the purpose. This was later replaced by a four-port sampler developed by ourselves which was battery operated, portable, and small enough to be accommodated wherever measurements were required. This equipment could have been adapted for the sampling of vapours, but for this purpose it was more convenient to use a device based on a piece of equipment developed by Little and Penketh.[8] This consisted simply of a thick-walled rubber bulb fitted with a fine bore capillary which allowed the gas to leak into the deflated bulb at a fairly constant rate: by suitably adjusting the dimensions of the capillary and fitting it was an adsorption tube containing tricresyl phosphate on Celite, the condensable components in

a sample representing a full 8-hour shift could be collected for subsequent chromatographic analysis. This apparatus was found liable to give spurious results on occasion; although it is now being replaced by a slow-sampling pump with carbon or silica gel adsorption tubes, it has given sufficient results to indicate fairly clearly what the general levels are of a number of toxic gases.

Epidemiological Study of Mortality on U.K. Plants

The first results of this monitoring programme were available in mid 1971 and showed immediately that the BSM values for the worst locations (larry car, oven lids, ram car, and guide car) were uniformly higher than the TLV. The values were, in fact, generally similar to those reported for U.S. plants, and it was concluded therefore that there was the same presumptive risk of an increased incidence of lung cancer on the U.K. plants as Lloyd had found for plants in the United States. This risk was likely to be lower than that of the most exposed men studied by Lloyd[9] because men in U.K. plants were now working for much shorter periods (1–2 years at most) on the oven tops than those studied by Lloyd, and their total ingestion of BSM, which appeared to determine the risk, was likely to be much lower. Because of this and other differences between conditions in the United States and in the United Kingdom, the U.S. results could not be accepted altogether as applicable to the United Kingdom, and it was decided therefore to undertake an epidemiological investigation, with the assistance of Dr. M. Jacobson of the Institute of Occupational Medicine. This was a retrospective-prospective study and, to increase the size of the study group and hence the reliability of the findings, it was undertaken collaboratively with the British Steel Company and included their somewhat larger workforce. The full results of this investigation are not expected until 1979, but the preliminary findings are available and are commented on below.

Engineering Modifications to Control Emissions

Although the precise relevance of Lloyd's work to United Kingdom plants was not known, there was strong evidence that the risks were real and significant and that whatever could be done to reduce emissions by present technology should be done. Accordingly, as soon as the first measurements became available, steps were taken to modify the charging procedure to reduce the emissions during this particularly dirty operation by the introduction of the so-called 'sequential charging' already mentioned. Plans were made to reduce the exposures of men working on the larry car, guide car, and ram car by enclosing the driving cabins and supplying them with filtered, air-conditioned air. Besides this, a much closer watch was to be kept on sources of emissions (leaking doors and oven lids in particular) and efforts made to reduce them as much as possible by better housekeeping and maintenance. To assist plant managers to determine how effective these measures were, a system of Smoke Leakage Assessment was instituted by which the magnitudes of the leaks from each of these sources were assessed on an arbitrary scale from 0–5 and summed to give an overall factor representing the overall emission of the plant. Although the method was a subjective one, depending on individual assessment of the magnitude of each leak, subsidiary investigations showed that it was surprisingly reproducible and that, after a short period of training, different observers would agree quite closely in their estimate. To assist in giving practical meaning to the results a further programme was carried out in which the amount of leakage corresponding to each of the steps on the arbitrary scale was measured quantitatively by drawing off the emission through a duct and sampling. In this way, a mass emission factor could be calculated for the plant as a whole. Measurements of this kind were incorporated in the routine of management of the plant and served to keep a constant check on the efficacy of the operators in keeping leaks to a minimum.

Personal Protection for Oven Top Workers

There were a number of working positions (e.g. door cleaning, oven lids) where the men's exposures were still likely to exceed the TLV even after the best technical means had been applied and after every possible improvement had been made in house-keeping and maintenance. Until better means were available, it was accepted that these men would need to depend on respirators to bring their exposures down to the required level. The Institute of Occupational Medicine has long expertise in the testing and development of respirators for use in coal mining, and with their assistance the respirators available were kept under constant review. The present state of development in this field is outlined briefly in a later section.

PRESENT STATE OF DEVELOPMENT OF ENVIRONMENTAL CONTROL

The programme of monitoring outlined above is a simple one, and there have been few technical difficulties in carrying it out. Systematic measurements have now been made for a number of years, and there is now enough data to assign typical values to all the principal working locations on NSF plants, in terms of personal exposures to dust, BSM, benzopyrene, and various toxic gases and vapours. Regrettably, although some improvements have been made in these levels, it cannot be claimed that they have yet been brought below the TLV nor even that the technical means are available yet that would allow this to be done. The whole matter is still in active development, as may be seen from the following brief account of the present position.

Dust and BSM Exposures

The routine monitoring programme has produced a vast mass of data on exposure levels for typical working locations on the plants: these show fairly marked differences from plant to plant, according to

250   R. M. Archibald, J. L. M. Launder, and J. D. Watt

the age, condition and state of maintenance, and marked differences also from location to location in any one plant. Table 1 gives a general broad idea of the levels observed and shows clearly that, except on occasion for the coke car, the concentrations of BSM have not so far been brought down to the required ACGIH level at any working location in any plant.

Table 1. Personal exposures to benzene soluble matter (BSM) in the atmosphere at various working locations on coke ovens

Working location	BSM exposure (mg/m³)	
	Best result	Worst result
Larry car	0.8	3.1
Oven lids	0.75	4.0
Ram-side doors	0.38	3.1
Guide-side doors	0.29	2.3
Guide car	0.46	1.7
Coke car	0.19	0.34

It was apparent from the preliminary measurements made before the full monitoring programme was developed that the measurements made by the personal samplers varied greatly from shift to shift, as shown by the typical measurements presented in Table 2. Some of the more extreme variations might be discarded as arising from mishandling of the sampler, intentionally or not, but most are inherent in the situation and arise presumably from day-to-day variations in the operation of the plant and variations in meteorological conditions. The standard deviation of a set of four or five shift observations, the number normally made at each working location in a plant visit, is such that the measurements can detect only large changes in working conditions. It is possible to

distinguish between the best and worst of the U.K. plants and between the best and worst working locations at a given plant, but the effect of engineering modifications made to improve the working conditions would not be detected readily unless they were very effective indeed. Although the personal sampler measurements are a valid measure of the conditions of a man's exposure and of conformity to the health standard, they may not, however, be a valid measure of the overall plant emission nor of the effectiveness of engineering modifications designed to reduce it.

Exposure to Toxic Gases and Vapours

The general levels of concentrations of certain of the gases monitored in this programme are shown in Table 3. It will be seen that they are uniformly low but may not be negligible: if their effects were additive, and if the analysis were widened to include the many other gases and vapours not represented in Table 3, the total concentration might approach or even exceed the TLV. The effect of such a combination of gases as this cannot, however, be estimated on present information: the usual formula in which each gas is weighted according to its TLV and the proportion of it present cannot be used for such a diverse mixture. The difficulty becomes greater when it is remembered that these figures not only represent 8-hour averages but also that the operation of a coke oven is such that the concentration a man is exposed to varies from minute to minute, reaching particularly high values during charging. In these circumstances it is necessary, before his conditions of work can be approved, to know whether these concentrations fall within the short term exposure limit of ceiling values laid down for some components

Table 2. Variations in dust concentrations measured by personal samplers during four shifts at various working locations on a typical coke oven

Working location	Dust concentrations (mg/m³)					Standard deviation (%)
	Shift No. 1	Shift No. 2	Shift No. 3	Shift No. 4	Mean	
Larry car	11.8	5.2	6.2	2.9	6.5	57
Oven lids	6.8	3.8	1.0	14.8	6.6	88
Ram car	1.2	2.0	2.7	1.3	1.8	38
Ram-side doors	6.2	2.6		0.54	3.1	91
Guide car	3.2	2.4	5.0	6.6	4.3	43
Guide-side doors	3.1	4.0	6.7	4.0	4.45	34
Coke car	1.4	1.8	5.1	2.1	2.6	65
Wharf	2.0	5.3	4.0	4.0	3.8	35
Heater	3.3	9.6	13.1	1.7	6.9	77
Tar main	3.6	4.9	3.3	2.6	3.6	26

Table 3. Concentrations of selected gases and
vapours in the larry car atmosphere

Substance	Concentration		
	Threshold limit vaue	Mean of all shift means	Range of shift means
	Vapour (ppm)		
Benzene	10	1.0	0.07–7.2
Toluene	100	0.95	0.04–9.9
Xylene	100	4.8	0.07–91
Naphthalene	10	3.4	0.13–31
	Gas (10^3 ppm)		
NH_3	25,000	144	10–820
HCN	10,000	33	1.1–180
SO_2	5,000	398	7–1400
NO_x	5,000	37	10–136

such as benzene, hydrogen sulphide, hydrogen cyanide, etc. There is no way of establishing this for the full range of compounds present, but an infrared analyser provides a ready means of determining the minute-to-minute concentrations of one gas, conveniently carbon monoxide, and if it is assumed that the composition of the pollution cloud as a whole is constant (that is, the amount of CO is constant in proportion to the amounts of the other components), then some guess can be made at the possible peak values of these others. This has been done with some success, and the peak values of some gases have been estimated and found to exceed the TLV by a wide margin. Nothing can be said about the physiological significance of this excess at present. For most of the gases and vapours, no ceiling values have been imposed, and brief exposures even to such high levels as this may be unobjectionable; for others the position is not clear. In the absence of data for interpreting such measurements, it may be necessary to acknowledge that the hazard cannot be assessed by predictive measurements. The hazard either may be tested for directly by frequent medical checks, or it can be assumed to be present and the appropriate remedial action taken by engineering measures if possible, by respiratory protection if not.

The Effect of Engineering Measures to Control Pollution

It was hoped that the introduction of sequential charging, which, as already mentioned, is strikingly effective in reducing emissions during the charging operation, would bring about a marked reduction in personal exposures. In fact, however, detailed statistical study of the measurements made over a period of five years or more fails to show any significant reduction in personal exposures, even though the reduction in smoke levels for the oven as a whole is markedly apparent to the eye. It is strongly suspected that one reason for this is that the man's exposure is determined not only by the level of plant emissions, but also by the man's movements through the field of pollution, which varies locally from point to point and from moment to moment. These movements, in turn, are determined partly by the nature of his duties and partly by his own attitudes. Watching men at work makes it apparent that there are some who consciously avoid the smokey areas and some who do not. It can be suspected that, for those who do not, the effect of any engineering improvements that limit the pollution will be partly nullified by the man's taking even less avoiding action himself in the less polluted atmosphere: there is direct evidence of this in the men's response to powered respirators. The men's personal exposures therefore may not be a trustworthy guide to the effectiveness of sequential charging in reducing total emissions.

It is not suggested that this is the only reason for the disappointing results of this remedial measure, although it is of interest enough to justify inclusion in the ergonomic investigation now in progress in NSF in which, among other things, the men's work habits will be studied in relation to the recorded expsoures. Another factor, probably more significant, is that, as shown by a subsidiary investigation, a large part of the men's exposure arises not from the charging operation but from leaking doors and other sources that contribute to the general haze around the plant and are not reduced by sequential charging.

Another important engineering modification that was hoped to reduce the exposure of the larry car operator to the approved level was the introduction of air conditioned larry cars. In principle, the larry car operator in these machines is exposed only to the air drawn in through a high-efficiency filter and cooled. At best, this equipment should be able to achieve the results expected; in practice it has proved difficult to reduce the number of occasions on which the man must leave the car to attend to duties elsewhere, and the reduction in personal exposures achieved in practice has therefore not been as great as hoped. The problem of how to combine the effective operation of these cars with the man's necessary duties is another aspect of the ergonomic programme now in progress.

Less important but still significant are the sanctuaries which have been built at several locations on a number of our plants: these again provide the man with cleaned and filtered air and provide him with at least a temporary respite from the conditions on the oven. Since he can spend only a small part of his working shift in these enclosures, the reduction in personal exposure is not large, and such installations could not be justified on these grounds alone. They have proved to be of considerable psychological value, however, and have been accepted, on some plants at least, as a tangible demonstration of the plant management's concern with the problem of pollution and of a willingness to do what can be done to relieve it.

Respiratory Protection

Until more effective means have been developed for reducing emissions, the only certain way in which a man's exposure can be brought within the TLV is to equip him with a respirator. The importance of respirators, at least as a temporary measure, has been recognised in the United States also, and both here and in the United Kingdom strenuous efforts have been made to improve on existing types and to make them as efficient and tolerable to wear for long periods as possible.

Under the working conditions of a coke oven, powered or air-supplied respirators of existing types could not be worn without exposing the man to a more serious hazard from accident than the toxic hazard from dust, but it was thought that a respirator of the oro-nasal type could give the required protection if properly designed. The Institute of Occupational Medicine, an organisation that forms part of the National Coal Board, already had considerable expertise in testing this kind of equipment, and with their assistance two were eventually selected for routine use from the many available on the U.K. market. These, under test, were found to be quite capable of reducing the inhaled concentrations to below the prescribed level of 0.2 mg/m^3, but the men clearly did not find them comfortable to work in and they were not regularly worn. It was looked on as a very important development, therefore, when in about 1973 a worker at the Safety in Mines Establishment, Dr. Greenhough, working almost privately, developed a respirator that was much lighter than the usual types and promised to be much more acceptable (Figs. 10 and 11). This, now marketed in the U.S. by Messrs.

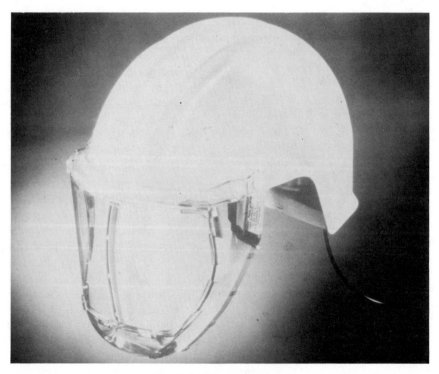

Fig. 10. "Airflow" helmet.

Fig. 11. Worker wearing helmet.

Racal as the 'Airflow' helmet, consists of a plastic helmet or 'hard-hat' incorporating a filter cloth as a lining; a motor in the back of the helmet, powered by a small battery at the waist, drives a stream of air at 250 litres per minute through the filter and down under a plastic visor over the man's face (Fig. 12).

Tests under working conditions have shown the helmet to be very effective, retaining some 95% of the dust admitted, and to be generally very acceptable in use. In effect the man is able to work, see and breathe normally with the advantage of having hard-hat protection to the head and a stream of cooling air over the face (Fig. 13). This is not to say that the device is without disadvantages: rain on the visor interferes with vision, and the plastic material of the visor is easily scratched when the wearer brushes it with a glove to remove coke dust. Other difficulties were unexpected and arose from the helmet's very effectiveness. It was found for example that the cooling effect of the stream of air on the face was so marked that men cleaning oven doors would expose themselves to much higher levels of radiation than normal, to the point at which the helmet itself had distorted: this has now been guarded against by providing the helmet with a reflecting surface to

reduce the rate of rise of temperature. Similarly, men cleaning the ascension pipes—a very dirty job from which a man not wearing a helmet usually stands well back—have got much closer to the work than they usually would, being protected from the smoke, and have therefore been exposed to much higher levels of toxic gases which are not trapped by the helmet. In other situations (e.g., handling pitch) where these complications are not met, the helmet has been found very acceptable, so much so that when only a few were available, it was very difficult to recover the helmets from the men after they had been issued for test.

INDUSTRIAL RELATIONS

One of the many problems which arose during the early years in which scientific results were accumulating and before our own epidemiological study was complete was how to interpret these extremely variable figures to the work people in a way that could be meaningful. Yet it was very necessary if their support in achieving better operation conditions was to be effective that they be informed on the necessity for this. A Joint Environmental Working Party of Management and Trade Unions was accordingly established as the most suitable mechanism for disseminating information. Training courses for all operatives were organised together with teaching for management and supervisors.

MORTALITY STUDY

The epidemiological study referred to above was carried out by Dr. M. Jacobson of the Institute of Occupational Medicine in Edinburgh, Scotland. I cannot do better for summary than quote from his work.

This paper describes mortality over a nine-year period among workers employed at two coal tar distillation factories, one patent fuel factory involving some coal carbonisation and 13 coke works. Vital statistics at 31st December 1975 were established for 98.3 percent of the 4836 men identified as eligible for the study.

The numbers at risk at the two tar distillation works are too small to justify any clear conclusion at this stage. It is noted that at one of them there is a tendency for death rates to be higher than for all men in England and Wales.

General mortality at the patent fuel factory in Wales was close to that which might be expected among similarly aged Welsh industrial workers. The proportion of deaths attributed to cancers of various anatomical sites was unusually high.

On average, there were fewer deaths among men in the 13 coke works than expected for all men in England and Wales; but coke workers aged less than 55 had higher death rates,

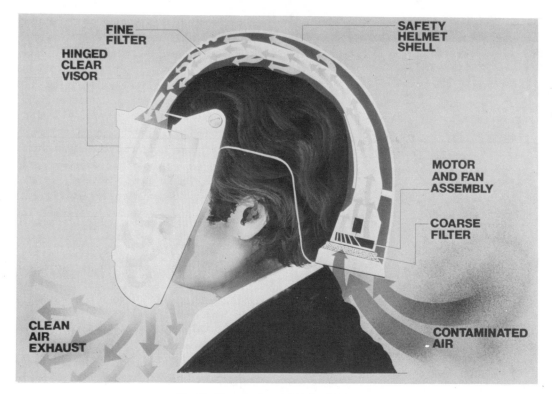

Fig. 12. Cross section of "Airflow" helmet.

Fig. 13. "Airflow" helmets in use at coke plant.

particularly as a result of ischaemic heart disease and malignant neoplasms, while their older colleagues had correspondingly lower mortality.

Lung cancer death rates for the 3962 coke workers were higher than among men of similar age in England and Wales. The excess occurred in nine of the 13 works considered. The effect is particularly marked among men aged 45 to 54 years. At older ages there was a small excess compared with all men in England and Wales and a deficit compared with male industrial workers. The proportion of deaths attributed to lung cancer was unusually high at all ages. There is no evience that the effect is associated exclusively or primarily with coke ovens.

Smoking habits at four of the coke works and at the patent fuel factory were similar to those of men outside the industry.

Some suggestive patterns are noted relating mortality from cancer of the lung and of the digestive organs to age and to period and type of employment in the industry. These will be studied further as the project continues.

Since then a similar survey has been carried out at the coking plants of the British Steel Corporation, the results of which bear a marked similarity to those of the earlier survey.

Both surveys continue, and a report covering the continuing experience of both companies is expected in 1979.

CONCLUSION

In this first review of the environmental health problems of the British coking industry, an attempt has been made to demonstrate how the industry has approached the problems of measurement of pollution and of studying mortality. Much still remains to be done, but at least the problems have been defined.

REFERENCES

1. E. L. Kennaway and N. M. Kennaway, "Study of Incidence of Cancer of Lung and Larynx," *J. Hyg.* **36**, 236–67 (1936).

2. E. L. Kennaway and N. M. Kennaway, "Further Study of Incidence of Cancer of Lung and Larynx," *Br. J. Cancer* **1**, 260–98 (1947).

3. R. Doll, "The Causes of Death among Gas-Workers with Special Reference to Cancer of the Lung, *Br. J. Ind. Med.* **9**, 180–85 (1952).

4. R. Doll et al., "Mortality of Gasworkers with Special Reference to Cancers of the Lung and Bladder, Chronic Bronchitis, and Pneumoconiosis," *Br. J. Ind. Med.* **22**, 1–12 (1965).

6. S. Kuroda and K. Kawahata, "Uber die gewerbliche Entstehung des Lungenkrebses bei Generatorgasarbeiten," *Z. Krebsforsch.* **45**, 36–39 (1936).

7. D. D. Reid and C. Buck, "Cancer in Coking Plant Workers," *Br. J. Ind. Med.* **13**, 265–69 (1956).

8. D. D. Walker, R. M. Archibald, and M. D. Attfield, "Bronchitis in Men Employed in the Coke Industry," *Br. J. Ind. Med.* **28**, 358–63 (1971).

9. J. W. Lloyd, "Long-Term Mortality Study of Steelworkers, V: Respiratory Cancer in Coke Workers," *J. Occup. Med.* **13**, 53–68 (1971).

5. R. Doll et al., "Mortality of Gasworkers–Final Report of a Prospective Study," *Br. J. Ind. Med.* **29**, 394–406 (1972).

DISCUSSION

Dick Gammage, Oak Ridge National Laboratory: Dr. Archibald, you mentioned that when this study started there was a reluctance in the United Kingdom to accept the U.S. standard of 200 $\mu g/m^3$ of air for benzene-soluble matter. Was the standard felt to be too high or too low, or was it perhaps believed that benzene-soluble matter is an inadequate parameter? Now that you report epidemiological data that seem to conflict with similar data in the United States, is there again reluctance to accept the benzene-soluble standard?

R. M. Archibald: I must be careful to be diplomatic in my answer. The reluctance at first was due to a feeling that the standard was perhaps, and I say this with all due humility, fairly hastily put together and not based on scientific data as sound as we would have liked. I think the broad answer is that seven or eight years later, we accept it. Government accepts it, industry accepts it, and we work toward it. So, the answer today is that it is acceptable. We haven't anything better to put in its place.

Jim Smith, SRC: Here is a two-part question. First of all, do you have a clothing protection program for the employees, and do you require showers or even provide them in the coking plants?

R. M. Archibald: Yes, protective clothing is provided and showers are available and used at all plants. I'm not myself completely satisfied with the protective clothing situation. We are, at this very moment, engaged with the unions in producing complete work clothing for every man.

MEDICAL SURVEILLANCE IN THE UNITED STATES AND ABROAD

Ralph E. Yodaiken*

ABSTRACT

A review of medical surveillance was undertaken. The World Health Organization (WHO) categories of occupational disease syndromes provide a basis for discussing the subject. Methods of data collection in this country, Europe, and the United Kingdom are unsatisfactory, and a simple plan for industry-wide surveillance needs to be established. A method for establishing priorities and areas of emphasis is suggested.

Surveillance means "watching over" and may be applied to a person or to a disease. It has been defined in various ways. For instance, the World Health Organization (WHO) defined disease surveillance as "a concerted attempt to keep under continuous observation all the factors that contribute significantly to the occurrence of disease in human populations."[1] Benenson,[2] in describing surveillance of disease, included the important words "effective control." He said, "surveillance of disease is a continuing scrutiny of all aspects of occurrence and spread of disease that are pertinent to effective control." A surveillance program must begin with the identification of health problems, and this includes a process of classification. With identification comes the need to assess the magnitude and potential impact of the problem. Implicit in this is the accurate recording and reporting of health effects and exposures to hazards. The collection of this raw data needs collation so that an overall picture of the situation emerges, followed by analysis and evaluation or interpretation of the data. Recommendations must flow from this data collection and must be acted upon at a local level, state level, and ultimately federal level. Finally the recommendations revert back to the source in the form of promulgated standards (Fig. 1).

WHO has suggested four categories of occupational disease syndromes (Fig. 2):[3]

1. only occupational in origin,

2. occupation as one of the causal factors,

3. occupation contributes to the occurrence of disease, as one of a complex of exogenous factors, and

4. occupation determines the course of preexisting disease.

The first of these categories is straightforward enough, but, nevertheless, constitutes a challenge. The classical example of an occupational hazard, studied in the eighteenth century by Percival Pott (Fig. 3), is the association of scrotal cancers and sweeping chimneys.[4] Present-day standards would require more than a crude statement of cause-effect relationship. The source of the coal, the precise chemical characterization of the soot, and a list of possible contaminants or confounding variables found in the walls of the chimney would have to be known. Even though the sex of the workers might have been clear, the age distribution, the ethnic group, the educational and socioeconomic status, (if that were in doubt), previous occupational exposures, and the latency period would be essential information. In the case of the chimney sweeps, the route of absorption might have been obvious, but today the metabolic pathway would need to be worked out, as it must be for vinyl chloride, which was discovered to cause angiosarcomas of the liver in the early 70s. Despite the difficulty of identifying biochemical pathways of these toxicants and carcinogens, the

---

*National Institute for Occupational Safety and Health.

ORNL-DWG 78-19861

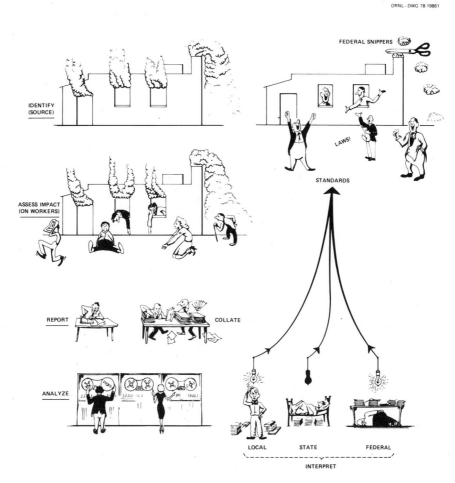

Fig. 1. The impact feedback.

problem of surveillance in known direct cause-effect relationships is relatively simple.

The second WHO category, however, is more complicated. The best example is that of asbestos and smoking: the former, a known occupational hazard; the latter, a social hazard; the combination of the two in 90% or more of cases being lethal. Even though the cause-effect relationship between asbestos and disease was identified many years ago, the synergistic effects of tobacco smoke took longer to assess. Even now the full impact of the hazard, involving perhaps hundreds of thousands of insulation workers and

pipefitters, is not known. This case brings home the need to recognize, classify, report, collate, analyze, and interpret data in a systematic and orderly fashion from the beginning.

The third category embodies the concept of "a total understanding of the interlocking influences on health of environmental stress at home and in the community as well as on the job."[5] A caveat to the "healthy worker" concept is that members of the work force are healthier than the general population because they are accepted into the labor force. The worker is exposed to the same environmental factors

ORNL–DWG 78-19860

OCCUPATIONAL ONLY

OCCUPATIONAL +
OTHER FACTORS
e.g. SMOKING AND DRINKING

OCCUPATIONAL +
AMBIENT ENVIRONMENT

OCCUPATIONAL +
UNRELATED HEALTH PROBLEMS

ONE FOOT IN GRAVE

Fig. 2. WHO's occupational disease syndromes—whose indeed!

as everyone else who lives, let us say, in the valley; but additionally, because he or she is exposed to higher concentrations of a variety of noxious substances at the work place, the healthy worker may soon enough become the unhealthy victim of substances that either interact with or have an additive effect on toxins absorbed from the ambient environment.

The fourth category involves the concept of preemployment examinations and a host of complex evaluations that come with such a procedure (Fig. 4). Will exposure to carbon monoxide, for instance,

exacerbate a preexisting atherosclerotic cardiac disease? Will exposure to lead promote subclinical hypertension? Should a man who smokes be permitted to work in an asbestos factory? In other words, does job classification or even exclusion from the work force start at the gates of the factory?

For some, surveillance means "watching over" the working population in its broadest sense and reporting disease patterns and hazards. It does not mean searching out disease. For those who interpret surveillance as the former, medical monitoring is best

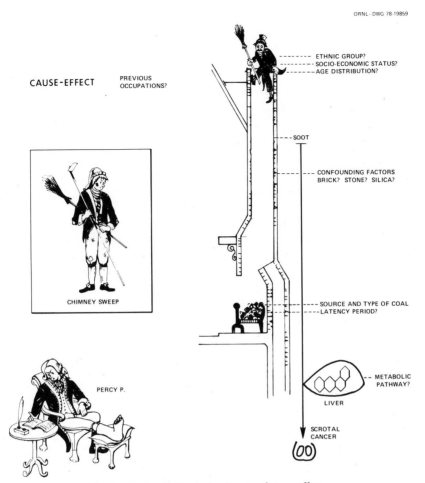

Fig. 3. Percival Pott—A simple case of cause-effect.

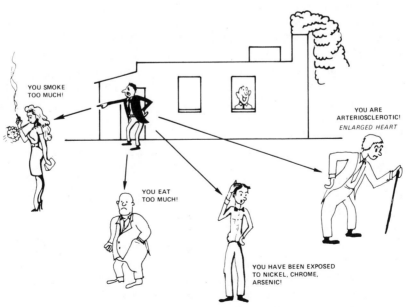

Fig. 4. Preemployment physicals.

limited to physical examinations with a minimal number of laboratory tests or even no tests at all. To others, surveillance means actively seeking out medical problems—in some ways a "fishing expedition" and in other ways a deliberate attempt to pick up specific disease processes *thought* to be related to a particular occupation. Speculation is introduced deliberately. Many disease entities are best known through animal experiments and have yet to be documented in human populations. Acrylonitrile is an example. Experimental work has shown acrylonitrile to be carcinogenic in rodents, but epidemiological proof of carcinogenicity in man is yet to be confirmed. Because experimental data indicates that central nervous system neoplasms occur in experimental animals fed acrylonitrile in the drinking water and, in addition, neurotoxic symptoms have been reported in a number of scattered populations and case reports, it was reasonable to suggest under "Medical Surveillance" in OSHA's Proposed Rules governing the use of acrylonitrile[6] that attention be given to the central nervous system. However, on the basis of reported experimental gastrointestinal tumors, OSHA proposed that proctosigmoidoscopy be included as part of the surveillance program offered to workers. Bringing complicated screening procedures into a surveillance program may not be practical.[7] In this particular case, the procedure is uncomfortable and time consuming. A doctor must be present, because this is not a procedure that can be carried out by paramedical personnel. Most important, it is of doubtful therapeutic-diagnostic value.[8] Is this indeed what is meant by medical surveillance? Should surveillance include difficult-to-interpret screening procedures? Even if there is little prospect of follow-up and cure? The latter question would be particularly true in seeking out *cancers* of the lung, in which the 5-year survival rate in most reported series is dismal.[9] Since the need to know is urgent, the need to develop reliable screening techniques is equally urgent. Furthermore, a protocol that points the doctor and, more particularly, paramedical personnel to specific target organs does not always serve its purpose. In the process of dwelling on one or another organ or system, the emphasis may be taken away from a complete medical history and thorough physical examination (Fig. 5).

Section 8(C)(2) of the Occupational Safety and Health Act requires the Secretary of Labor to prescribe regulations requiring employers to maintain accurate records of, and to make periodic reports on, work-related deaths, injuries, and illnesses other than minor injuries requiring only first aid treatment, which do not involve medical treatment, loss of consciousness, restriction of work or motion, or transfer to another job.[10] The responsibility for implementing programs for recording and reporting occupational injuries and diseases was given to the Bureau of Labor Statistics. Occupational illnesses here include acute and chronic illnesses caused by inhalation, absorption, and ingestion or direct fatalities, lost work days or cases without lost work days, including transfer of jobs, termination of employment, restriction of work or motion. All employers are required to make a one-line entry on the log of occupational illnesses and injuries and a more detailed report on a supplementary record. It is important to note, particularly in the light of surveillance programs in other countries (which will be discussed briefly below), that the requirements are for reporting of minor injuries and illnesses as well as major occupational health problems.

The average number of cases per 1,000 recorded in the United States is higher than that for most European countries (it works out to be 1.7 times higher), because most cases reported in this country would not fall under Workmen's Compensation systems in other countries.[11] In 1972 NIOSH implemented a pilot study designed to estimate inaccuracies in the reporting system.[12] Workers exposed to occupational health hazards were examined by qualified health personnel and the cases divided up into five categories: (1) probable occupational disease, (2) doubtful, (3) suggestive history (no objective findings), (4) cannot be evaluated, and (5) probably not occupational (Fig. 6).

In the study 1,100 medical conditions were found among 908 employees, 31% of which were regarded as occupational, falling into the following categories: 28% hearing, 18% skin, 14% lower respiratory tract, 14% low-grade toxic, (e.g. raised blood lead), 11% upper respiratory tract, 9% eye conditions (e.g. conjunctivitis), and 6% miscellaneous. Employers' logs, however, reported only 11 occupational diseases. Workers' compensation files yielded an additional 9 so that 20 of 346 or about 6% of cases were actually identified as occupational. It has been stressed that only claims paid are reported and that the reporting concentrates on financial matters and ignores the breakdown by cause or condition.[13]

How do the surveillance systems in this country compare to those in other countries? In 1976 the Director General of WHO noted that schedules of occupational diseases exist in at least 80 countries.[14]

ORNL–DWG 79-19857

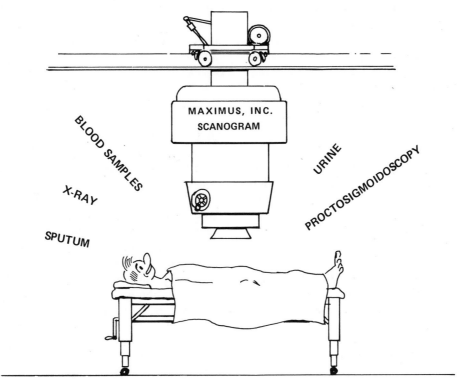

Fig. 5. The workers' couch.

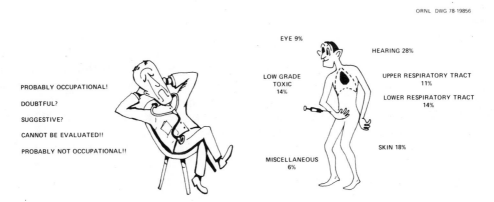

Fig. 6. There has to be a better way!

One of the major drawbacks of scheduled systems is that the schedules do not contain certain diseases known to be occupationally caused, for example, hearing loss. In France, wide underreporting of compensation is probably related to the 48 conditions that appear in the schedule. Britain has 47 categories of occupational diseases plus pneumoconioses and allied chest disorders. In addition, the

Industrial Injuries Advisory Council provides advice to the Secretary for Social Services as to whether a particular occupational risk is limited to an industry or is common to all persons. And the Factory Act requires 17 occupational diseases to be reported. But, as in this country, there is a considerable divergence between the number of industrial injury claims and the number of injuries reported to the safety authorities.[11] In the Federal Republic of Germany, each physician and dentist is required to report observations of scheduled cases to the National or State Physicians or the Occupational Protection Physician of the Federal Railway. Factors taken into consideration are that the disease be specific to the occupation and that the workers be exposed to a higher level of the specific hazard than the general population. Once the cause of the relationship has been established and it has been decided that this is a severe or recurrent disease or that the worker has been forced to give up his activity, compensation requirements will have been met. Quite clearly under these circumstances, minor illnesses will not appear in the records.[11]

Turning now specifically to fossil fuels surveillance programs, at the end of 1977 NIOSH visited a number of commercial coal gasification plants: Sasol, the South African Coal, Oil, and Gas Corporation, and The Koppers-Totze Coal Gasification Plant of the African Explosives and Chemicals Inc.—both in or near Johannesburg, the RURGAS in Dorsten, Germany, and the Westfield Development Center in Fife, Scotland. Our major interest was centered in the plant at Sasolburg, which had been in operation for over 20 years and where we were able to review the coal gasification process thoroughly and collect information on process engineering, industrial hygiene, work practices, engineering controls, and health problems. We were able to identify hazard areas in coal gasification with respect to chronic low-level exposures to substances such as carbon monoxide, hydrogen sulfide, and polynuclear aromatic hydrocarbons. The Division of Criteria Documentation has put out a criteria document based on this information and the usual process of collecting, assessing, and evaluating the world-wide literature and experiences of others on a national and international basis which covers a wide range of potential health hazards at various points in the coal gasification process and a general recommendation for medical surveillance.[15] This document includes a comprehensive preplacement medical examination. Primary lung function tests, cytology, urine analysis, and a variety of other tests are suggested. Our major concern is to anticipate health hazards rather than collect the data in a retrospective fashion when it is too late to do anything for current workers. Perhaps one of the most disappointing aspects of our visit was the failure in all of these areas to retain adequate follow-up records. Although we have a relatively clear picture

of acute hazards and their consequences, the long-term follow-up data is missing. Without the benefit of a research survey we have *only* anecdotal information that superficially suggests that chronic disease may not play a significant role among the worker population, provided good technology control and work practices are in place. It is essential to confirm this impression.

To ensure that medical surveillance is properly applied throughout the fossil fuel industry, certain recommendations are appropriate (Fig. 7). The first, of course, is adequate control technology. This is the primary requirement. The minimization of leakage, the surveillance of collection systems, attention to valves, and all potential leak sources that go with engineering controls should be rigorously applied.

Good work practices are dependent on the training of personnel. Untrained personnel provide the statistics for injuries and acute illnesses. For instance, among the reported acute injuries that we noted during our visits to the coal gasification plants, we were struck by steam burns and gassings. Fatal cases have occurred in at least three of the plants we visited. Other problems we noted that fall under the second WHO category were the social problems of smoking and drinking. It would be inappropriate for any ongoing medical surveillance program to ignore attention to these details. Education on smoking is essential not only because of the inhalation hazard but also because of the ingestion problem that goes hand in hand with smoking.

Medical monitoring should be adequate but simple. Laboratory tests should be limited to those that are meaningful. Collecting data for the sake of collecting data is a futile exercise. On the other hand, tests that are specific to a particular industry should not be overlooked. For example, many aspects of coal gasification are extremely noisy, and the omission of audiometric examinations would be detrimental to an evaluation of the health problem.

Finally, where should the emphasis be placed? I want to draw attention to the recommendations for a National Strategy for Disease Prevention, a report put out by Communicable Disease Center through an Ad Hoc Advisory Committee in June of this year.[16] It provides a mechanism for pinpointing priority areas for surveillance. These may be classified as follows:

1. *High priority* means that if appropriate measures are implemented a *large* decrease in *mortality or*

ORNL-DWG 78-19855

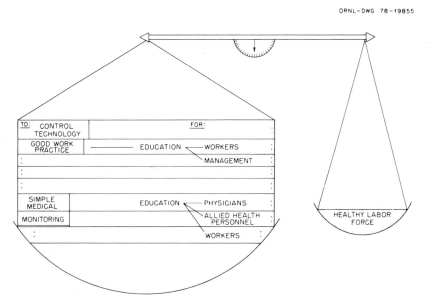

Fig. 7. Surveillance.

morbidity could be *predicted*, but a failure to control the problem would bring about an unacceptable increase or persistence of the problem.

2. *Potentially significant problem* means that a *moderate* decrease in *morbidity* would be *probable* if appropriate measures were implemented, whereas an undesirable increase or persistence will be probable if the problem is not addressed.

3. *Low priority* means that appropriate measures would correct a relatively *small* but significant increase in morbidity.

A *numerical system* was devised for evaluating the relative importance of each of these categories, each factor (Prevalence, Mortality and/or decreased Quality of Life, Technical Feasibility) being assigned a value of 0-4. The resulting factors are multiplied by each other and the product classified as follows: 28-64, high priority; 9-27, potentially significant; 1-8, low priority; 0, no priority. An example follows:

Let us assume the problem is noise (Fig. 8).

- *Prevalence* (0-4)—A large percentage of the population is affected by noise in the coal gasification process; therefore, this would be given a value of 4.

- *Mortality* and/or *Decreased Quality of Life* (0-4)—Those affected by significant hearing loss

suffer a disability and possible economic loss caused by this disability. The quality of life is significantly decreased, and, therefore, the value is 4.

- *Technical Feasibility of Successful Intervention* (0-4)—The problem can be prevented or controlled by available technology. Therefore, the technical feasibility of successful intervention is given a value of 4.

- $4 \times 4 \times 4 = 64$—A high priority category.

An annual review of priorities by the entire industry with labor, state, and federal input would place the priorities in perspective.

To sum up: interpreting surveillance depends on where one stands—the blind men and the elephant situation. To the idealist, medical surveillance fulfills the preventive medicine dream—anticipating occupational hazards or aborting induced diseases or at least effecting cures in good time. To the skeptic it is a routine and sometimes costly task which has doubtful benefits but is mandated by law or public pressure. To the legislator it is a prayer—a hope that it will prevent or abort disease and that the doubts about its effectiveness will be proven wrong. As long as technology control is imperfect however, medical surveillance is essential—at best as a preventive measure, at worst as a record of the actual health effects of the myriad of occupational hazards. It

ORNL—DWG 79-10417

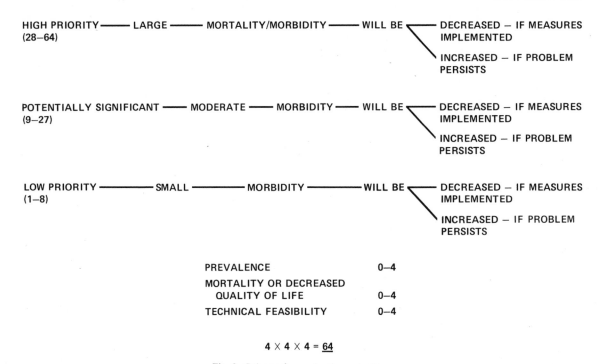

Fig. 8. Scheme for evaluating priorities.

should be simple, incorporate an accurate reporting, storage, and retrieval system, and be constantly under review and revision.

REFERENCES

1. World Health Organization, "The Surveillance of Communicable Diseases," *WHO Chronicle* **22**, 439–453 (1968).

2. A. S. Benenson, *Control of Communicable Diseases in Man*, 12th Ed. Am. Preventive Health Assn., Washington, D.C., 1975.

3. World Health Organization, *Early Detection of Health Impairment in Occupational Exposure to Health Hazards*, Technical Report Series, No. 571, WHO, Geneva, 1975, p. 78.

4. Percival Pott, *Chirurgical Observations Relative to the Cataract, the Polypus of the Nose, the Cancer of the Scrotum, the Different Kind of Ruptures and the Mortification of the Toes and Feet*, Hawks, Clarke, and Collins, London, 1875.

5. T. F. Hatch and H. E. Williamson, "Priorities in Preventive Medicine," *Arch. Envir. Health* **22**, 52–55 (1974).

6. "Proposed Rules," Dept. of Labor, *Federal Regist.* **43**(11), 2616 (1978).

7. R. E. Yodaiken, Statement before Dept. of Labor, OSHA Public Hearing in Occupational Exposure to Acrylonitrile, 1978.

8. State of the Art Conference: *Screening and Early Detection of Colo-Rectal Cancer*, Division of Cancer Control and Rehabilitation, NCI, June 1978. (Summary to appear in *J. Am. Med. Assoc.*)

9. *Consensus Conference on Screening for Lung Cancer*, NCI, September 1978. (Summary to appear in *J. Am. Med. Assoc.*)

10. *Record Keeping Requirements under the Occupational Safety and Health Act of 1970.*, U.S. Dept. of Labor, Occupational Safety and Health Administration, 1975.

11. V. C. Rose, *Reliability and Utilization of Occupational Disease Data*, U.S. Dept. of Health, Education, and Welfare, National Institute for Occupational Safety and Health Publication No. 77-189, 1977.

12. D. P. Discher, G. D. Kleinman, and F. J. Foster, *Pilot Study for the Development of an Occupational Disiox Surveillance Method*, National Occupational Hazard Survey, U.S. Dept. of Health, Education, and Welfare, National Institute for Occupational Safety and Health Publication No. 75-165.

13. R. Spirtas, D. S. Sunden, J. B. Sestito, V. J. Behrens, and J. French, *A Conceptual Framework for Occupational Health Surveillance,* U.S. Dept. of Health, Education, and Welfare, National Institute for Occupational Safety and Health Publication No. 78-135.

14. *Occupational Health Programme,* Report by the Director-General, World Health Organization, Twenty-ninth World Health Assembly, Geneva, 1976.

15. *Criteria for a Recommended Standard—Occupational Exposure in Coal Gasification Plants,* U.S. Dept. of Health, Education, and Welfare, National Institute for Occupational Safety and Health Publication No. 78-191, 1978 (in press).

16. *Recommendations for a National Strategy for Disease Prevention,* Report to the Director, Communicable Disease Center, by the Center for Disease Control Programs and Policies Advisory Committee, U.S. Dept. of Health, Education, and Welfare, Communicable Disease Center, Atlanta, 1978.

DISCUSSION

Jim Smith, SRC: Did I understand you to say that there are asbestos workers now that are denied employment on account of past occupational exposure and habits? If so, would you care to elaborate?

Ralph Yodaiken: I believe Johns-Manville is demanding sputum tests for people who apply for work, and where the sputum test is positive on two separate occasions, then either the man is not admitted to work or he is located at a point where he is not potentially exposed to asbestos. I am not quoting precisely because this was information given to me at a conference that I recently attended. I don't know how widely we could apply this sort of philosophy. There are such a lot of factors involved in screening people in a preemployment situation. If it were true that, on the basis of sputum cytology, we could anticipate a high cancer risk with any degree of certainty, we might have a different approach. Our techniques, however, give results of questionable value, so there are reasons why a positive sputum cytology result in itself is not sufficient reason to exclude a man from work.

COMPARATIVE INDUSTRIAL HYGIENE ASPECTS FOR COAL GASIFICATION AND LIQUEFACTION, OIL SHALE, AND TAR SANDS PROCESSING

James O. Jackson*

ABSTRACT

Occupational health programs for synthetic fossil fuel technologies are in various stages of development today. Although it is not possible to compare industrial hygiene programs for coal gasification, coal liquefaction, oil shale and tar sands now for several reasons, it is possible to investigate each technology for its potential occupational health problem areas. This evaluation is called a process profile, and a preliminary version is developed for each of the four technologies described here. Areas for future research are considered as well as several factors that are fundamental to any synthetic fossil fuel industrial hygiene program.

When Dr. R. B. Gammage asked me to address this symposium on comparative industrial hygiene programs for coal gasification, coal liquefaction, oil shale, and tar sands, I welcomed the opportunity to delve into such an analysis. Pittsburg & Midway Coal Mining Company (P&M), a subsidiary of Gulf Oil Corporation, is presently operating a 50-ton-per-day Solvent Refined Coal (SRC) plant at Ft. Lewis, Washington. Because of my involvement at Ft. Lewis, I knew that few literature references were available describing industrial hygiene programs for coal gasification or liquefaction; however, I have subsequently learned that even fewer references are available on oil shale or tar sands processing.

While published information is not available to an appreciable extent, I am gratified professionally to find that industrial hygiene programs do indeed exist for each of these four synthetic fossil fuel industries. A thorough comparative analysis of industrial hygiene programs in these four areas is not possible at this time for several reasons including (1) the lack of published and/or available program protocol, (2) the absence of a comparative data base, and (3) the fact that several programs in key areas are now just beginning. With this in mind, another approach can be useful today in comparing these four industries. This approach can be described as a process profile, that is, a systematic effort to carefully

scrutinize a specific process and answer the who, what, where, why, and when of occupational and environmental health and safety. This morning we can explore a miniversion of a process profile for each of these four synthetic fossil fuel technologies and, at the same time, look for comparisons as they exist. Then, let us examine some of the industrial hygiene data from the P&M facility at Ft. Lewis, and, finally, allow me to present a series of recommendations that must be addressed by responsible management in the synthetic fuels industry.

HEALTH EFFECTS OF SYNTHETIC FOSSIL FUELS

This symposium has already considered the general area of health effects of synthetic fossil fuels and, to some extent, the specifics as well. The following discussion elaborates these specifics in light of the background needed for this presentation.

Coal Conversion

Coal gasification and liquefaction will be considered together because of the nature of the literature

---

*Gulf Science and Technology Company, Medical and Health Resources Division, Industrial Hygiene Laboratory, Pittsburgh, Pennsylvania 15230.

citations. The Union Carbide study from the 1950s at Institute, West Virginia,[1-4] is most often mentioned whenever discussing employee exposure at coal gasification or liquefaction facilities. Dr. Alan Palmer of Stanford Research Institute is here today discussing follow-up activities for some of the employees from that study. Enviro Control, Inc.,[5] prepared a document recommending health and safety guidelines for coal gasification pilot plants for the National Institute for Occupational Safety and Health (NIOSH). NIOSH will also soon release a criteria document on commercial coal gasification facilities which was likewise prepared by Enviro Control. Mr. Jim Evans, our session chairman today, played a leading role in both of these publications. Gruhl[6] has published a novel approach to developing a method to evaluate cancer risk to the public posed by atmospheric pollutants from the coal industry. He also includes an econometric discussion where preliminary estimates are given for the cost impact on industry in providing protection against exposure to carcinogens. Gulf's industrial hygiene program for both the P&M Ft. Lewis SRC plant and the Harmarville research facility has been reported previously.[7-9] Further, the medical surveillance, training, toxicological and industrial hygiene efforts associated with the Ft. Lewis Energy Research and Development Administration/Department of Energy (ERDA/DOE) contract have been and continue to be published.[10,11] NIOSH contracted with Bendix Corporation to perform an industrial hygiene characterization of coal gasification plants, and at last year's symposium, Phillips[12] reported on the preliminary investigation of polynuclear aromatic hydrocarbon (PNA) emissions from HYGAS. The comprehensive industrial hygiene and medical program for health protection in coal conversion technology developed at Oak Ridge National Laboratory has been described[13] and also presented at last year's symposium.[14]

Oil Shale

Weaver and Gibson[15] have recently described the American Petroleum Institute's (API) research program for oil shale. A variety of acute and chronic toxicity studies are underway or planned and include carcinogenesis, mutagenesis, and teratogenesis. Further, the API, the Department of Energy, and industry[16] are jointly sponsoring an industrial hygiene study of the Paraho oil shale facility. The Los Alamos Scientific Laboratory is performing

this project where completion is anticipated for later this year.[17] I am pleased to see this cooperative effort, particularly where industry and government can work together in determining the relative health risk with a specific technology. Coomes[18] presented a paper in Denver describing the TOSCO II bioassay program and, in general, reviewed the carcinogenic aspects of oil shale. The industrial hygiene impact of producing shale oil has been discussed by Bachman, who developed an industrial hygiene program for the Colony Development Corporation.[19] Another cooperative program is beginning at the Rio Blanco Oil Shale effort in Colorado, where Gulf is developing the industrial hygiene program.

Tar Sands

I am not aware of any published or ongoing toxicology studies with tar sands, process streams or products. It is hoped the Canadian operations will fund a joint bioassay program in the near future. Meanwhile, it is advisable that we assume an attitude of due caution when dealing with tar sands processes.

PROCESS PROFILES

Let us now begin to examine the four synthetic fossil fuel technologies that are of interest to us today. Again, these are very brief process profiles designed to point out areas within each process where exposures may and can occur and, further, to mention what some of these obvious exposures are. This is not intended to be an exhaustive study since we do not have sufficient time today, and an adequate data base to prepare the definitive study on each of these four industries does not exist at this time. We will begin with consideration of tar sands, move on to oil shale, and then to coal gasification and liquefaction.

Tar Sands

Tar sands, oil sands, and bituminous sands are all commonly used terms to describe the same bitumen-impregnated sands of Alberta. There is no technical difference in the choice of these terms, and therefore, I will refer to them as tar sands for simplification.

Typical tar sand from Alberta consists of a mixture of bitumen, sand grains, and water.[20] Bitumen, which is the viscous, dense petroleum substance that impregnates the sands, gives a black color to the tar sands and surrounds the water layer covering the sand particles (Fig. 1). Sand grains constitute about 83% of the weight of the tar sands, with bitumen and

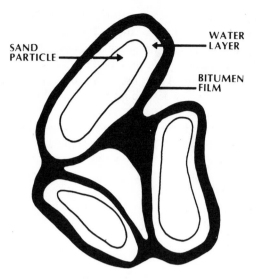

Fig. 1. Typical arrangement of tar sand particles.

water being the remaining 17%. A health hazard potential exists from these sand grains because they are in the form of quartz. Elemental analysis of the sand grains indicates that silicon is 40–50% of the mineral content.

Although tar sand deposits were first noted along the Athabasca River in 1788, commercial plants used for processing Alberta tar sands are few. The Great Canadian Oil Sands Limited plant (GSCOS) has been in operation since 1968, and the Syncrude Canada Limited facility is coming on stream at this time.[21] Other facilities are planned for construction and have been approved or are awaiting application approval.

A simplified process flow diagram for a typical commercial plant is shown in Fig. 2. Health hazards are presented by following the various steps of tar sands processing.

Mining

Mining of tar sands presents the gamut of hazards traditional to open-pit or strip mining as found in other industries. Usually one envisions noise, dust, and, of course, safety as the major concerns here; however, the northern Alberta location of the tar sands mines adds another dimension to these conventional hazards—extreme cold stress.

Noise sources include the expected multitude of heavy equipment such as draglines, scrappers, front-end loaders, etc. Particulate-level concern also encompasses consideration of the associated α-quartz concentrations. Fortunately, this potential problem area is reduced because the tar sand itself is a tacky material which certainly diminishes the mining dust levels. When the tailings are returned to the mine, and during the subsequent reclamation effort, the bitumen has been removed from the tar sands and, for the most part, only the sand particles remain. In this situation, dust levels must be evaluated more critically because the sand particles are fine-grained and, as just mentioned, freed from their organic binder.

Excessive cold stress in northern Alberta tar sands mines poses unique occupational health problems. Work practice procedures must be developed very carefully for operating personnel as well as maintenance employees. Obviously, skin contact with metal

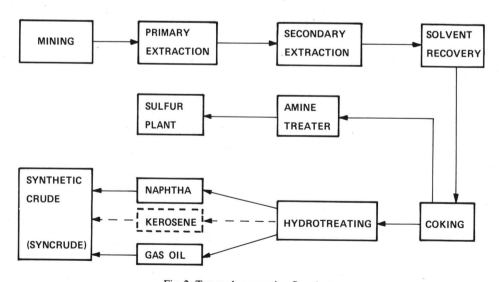

Fig. 2. Tar sands processing flowchart.

equipment at these subzero temperatures must not be allowed to occur. Provision must also be made to ensure that operators cannot be stranded too far from shelter, should equipment become inoperable.

Extraction

Noise, heat stress, silica, steam, and hot water are some of the potential health hazards associated with the initial hot water extraction process. Here, a rotating drum can be used to slough off successive layers of tar sand lumps until a pulp is formed, which is controlled at a pH of 8–8.5, then screened and heated again.

After conditioning, the screened pulp is pumped to separation cells where bitumen is freed from the sand grains by a froth flotation. Dilution centrifuging with naphtha is then used to dehydrate and demineralize the froth to provide an acceptable feedstock for the following bitumen upgrading. This latter froth cleanup procedure provides the opportunity for skin contact, particularly if the centrifuges plug. Exposures to naphtha vapors are possible here as well and during the subsequent solvent recovery operation. The aromatic content of the naphtha stream must be noted carefully.

Coking

Delayed or fluid coking processes can be employed to begin upgrading the separated bitumen. Since bitumen can be high in sulfur, the resulting desulfurization during coking can generate considerable hydrogen sulfide (H_2S). Heat, hot water, steam, H_2S, and some light hydrocarbons are potential occupational stresses associated with the upper deck of delayed cokers. Opening of delayed coking drum heads warrants careful attention also, particularly to hydrocarbons. Excessive coke dust, carbon monoxide, and PNAs can be further generated by this process.[22]

Amine treater–sulfur plant

H_2S is the obvious stress in this situation.[21]

Hydrotreating

Bitumen is further upgraded with hydrotreating where two or three cuts are made and then combined to form the desired synthetic crude or syncrude. Essentially, hydrotreating is the saturation of olefinic hydrocarbons and the removal of sulfur, nitrogen, oxygen, and halogens. Various catalysts are used in hydrotreating, as are pressures of up to 1000 psig. Potential health hazards for this process include the aliphatic and aromatic hydrocarbons, H_2S, noise, heat, and metal-catalyst carbonyls.[22]

Oil Shale

Oil shale processing is not a new industry, as it can be traced back to the latter part of the 19th century in both Scotland and the United States. The hazards of the oil shale industry are not necessarily new either, and the potential carcinogenicity associated with this industry is certainly recognized. Coomes[18] has observed that the shale oil itself is carcinogenic by animal bioassay, while the upgraded or hydrotreated shale oil is essentially noncarcinogenic.

Oil shale industry terminology is often misunderstood, particularly with the frequent interchanging of the terms oil shale and shale oil, for example. The following definitions are offered to simplify this discussion:

Oil Shale: fine-grained sedimentary rocks containing substantial amounts of organic material called kerogen.

Kerogen: organic material in oil shale which is insoluble in organic solvents.

Shale Oil: liquid oil product from the retorting of kerogen.

Spent Shale: solid material remaining after retorting.

Figure 3 presents a simplified schematic of an oil shale process.[20] Again, this is a generalized flow diagram and is intentionally brief. As with tar sands, each major processing step will be considered separately.

Underground mining

Oil shale can be mined underground and, when it is, the occupational exposures are similar to those experienced in other underground mining operations. The free silica content of the mined oil shale can approach 10 or 12%, and thus dust exposures should be minimized. Noise from mechanical and electrical equipment may be a hazard, as diesel exhausts may be, if not properly ventilated. Where underground blasting occurs, sufficient time must be allocated to clean the area of dust levels, nitrogen dioxide (NO_2), and probably carbon monoxide (CO); otherwise, overexposures can occur. Mucking and other mined-material transfer operations present the usual underground hazards, although, in general, safety problems are often a higher priority than

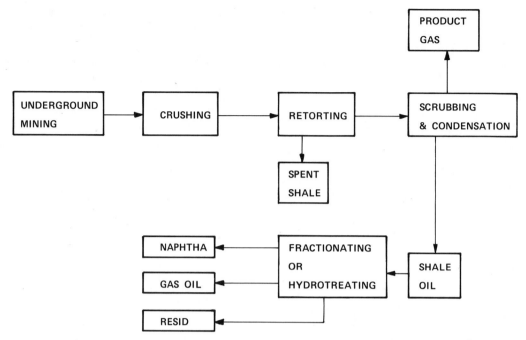

Fig. 3. Oil shale processing flowchart.

industrial hygiene concerns because of immediate life-threatening situations.

Crushing

The mined oil shale is transferred to the crushers for subsequent pulverization. This transfer can occur by a variety of methods; however, exposure patterns will be determined by the method of conveyance. The crushing operation will naturally result in noise and particle levels needing evaluation of occupational exposure. The extent of a free silica hazard, depending upon the degree of crushing required, should be investigated

Retorting

Following crushing, the oil shale is transferred to the retort and can be loaded from either the top or bottom of the retort, depending on the type of process. Retorting can be a continuous or a batch process. The oil shale moves through a series of zones in the retort or kiln. As the oil shale is heated, kerogen is pyrolyzed to shale oil vapors, shale gas, and organic residue. The organic residue is then burned (perhaps with recycle) to release additional pyrolysis products. The potential for skin contact is high in the retorting process where PNAs, heterocyclics, and other potential carcinogenic compounds may be

encountered. Heat stress, steam, dust, CO, carbon dioxide (CO_2), H_2S, and methane (CH_4) are additional concerns.

Scrubbing and condensation

Usually a cyclone by itself or in series with an electrostatic precipitator is used to collect the shale oil mist droplets and any carry-over particulate matter in the off-gas stream. Further processing with condensation removes the remaining shale oil. Again, skin contact with the constituents listed above is a possible source of exposure in this part of the process, although to a lesser degree than in the retorting stage.

Product gas

The resulting product gas, if used, must be stripped of sulfur compounds prior to further processing. Hydrogen sulfide and other sulfur compounds are the principal hazards from sulfur recovery operations.

Fractionation or hydrotreating

Upgrading of the shale oil to usable products could be accomplished with either fractionation or hydrotreating. If hydrotreating is used, arsenic removal is necessary to avoid poisoning the catalyst. The hazards associated with hydrotreating have just been discussed for tar sands; those for fractionation are not substantially different.

Spent shale

The spent shale is removed from the retort, cooled, and disposed of as tailings. Dust, associated metals (and perhaps free silica), and some volatile material are potential exposures, especially to bulldozer and shiploader operations when transporting and dozing the spent shale.

Coal Gasification

Coal gasification can produce a low-Btu product (125–500 Btu/scf) with resulting oils and tars if bituminous or lower-ranked coals are used or no tars if anthracite coal or high-temperature gasification is used. Coal gasification can also produce a high-Btu product (900–1000 Btu/scf) where oils and tars result as well.[5] Because of the complexities of these differing processes, let us propose a simplified coal gasification process for discussion that will include the salient health stress features of interest (Fig. 4).

Coal preparation and feeding

Coal is delivered from the mines and stored until used when it is dumped into bins and transferred by means of feeders and conveyors to primary and secondary mechanical crushers. From storage bins, the pulverized coal is fed into the gasifier through a pressurized lock hopper system. Occupational exposures obviously include heavy equipment noise

and fumes, dust and the associated α-quartz potential stress, along with trace elements in the coal and spontaneous combustion in storage pits. Gasifier fumes (up to 20% CO) can present potential health hazards as well.

Gasification

Gasification, the center of the entire process, is generally performed in a fluidized bed system. Coal is injected into the gasifier from the lock hopper. Simultaneously, fluidizing gas and/or steam and oxygen are fed in through the bottom of the bed. Product gas passes through a cyclone to separate the entrained solids. The resulting ash is removed from the bottom of the bed, quenched with water, cooled and disposed of properly. Noise and heat; dust; gasifier gases; including CH_4, CO, and products of incomplete combustion; phenols; tars and oils; high-pressure steam and oxygen; and trace elements are some of the potential health hazards that may occur with gasification and the subsequent ash removal and disposal.

Quenching and scrubbing

The gasifier off-gas still contains solids and hydrocarbon material that is now removed by quenching and scrubbing processes. For example, a venturi water scrubber would do both desired actions where the resulting gas is ready for drying if a low-Btu product is needed, or shift-conversion if a high-Btu gas is required. The resulting condensate or quench water which results from quenching and scrubbing contains substantial amounts of hydrocar-

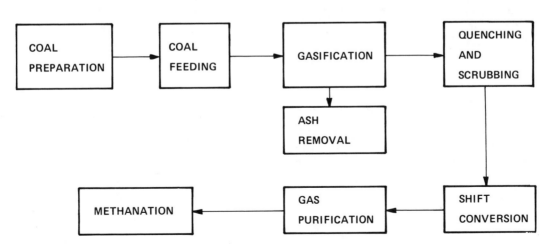

Fig. 4. Coal gasification processing flowchart.

bon material as well as trace elements. The subsequent settling and separation procedures provide opportunities for potential occupational hazards from various hydrocarbon and tarry materials, along with many metals. Exposures are also possible to H_2S, CH_4, CO, CO_2, volatile hydrocarbons, tars, and noise. Maintenance personnel in this area should be particularly cautious of the tar build-up within the cooling and scrubbing equipment.

Shift-conversion

This unit focuses on the catalytic reaction of CO + H_2O converted to CO_2 + H_2 to ensure that the H_2:CO ratio is correct for subsequent methanation. Potential exposures are possible here to hydrogen cyanide (HCN), CO, CO_2, high-pressure steam, and any residual hydrocarbons.

Gas purification

Sulfur compounds must be removed from the gas streams before methanation occurs; otherwise the methanation catalyst could be deactivated or poisoned. Also, CO_2, H_2O, and some remaining hydrocarbons are selectively removed. Several systems can be used for this clean-up including the Benfield process where a hot potassium carbonate solution with diethanolamine (or diglycolamine) is used. With a Lurgi gasifier, a Rectisol process is used here, whereby the gas is chilled and washed with cold methyl alcohol for the desired removal. Again, H_2S and other sulfur compounds are an obvious potential hazard as is the product gas, i.e., CO, CO_2, and CH_4. While additional exposures can occur, they usually depend on the purification process. Subsequent sulfur treatment could employ a Claus or Stretford unit and tail gas desulfurization; employee exposure could involve H_2S and other sulfur compounds as well as exposures to the feed gas.

Methanation

To upgrade the purified product gas and to reduce its CO content, the following reaction is produced: CO + $3H_2 \rightarrow CH_4$ + H_2O. This reaction uses a nickel catalyst in combination with pressure and temperature. Potential occupational stresses here are CO, nickel carbonyl [$Ni(CO)_4$], iron pentacarbonyl [$Fe(CO)_5$], CH_4, H_2, high-pressure steam, and product gas and heat.

Coal Liquefaction

There are a number of coal liquefaction schemes which can be separated into two processes: pyrolysis (carbonization) and dissolution.[11] Dissolution processes can be further distinguished as those that use (1) no catalyst or hydrogen, (2) both a catalyst and hydrogen, or (3) hydrogen but no catalyst. This last dissolution technique using hydrogen without an added catalyst is the basis of the Solvent Refined Coal process. P&M uses this technology in the operation of the 50 ton-per-day facility at Ft. Lewis, Washington. Because of my familiarity with this facility, I shall concentrate on SRC as the example of coal liquefaction.

The SRC I process (Fig. 5) produces a high-Btu, clean-burning, low-sulfur, and low-ash product. Alternatively, the SRC II process (Fig. 6) gives a liquid distillate that can be fractioned into several products, such as fuel gas, naphtha, fuel oil, and a heavy bottoms cut.

SRC I—coal preparation

Coal is moved from storage bins to the preparation area. Here, the coal is pulverized and sized, then mixed with the solvent to form the 25–40% coal slurry. Anticipated potential employee exposures are the usual noise and dust (including the possibility of α-quartz) problems of moving, storing, and subsequent pulverizing of coal. At Ft. Lewis, the inert gas contains CO, due to incomplete combustion at the inert gas generator. The hazard potential of coal slurry can vary up to and including carcinogenicity. For this reason, work practices and personal protection must be evaluated carefully to ensure employee health. Of course, skin contact is the main occupational concern with the coal slurry, although aerosolization is a possibility as well.

Preheater and dissolver

The slurry is mixed with hydrogen and fed through a slurry-preheater to a reaction zone defined as the dissolver. At this point, about 90% of the organic material in the coal is dissolved in approximately 20 minutes at about 800°F and 1500 psig. Again, skin contact is of concern here, should spills or leaks occur.

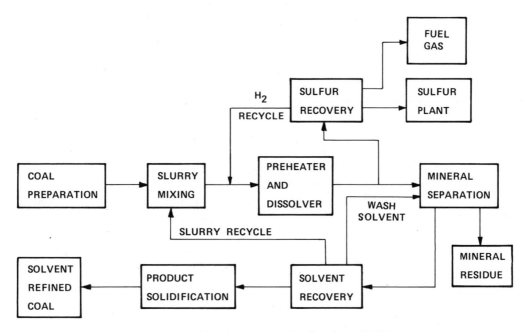

Fig. 5. Coal liquefaction processing flowchart—SRC I.

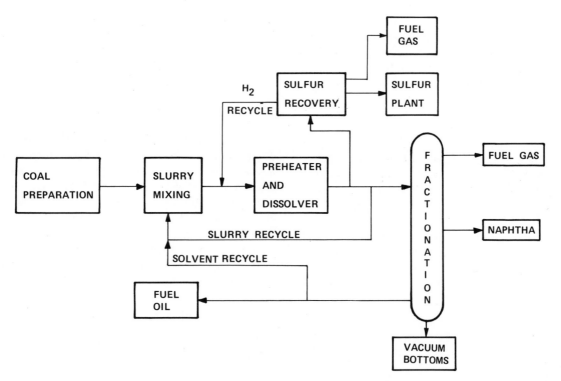

Fig. 6. Coal liquefaction processing flowchart—SRC II.

Sulfur recovery

This process and the subsequent hazards were described earlier in this review.

Mineral separation

The product stream is separated to remove light gases, and the slurry of liquid and undissolved solids is filtered. The excess hydrogen is stripped of its H_2S content by acid gas absorption and then recycled to the slurry preheating system. Following filtration, the undissolved solid or mineral residue is washed and then dried. This mineral residue can be hazardous because it is the concentrate of the undissolved trace metals and minerals, such as α-quartz present in the feed coal. Caution is necessary if these materials are present in the filter pre-coat preparation area. Enclosure of the pre-coating operation is essential because of the nature of these materials. Noise again could be a problem in this area because of the extensive mechanical equipment involved.

Solvent recovery

Vacuum flash distillation recovers the solvent from the liquid filtrate. The overhead is fractioned into three parts: a light liquid by-product, a wash solvent for the filter just described, and a process solvent for recycle to the slurry-mixing area. Again, skin contact and noise are the principal sources of occupational stress.

Product solidification

The bottoms from the flash distillation is a heavy residual oil called solvent refined coal. Upon cooling, it is a pitch-like material with a melting point of 300°F to 400°F and a heating value of about 18,000 Btu/lb. It contains less than 0.1% ash and less than 0.8% sulfur. Employee exposure to noise and SRC particles can occur here, especially in the immediate area of solidification of the SRC product. The SRC material flows onto a stainless steel belt for cooling; it is at this point in the process that noise and particle levels can be excessive.

Miscellaneous

Other areas of occupational stress can be associated with the subsequent handling and transfer of the SRC product where noise and dust levels again may be of concern. Industrial hygienists should note carefully that welders working on contaminated equipment may be exposed. In addition to traditional welding hazards, coal tar and PNAs can also be generated here. Phenols are present in many SRC process streams, so skin contact should be avoided.

SRC II

As previously noted, SRC I yields primarily a solid product while SRC II produces liquids. A comparison (Figs. 5 and 6) shows that the processes are similar in the coal preparation, the slurry mixing, preheating, and dissolving, and the desulfurization stage. Instead of filtration as in SRC I, the SRC II process slurry is fractionated. Distillate fractions include naphtha, fuel oil, and heavy distillate. The distillation bottoms contain the mineral residue, which is somewhat different in its toxic potential when compared to SRC I. The middle fraction yields fuel oil.

SRC I AND II MONITORING DATA

Over two years of operation in the SRC I mode has resulted in an industrial hygiene monitoring data base that is presently available.[10] Similarly, operation in the SRC II mode for nearly one year yields a second set of industrial hygiene data that will be available in the near future.[11] A comparison of the two data sets is also in progress where exposures in several areas of the facility should show lower concentrations for SRC II operations.

Skin Wipe Program

Because of the hazards associated with skin contact and certain streams in the SRC process (such as the coal slurry or the recycle solvent), we are attempting to develop an index of skin exposure to distinguish skin exposures quantitatively by job classification and SRC operating mode as well as to indicate the extent of exposure. We anticipate using several parameters to characterize this index from an organic extract of the skin wipe and subsequent PNA determination.

PNA MONITORING

An overriding theme throughout this presentation has been the concern for exposure to PNAs. The SRC skin wipe program was just discussed above, and Phillips' efforts at evaluating PNA emissions at HYGAS have been mentioned.[12] Jackson and Cupps[23] have developed an alternative procedure for collecting PNAs by both personal and area sampling.

They reported that, with the traditional glass fiber/silver membrane filter system, collected PNAs can sublime from the filters. The alternative developed consists of backing up the traditional filter system with a solid sorbent section.

SUMMARY

Today we briefly examined several synthetic fuel processes to determine their obvious industrial hygiene problems. We have looked at some data from one process, solvent refined coal. During preparation of this paper, it occurred to me that one general industrial hygiene program might be sufficient to encompass the four process types described. If such an approach is possible, the program outlined by Cheng and Jackson[7-9] at last year's ORNL seminar might be used. This approach considers coal-derived materials only, although it could certainly embrace oil shale processing and possibly tar sands. The principal features of this industrial hygiene program emphasize personal protection and personal hygiene practices, employee health education, identification and measurement of known and suspected hazardous agents, and procedures to reduce employee exposures to effectively safe levels.

Finally, there are at least eight considerations to be included in planning any industrial hygiene program for synthetic fossil fuels:

1. Barrier creams can provide a false sense of security and protection; however, they do indeed assist in washing stain-producing materials from exposed skin.
2. A procedure is needed to evaluate skin exposure to potential carcinogenic material. The skin wipe index and the reflectance fluorometer being developed at ORNL are steps in this direction, but further development is desirable.
3. Personal monitoring procedures must be refined to reflect the variety of exposures found in synthetic fossil fuel facilities. The emphasis thus far has been on PNAs, and while this is certainly worthwhile, information is also needed on exposures to a number of other chemicals, such as heterocyclic compounds and aromatic amines.
4. Employee exposure criteria for synthetic fossil fuel facilities must be developed for the specific industry of concern. While the resulting Threshold Limit Values (TLVs) may indeed be similar to or even the same as those observed in other natural or synthetic fossil fuel industries, the unique characteristics of each industry must be evaluated individually.
5. One valuable input for setting exposure criteria is toxicology data. Animal bioassay data supplemented with chemical analyses of process streams is needed for each synthetic fossil fuel industry.
6. Detailed profiles should be developed for the various processes now in the pilot plant stage. This effort will be invaluable when large demonstration or full-scale commercial facilities are built.
7. Employee training at all levels must be started with periodic review sessions. Workers today are interested in their health and well-being, and it is management's responsibility to see that employees are aware of hazards associated with their workplace.
8. Epidemiological programs must be developed and used to follow the work force at these synthetic fossil fuel facilities.

These are just a few of the concerns I have about the general health aspects of synthetic fossil fuel processes. Perhaps this effort will stimulate the further publication of industrial hygiene programs and data in these areas which will subsequently aid in resolving a number of issues that have been raised.

REFERENCES

1. R. J. Sexton, "The Hazards to Health in the Hydrogenation of Coal I: An Introductory Statement on General Information, Process Description, and a Definition of the Problem," *Arch. Environ. Health* 1, 181–86 (1960).
2. C. S. Weil and N. I. Condra, "The Hazards to Health in the Hydrogenation of Coal II: Carcinogenic Effect of Materials on the Skin of Mice," *Arch. Environ. Health* 1, 187–93 (1960).
3. N. H. Ketcham and R. W. Norton, "The Hazards to Health in the Hydrogenation of Coal III: Industrial Hygiene Studies," *Arch. Environ. Health* 1, 194–270 (1960).
4. R. J. Sexton, "The Hazards to Health in the Hydrogenation of Coal IV: The Control Program and the Clinical Effects," *Arch. Environ. Health* 1, 208–31 (1960).
5. *Recommended Health and Safety Guidelines for Coal Gasification Pilot Plants,* DHEW (NIOSH) Pub. No. 78-120, January 1978.
6. J. Gruhl, *Review of Methods for Assessing the Carcinogenic Hazards from Coal-Using Energy Technologies,* NTIS Pub. No. PB 270-682, September 1976.

7. R. T. Cheng, "Industrial Hygiene Programs and Experience for a Fifty Ton per Day Solvent-Refined Coal Plant," in *Proceedings of the Second ORNL Workshop on Exposure to Polynuclear Aromatic Hydrocarbons in Coal Conversion Processes,* CONF-770361, Oak Ridge National Laboratory, Oak Ridge, Tenn., 1977, pp. 39–46.

8. R. T. Cheng, J. O. Jackson, and J. F. Smith, "Industrial Hygiene Monitoring Programs for a Solvent-Refined Coal Pilot Plant," presented at the 1977 American Industrial Hygiene Conference, New Orleans (May 1977).

9. J. O. Jackson, "Industrial Hygiene Program and Experiences with Small-Scale Coal Conversion Pilot Plants," in *Proceedings of the Second ORNL Workshop on Exposure to Polynuclear Aromatic Hydrocarbons in Coal Conversion Processes,* CONF-770361 Oak Ridge National Laboratory, Oak Ridge, Tenn., 1977, pp. 31–38.

10. *Solvent Refined Coal (SRC) Process: Health Programs,* Research & Development Report No. 53, Interim Report No. 24, Volume III—Pilot Plant Development Work, Part 4—Industrial Hygiene, Clinical and Toxicological Programs, FE/496-T15, January 1978.

11. *Solvent Refined Coal (SRC) Process: Health Programs,* Research & Development Report No. 53, Volume II, Industrial Hygiene, Clinical and Toxicological Programs, in preparation.

12. R. D. Phillips, "Preliminary Investigation of Airborne PNA Emissions from the HYGAS Pilot Plant," in *Proceedings of the Second ORNL Workshop on Exposure to Polynuclear Aromatic Hydrocarbons in Coal Conversion Processes,* CONF-770361, Oak Ridge National Laboratory, Oak Ridge, Tenn., 1977, pp. 49–58.

13. N. E. Bolton et al., *Occup. Health Safety,* **46,** 30 (1977).

14. N. E. Bolton, "An Industrial Hygiene Program for Coal Conversion Technology," in *Proceedings of the Second ORNL Workshop on Exposure to Polynuclear Aromatic Hydrocarbons in Coal Conversion Processes,* CONF-770361, Oak Ridge National Laboratory, Oak Ridge, Tenn., 1977, pp. 13–18.

15. N. K. Weaver and R. L. Gibson, "The U.S. Oil Shale Industry. A Health Perspective," presented at the 1978 American Industrial Hygiene Conference, Los Angeles (May 1978).

16. N. K. Weaver, personal communication (September 8, 1978).

17. E. E. Campbell, personal communication (August 25, 1978).

18. R. M. Coomes, "Carcinogenic Aspects of Oil Shale," presented at the American Nuclear Society Environmental Aspects of Non-Conventional Energy Resource II—Topical Meeting, Denver (September 1978).

19. J. A. Bachman, "The Industrial Hygiene Impact of Producing Shale Oil, A New Energy Source," presented at the Southern California Section of the American Industrial Hygiene Association (January 1974).

20. T. A. Hendrickson, *Synthetic Fuels Data Handbook,* Cameron Engineers, Inc., Denver, Colo., 1975.

21. J. A. Dallinger, personal communication (August 25, 1978).

22. *Industrial Hygiene Monitoring Manual for Petroleum Refineries and Selected Petrochemical Operations,* American Petroleum Institute, Washington, D.C.

23. J. O. Jackson and J. A. Cupps, *Carcinogenesis: Polynuclear Aromatic Hydrocarbons, Vol. 3,* Raven Press, New York, 1978, pp. 183–91.

DISCUSSION

Otto White, Brookhaven National Laboratory: What is the extent of the industrial hygiene surveillance? Are you currently monitoring a certain percentage of your production workers, or have all the industrial hygiene surveys been just that, surveys of a small number of production workers to get some trends?

James Jackson: We have a schedule that has just been revised. We are running in excess of 50 personnel samples a month. The data are being reviewed and evaluated on a statistical basis to see if we have a sufficient data base to work with. We will make sure that we are not missing anything in the maintenance of our industrial hygiene programs.

NEW DEVELOPMENTS IN MONITORING INSTRUMENTS*

Alan R. Hawthorne†

ABSTRACT

The need for a real-time area monitor for toxic chemicals at coal conversion facilities is reviewed. A brief description of some new monitoring instruments capable of giving multicomponent analyses is provided. These instruments include portable mass spectrometers, a computerized infrared analyzer, a derivative ultraviolet-absorption spectrometer, and a miniature gas chromatograph for trace gas analysis. Instruments capable of measuring polynuclear aromatic compounds on surfaces and as respirable particulate matter are also discussed.

INTRODUCTION

Large-scale energy planning and development in the U.S. must be predicated on utilization of the extensive coal resources available in this country to meet our increasing energy demands. With limited U.S. and world supplies of petroleum resources, the Department of Energy is committed to producing synthetic fossil fuels to substitute eventually for dwindling petroleum reserves. Concurrent with this effort is a commitment to produce alternative energy sources with minimal adverse environmental and health effects. Coal conversion technologies pose a challenge to control engineers, life scientists, and industrial hygienists because, from the wide range of compounds found in process streams, many may prove to be detrimental to both health and environment.

The potentially hazardous environment at a conversion facility places significant responsibility on the industrial hygienist in that many of the chemicals either are acutely toxic or are carcinogenic, thus requiring exposures to be kept to a minimum. Exposure to potential carcinogens requires a special commitment to worker protection, since there are often no detectable warning signs of exposures that may result decades later in skin, lung, or other cancers.

Given this commitment of protecting both health and environment from deleterious effects of coal conversion technologies, a concerted effort will be required of many people—the analytical chemist, the life scientist, the environmental scientist, the control engineer, and the industrial hygienist—to meet this goal. This paper discusses one aspect of the role to be filled by the industrial hygienist: the problem of area monitoring using new, real-time, or near real-time instrumentation.

Not addressed is the problem of providing detailed laboratory analysis of samples, where powerful techniques such as gas chromatography–mass spectrometry (GCMS) applied to fractionated samples have given much detailed and valuable information. Also outside the scope of the present paper is personal monitoring using passive dosimeters such as charcoal tubes or filter samplers that require subsequent laboratory analysis to yield data on worker exposures. These monitoring techniques are of vital importance to an overall industrial hygiene

---

*Research sponsored by the Division of Biomedical and Environmental Research, U.S. Department of Energy, under contract W-7405-eng-26 with Union Carbide Corporation.

†Health and Safety Research Division, Oak Ridge National Laboratory, Oak Ridge, Tennessee 37830.

program and deserve full attention. The subject of the present work is confined to area monitoring; examples are presented of new instruments recently developed or under development. This list is not meant to be definitive but only to give a sampling of the types of work being done.

MONITORING NEED

Synthetic fossil fuel technologies provide potential exposure to a wide range of toxic chemicals (Table 1), many of which are known to be carcinogenic.[1,2] Many more are suspected of being either primary or secondary carcinogens, or of having cocarcinogenic or promotional activity. Effects of worker exposure to the toxic chemicals present in various process streams of coal conversion facilities have been established most often when the toxicant under investigation was the only harmful compound present. Conversion facilities present the hazard of exposure to a complex mixture of chemicals, where the possibility of severe synergistic effects may greatly enhance the result normally expected if each compound were considered individually.[2] A striking example is the 10,000-fold decrease in threshold dose for mouse skin tumors due to benzo (a) pyrene (BaP) in the presence of dodecane.[3]

This potential for synergistic exposures heightens the need for a monitoring and record-keeping program that will measure a variety of chemicals, permitting correlations to be made between the exposure source terms and the ultimate insults to human health. Certainly much of the initial work for such a monitoring program must be viewed as a research and development effort to derive an appropriately selected list of compounds to monitor, from the thousands of compounds that exist in process streams. Such a list, when correlated with more completely characterized samples, could be used to estimate occupational risk to the worker. It is overly optimistic to expect that this list will be short, as the relative amounts of chemicals vary considerably from process to process, as well as within a given process, due to varying operating parameters and feedstocks. The industrial hygiene program, charged with protecting workers from chronic and accidental exposures to such a variety of compounds, should have at its disposal instrumentation capable of rapid, sensitive, and selective measurement, while at the same time being easily operated by a chemical or industrial hygiene technician.

An important group of chemicals is the polynuclear aromatic (PNA) compounds, both heterocyclic and homocyclic, which, because of the proven carcinogenic potency of many members of the group, deserves special consideration in health protection practices. For these compounds a particularly thoughtful approach to worker protection must be pursued because a small exposure, either accidental or chronic, may result in carcinogenesis many years later with no immediate warning signs to the exposed person. Polynuclear aromatic compounds are also a problem because they may pose a risk through multiple pathways. Exposure may occur from contaminated surfaces, respirable particulate matter, and vapors. Because the more potent PNAs have high boiling points, they exist in the vapor phase only at extremely low levels and probably pose little hazard through vapor inhalation. However, some of the lower-ringed PNAs such as methylnaphthalenes and quinoline have variously been suggested as tumor promotors, inhibitors or actual carcinogens,[4,5] whereas benzene and naphthalene are chemicals with specific Occupational Safety and Health Administration (OSHA) exposure limits of their own.

A wide gap now exists between the sophisticated laboratory-based instruments, such as GCMS, liquid chromatoraphy–mass spectrometry (LCMS), and low-temperature Fourier transform–infrared spectrometry (FTIR), and the crude techniques of weighing a filter sample extract to determine the benzene-soluble fraction of total particulate matter (BSFTPM). This latter technique, although inexpensive and simple, suffers from a lack of selectivity between PNAs, some of which are carcinogenic while others are not, and from lack of sensitivity, reproducibility, and accuracy due to taking the difference of two relatively large, nearly equal weights. The threshold limit value for BSFTPM is only 150 $\mu g/m^3$ (ref. 6). The high quality research instruments, on the other hand, can provide a much more complete analysis, yielding results on hundreds of PNA compounds. These results are expensive, often costing hundreds and even thousands of dollars per sample for a complete analysis, and time consuming, requiring weeks and perhaps months to get final results. Although this vast amount of data is valuable to the researcher and for the assessment of environmental and health effects, to assess worker exposure the industrial hygienist needs only a few (maybe 10, not hundreds) selected PNAs or groups of PNAs to correlate measured breathing environments and contaminated surfaces with more accurately characterized samples. An estimate is often desired in essentially real time.

The need for real-time analysis of work areas is based on more than a desire simply to offset delays in assessing a work zone until a laboratory analysis has been performed on samples taken from the area. The potential for exposure to acutely toxic compounds and to carcinogens requires that a worker be warned immediately when exposure levels rise above predetermined values. One will readily admit that under routine operations in conversion facilities, the likelihood of such an exposure is small. On the other hand, leaky valves, component repair, and other abnormal procedures offer considerable opportunities for high exposure levels. Portable instruments capable of detecting small leaks allow the industrial hygienist and operations engineer to prevent small problems from becoming large problems.

NEW MONITORING INSTRUMENTS

Having established the monitoring needs for a wide variety of toxic chemicals, one can produce a list of desirable attributes for a monitoring instrument. An ideal instrument would be:

1. portable
2. sensitive
3. selective
4. accurate and reproducible
5. real time
6. multicomponent
7. easily operated
8. inexpensive

To design an instrument possessing all of the above qualities would be very difficult, but this list does provide a goal to work toward.

Some new monitoring instruments that fulfill some of the above requirements are briefly reviewed below. Many of these instruments are still in the design and development stage and will require extensive field testing before they can be made available for routine use. Descriptions of some of the instruments are necessarily brief due to incomplete development and patent considerations. Because workers may be exposed to the toxic chemicals potentially present in conversion facilities through gas-phase exposure, through respirable particulate matter, and through skin contact with contaminated surfaces, instruments useful in monitoring exposures from each of these pathways are discussed.

Many of the instruments described share common attributes. Particular emphasis has been placed on instruments with a multicomponent capability. With such a large number of possibly hazardous compounds, many of which may be regulated, the need to monitor several compounds at one time exists in order to decrease sampling time and effort per monitored compound. One feature common to several instruments is the use of microcomputers for both instrument control and data analysis and display. This capability requires less detailed knowledge of the analytical procedure and often improves the reproducibility of instrumental measurements, permitting operation by industrial hygiene or chemical technicians.

In addition to handling the more mundane chores associated with taking measurements, the microcomputer allows the use of an interference matrix among the various compounds being measured. A set of inverse interference coefficients can be calculated for these compounds, and, when the matrix equations are solved by the microcomputer, analytical values corrected for interference are obtained. It is worth pointing out that increased power, increased speed, and reduced cost are characteristic of the new microprocessors, and continued improvement of these qualities is likely. Computer memory is becoming cheaper and faster. These advancements suggest that more and more computing power will become available for the improvement of monitoring instruments.

Portable Quadrupole Mass Spectrometer

One of the most important tools for the complete analysis and characterization of samples from complex process streams, effluent pathways, and environmental monitoring is the GCMS. These large and powerful laboratory-based instruments supported by extensive computer libraries of mass spectra have provided the analytical chemist with the capability of a detailed quantitative, multicomponent analysis on a variety of samples. Recently some of the power of mass spectrometry has been extended to real-time monitoring with the development of semiportable (requiring only 110-Vac power supply) quadrupole mass spectrometers for trace gas analysis.[7,8,9]

These instruments can monitor a wide range of gaseous compounds present at conversion facilities. In addition to the examples listed in Table 1, the portable mass spectrometers can also measure vapors from the more volatile PNA compounds.

Table 1. Occupational Safety and Health Administration standards for some airborne compounds that may be found in coal conversion facilities

Compound	OSHA standard, TWA[a] (ppm)	Monitoring technique[b]		
		IR	MS	DUVAS
Acetic acid	10	+	+	?
Ammonia	50	+	−	+
Aniline[c]	5	+	+	+
Benzene	10(1)	+	+	+
Carbon disulfide	20(1)	+	+	−
Carbon monoxide	50(35)	+	−	−
Coal tar pitch volatiles[d]				
Cresol[c]	5	+	+	+
Ethyl mercaptan	10[e]	−	+	?
Hydrogen chloride	5[e]	+	+	−
Hydrogen sulfide	20(10)[e]	−	+	−
Methyl mercaptan	10[e]	+	+	?
Naphthalene	10	−	+	+
Nitrogen dioxide	5(1)	−	+	+
Phenol[c]	5	+	+	+
Pyridine	5	+	+	+
Styrene	100	+	+	+
Sulfur dioxide	5(2)	+	+	+
Toluene	200(100)	+	+	+
Xylene	100	+	+	+

[a]Time-weighted average. Numbers in parentheses are NIOSH recommended standards.

[b]Techniques discussed in the text for analyzing airborne compounds: IR—infrared analyzer, MS—portable mass spectrometer, DUVAS—derivative UV-absorption spectrometer. A plus indicates the instrument is capable of monitoring that compound at the OSHA limit.

[c]Exposure may be through the skin.

[d]Coal tar pitch volatiles, as measured by the benzene-soluble fraction of particulate matter, include such polycyclic aromatic hydrocarbons as anthracene, benzo(*a*)pyrene, phenanthrene, acridine, chrysene, and pyrene. The instruments in this table measure vapors rather than particulates. Other techniques are required to detect particulates.

[e]Ceiling concentration.

A relatively small quadrupole mass spectrometer[7] has been developed by Analog Technology Corporation (ATC) under a contract administered by National Aeronautics Space Administration (NASA) and funded through an interagency agreement between NASA, the Department of Energy (DOE), Environmental Protection Agency (EPA), and various branches of the Department of Defense (DOD). The instrument is capable of making rapid, repetitive measurements with measurement times as short as a second or less per gas constituent and with results displayed directly in concentration units. The integral computer can correct for interference of up to 40 different mass peaks, provided the mass spectrometer has been calibrated for the interfering gases. This method, in essence, provides "separation" of interfering compounds by computer. The need for prior calibration of interfering compounds is one of the limitations of this method.

The mass range of the instrument is 2 to 200 mass units and has a detection limit of about 0.5 ppm for most gases. Figure 1 is a photograph of the mass spectrometer, which consists of the following major components: (1) a computer that controls the mass spectrometer, performs data manipulation and storage of spectra, and provides a cathode ray tube (CRT) display of data and operating parameters; (2) an electronic support package and power supplies; and (3) a vacuum envelope containing (a) the gas inlet components and ion source, (b) a 2-in. hyperbolic quadrupole mass filter, and (c) a continuous dynode electron multiplier. Attached to the base of the vacuum tee is a 30-liter/s ion pump, which maintains the operating pressure at approximately 10^{-3} Pa. The gas inlet line and ion source are mounted to one end flange, and the electron multiplier is attached to the other flange. Air to be sampled is pumped continuously through a capillary tube by a mechanical pump to provide partial pressure reduction; a pin-hole leak in a gold foil allows entrance of the sample into the low-pressure mass spectrometer.

The instrument has three modes of operation: (1) automatic analysis for monitoring selected peaks, (2) continuous analysis of a single mass peak, or (3) a conventional mass spectrum displayed at 1/20 of a mass unit per data point. The automatic analysis mode uses peak-switching to measure up to 40 mass peaks and gives a digital readout of concentration after interference corrections have been made. This mode allows rapid, continuous monitoring while also correcting for interfering compounds.

Performance evaluation of the mass spectrometer by the various agencies is continuing. The feasibility of using the instrument for monitoring for exposures to potentially hazardous gases from coal conversion facilities is currently being investigated by the Health and Safety Research Division of Oak Ridge National Laboratory (ORNL).

Another portable mass spectrometer, designed and built by Varian Associates, has been used during the past three years by the U.S. Army Environmental Hygiene Agency for industrial hygiene and air

Fig. 1. Field portable mass spectrometer developed by NASA for trace gas analysis.

pollution monitoring.[8,9] The instrument consists of two heavy-duty aluminum suitcases, one containing the control panel and display, computer, sample pump, and power supplies, while the other contains the quadrupole mass spectrometer, vacuum system, rf-dc power supply, electron multiplier, inlet system, and temperature controls (Fig. 2). The mass range is from 10 to 346 mass units with only 14 mass units being displayed on the CRT at one time. The computer is also capable of displaying a series of real-time peaks and their background intensities available from read-only memory as a bar graph. This mode

permits the operator to observe directly changes in concentration, which is quite useful in detection of point source leaks. The computer, however, does not have the capability of correcting for interfering compounds; thus one must choose judiciously the ion fragment to monitor for a given compound. Although designed mainly for qualitative analysis, the instrument does provide quantitative results within approximately $\pm15\%$.

Perhaps the most unique feature of this instrument is the use of a membrane inlet system that provides for up to a factor of 10^6 enrichment of many organic

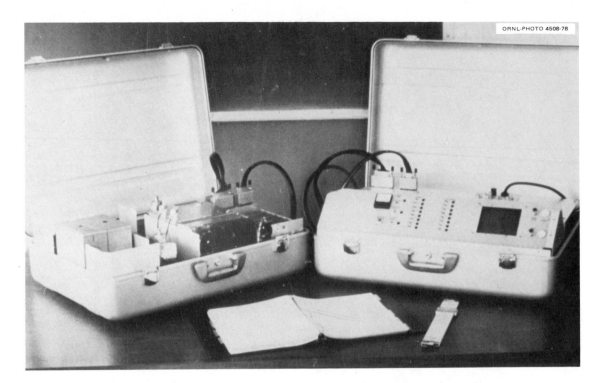

ORNL-PHOTO 4508-78

Fig. 2. The U.S. Army Environmental Hygiene Agency uses this portable mass spectrometer for industrial hygiene and air pollution monitoring.

gases relative to air. The system consists of two parallel, three-stage, dimethyl silicon polymer membranes that can be temperature controlled to improve the solubility of organic compounds of interest in the silicon membrane. One side of the inlet system remains at atmospheric pressure while the other side, containing the "enriched" mixture, is at the high vacuum of the mass spectrometer. Using this inlet system the instrument has a detection limit of a part per billion or so for many organics and is suitable for measuring most organic compounds with vapor pressures of 10^{-3} Pa or above.

A second-generation instrument based on improvements derived from experience with the prototype is being developed. The system will contain a magnetic cassette tape and an acoustic coupler for transmission of data back to a base laboratory for further analysis if needed. An optional inlet system for a gas chromatograph will give improved selectivity for complex mixtures.

Multicomponent Microcomputer-Controlled Infrared Analyzer

Infrared (IR) spectroscopy is an instrumental analytical technique capable of rapidly measuring a wide range of hazardous gases. Until recently, the use of IR spectroscopy for industrial hygiene monitoring was somewhat restricted because of the potential of severe interferences from the broad range of compounds that might be encountered at a coal conversion facility. A quantitative multicomponent analysis can now be provided by the Wilks MIRAN-80 computerized IR spectrometer that uses an integral microcomputer to correct for interferences[10,11] (Fig. 3).

Using a 0.5-m multipass sample cell with a total pathlength of up to 20 m and an air sampling pump, coupled with the spectrometer, one can obtain a sub-parts-per-million detection limit for many of the hazardous compounds listed in the American

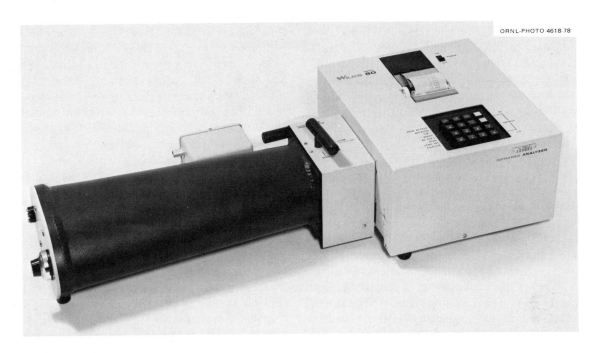

ORNL-PHOTO 4618-78

Fig. 3. The MIRAN-80 computing IR analyzer analyzes a mixture of vapors and prints out concentrations on a permanent paper tape.

Conference of Governmental Industrial Hygienists (ACGIH) threshold limit values.[12] The MIRAN-80 has a wavelength range of 2.5 μm to 14.5 μm with 0.4 μm accuracy and 0.05 μm reproducibility.[10] It is a single-beam instrument, with a LiTaO$_3$ pyroelectric detector and a Nichrome wire source, capable of detecting 0.0001 absorbance unit. The heart of the instrument is a microprocessor that controls the wavelength, records transmission data, and calculates concentrations from stored matrix coefficients that account for compound interferences. The stored coefficients must be obtained from a matrix inversion operation performed on calibration data by a computer other than the MIRAN-80. In routine operation a mixture of five compounds can be analyzed quantitatively in approximately 1 min. The computer is capable of handling up to 11 interfering compounds. The instrument has an operator keyboard and an integral dot-matrix printer to handle communication with the computer and give a permanent copy of the data analysis. The MIRAN-80 can also be used to analyze liquids and solids by exchanging appropriate sampling accessories.

Derivative Ultraviolet-Absorption Spectrometer

Ultraviolet(UV)-absorption spectrometry has good sensitivity for the measurement of PNAs, since most PNAs absorb strongly in the UV region of the spectrum. The principal shortcoming of UV spectrometry is a lack of selectivity between compounds in a mixture. A method of improving the selectivity of UV-absorption spectrometry is the use of derivative techniques to resolve overlapping absorption peaks.[13-16] Second-derivative UV absorption tends to enhance signals due to narrow band absorption while reducing signals due to broad band absorption. Compounds within a mixture having narrow band absorption have second-derivative peaks that are selectively enhanced. The technique is insensitive to sample opacity. Likewise, source light intensity variations and source energy distribution are inconsequential, and thus the noise level is reduced, as compared to conventional second-derivative UV-absorption spectroscopy.

By using a multipass air-sampling cell, a second-derivative UV-absorption spectrometer can measure part-per-billion levels of gas-phase PNAs such as naphthalene, methylnaphthalenes, and acenaphthylene, as well as benzene, toluene, pyridine, phenol, and cresols. When wavelength-modulated techniques are used to obtain the second derivative of the absorption signal, a selected peak may be monitored as a function of time to give a continuous measurement.

Continuous real-time monitoring of aqueous samples can be provided using a flow-through quartz solution cell. This mode of operation could give levels of phenol in wastewaters, for example.

The SM-400 second-derivative spectrometer,[13] manufactured by Lear-Siegler, Inc. (Fig. 4), uses a vibrating entrance slit to the monochromator to produce a 45-Hz sinusoidal displacement, resulting in a wavelength modulation of approximately ±1.5 nm. The second-derivative signal is obtained by detecting the second harmonic of the oscillation frequency (90 Hz) and normalizing this signal by the direct absorption signal. The second-derivative signal of a compound is directly proportional to the trace concentration of that sample in air. The SM-400 has both visible and UV light sources and is able to scan

through the spectrum at 2.5 or 10.0 nm/min. A prototype instrument is being developed at ORNL[17] in an effort to bring about a reduction in size and to add the power of a microcomputer to correct further for interfering compounds. The wavelength modulation is obtained by placing a vibrating mirror within a small holographic grating monochromator. The modulation frequency is 400 Hz, and the second harmonic is 800 Hz, allowing better filtering of 120-Hz noise from ac power supply ripple. The faster modulation also permits much faster scanning, with speeds variable from 30 nm/min to 240 nm/min.

The prototype instrument consists of two units (Fig. 5). The spectrometer box (10 in. × 16 in. × 6 in.) contains a UV lamp, the monochromator-modulator, a stepping motor, photomultiplier tube, and associated electronics. An integral microprocessor controls the taking of spectra and stores a 100-nm scan with a resolution of 0.25 nm per data point. The second unit of the system consists of a commercial personal microcomputer to provide data analysis, with a CRT for display and concentration readout, a keyboard for communication with the spectrometer and for program selection, and a magnetic tape

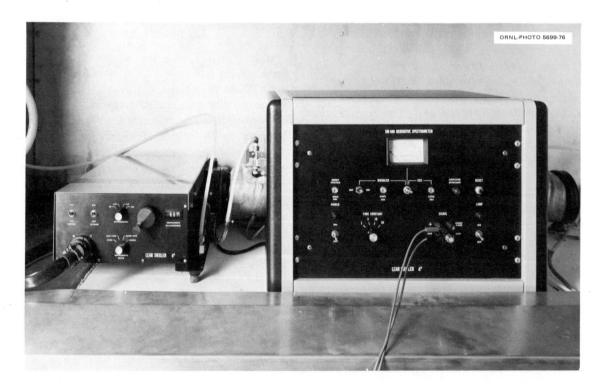

Fig. 4. SM-400 derivative spectrometer with the 1-m heated multipass air-sampling cell in the background.

Fig. 5. Computer-controlled derivative spectrometer developed by ORNL, shown here with liquid sample analyzer compartment instead of the 1-m multipass air-sampling cell.

cassette data-storage system. This prototype instrument will be evaluated for use as a real-time monitor at the University of Minnesota at Duluth low-Btu gasifier during the coming year.

Miniature Gas Chromatograph

One of the tools most used by the industrial hygienist in evaluating worker exposure to a wide variety of hazardous chemicals is the gas chromatograph (GC). When coupled with a personal sampler to collect an integrated sample, the GC is used to analyze the sample extract giving a measure of integrated exposure. Portable field chromatographs are also routinely available to give on-the-spot analysis of a wide variety of compounds.

A novel gas chromatograph is being developed by Stanford University[18] as a personal air monitor for toxic gases. This miniature GC is produced using integrated circuit technology. The chromatograph components are contained in a 5-cm silicon wafer. Etched into the silicon is a 0.5-m-long, open-tubular capillary column covered with a glass plate. The sample injection valve is a solenoid-controlled nickel diaphram with a valve seat etched into the silicon.

The thermal conductivity detector is a microbead thermister placed within a silicon cavity at the gas stream exit.

In addition to the miniature GC, the instrument contains a carrier gas supply and a microcomputer. The microcomputer controls the frequency of injection of air samples and measures up to 10 gas peaks. It also calculates and updates the time-weighted exposure for these gases. The instrument is expected to be small enough to be worn by an industrial worker without interfering with his movement, thus giving nearly real-time values for exposure to a mixture of toxic gases.

Laser Spectroscopy Techniques

Increased use of laser techniques for laboratory analysis is likely to extend more and more to field monitoring in the future. Having demonstrated the ultimate sensitivity of detecting a single atom among 10^{19}, researchers using multiphoton absorption techniques are optimistic that in the future a wide range of compounds can be detected with high selectivity using this technique.[19]

Another instrument, which is currently available, couples a laser with the relatively new technique of optoacoustic spectrometry.[20] A laser optoacoustic spectrometer developed by Gilford Instrument Laboratories is capable of detecting toxic gases in the parts-per-billion range. The instrument uses a tunable wavelength CO_2 laser to select from approximately 30 laser lines between 9.2 and 10.8 μm. The spectrometer has both a reference cell and a sample cell with sensitive electret microphones serving as pressure transducers to detect energy absorbed by the sample due to radiationless processes. The reference cell provides cancellation of signals from absorbed radiation in the interior surfaces of the cells. It can also be used to subtract the signal due to an interfering compound. A lock-in amplifier is used to detect the signal modulated by a mechanical chopper at a rate ranging between 30 and 100 Hz. An HP-9815A programmable calculator is interfaced to the instrument to provide control during operation and to perform calculations on an interference matrix allowing up to 13 interfering gases to be measured in a mixture.

Another laser technique has been suggested in the National Institute for Occupational Safety and Health (NIOSH) report *Recommended Health and Safety Guidelines for Coal Gasification Pilot Plants.*[2] The device is based on low-temperature fluorescence excited by a laser with 520-nm light. As presented, a 540-nm filter would be used to select fluorescence from BaP. At low temperatures, both increased sensitivity and selectivity for BaP are obtained. Figure 6 illustrates the concepts for the real-time monitor that would indicate changes in BaP concentration in air particulate matter.

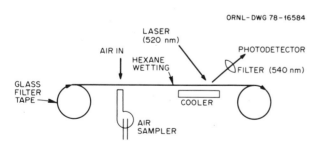

ORNL–DWG 78–16584

Fig. 6. Schematic of a proposed real-time BaP monitor.

Room Temperature Phosphorimetry

Recent developments in room temperature phosphorescence (RTP)[21] for spectrochemical analysis

show promise that this technique can yield semicontinuous, nearly real-time data on various PNA concentrations in airborne particulate matter. Adsorbing the sample onto an appropriate surface such as filter paper allows phosphorescence to be obtained at room temperatures without the use of difficult and tedious cryogenic techniques normally used for phosphorescence. Several techniques are being investigated that will improve the selectivity of RTP. These include selective use of heavy-atom perturbers such as NaBr and NaI to enhance the room temperature phosphorescence of some PNAs relative to others.[22] Time-resolved techniques can be used to take advantage of the differences in phosphorescence lifetimes among compounds in order to enhance selectivity. Improved resolution can also be obtained using synchronous scanning[23] and derivative techniques. Evaluation of these various techniques to enhance selectivity is continuing.

Design of an on-line PNA monitor at ORNL is still in the preliminary phase. Filter tape, impregnated with selected heavy atoms, is pulled across the entrance to a high volume air sampler. Air is made to flow through the filter tape for a time (10 min, for example); then the tape is advanced to a solvation and drying chamber. The sample is then dried using an infrared lamp. The preceding spot is then moved into the phosphoroscope for measurement. Data taking and analysis as well as tape movement, solvation, and spectrometer operation will be controlled by the instrument microcomputer. A goal is to produce a prototype device by the end of 1979.

Portable Survey Fluoroscopes

Since contact of the worker's skin with PNA-contaminated surfaces is an important exposure pathway, instrumentation is needed to provide at least a qualitative, and hopefully a semiquantitative, estimate of PNA levels on surfaces. The present crude method consists of shining a UV lamp onto surfaces suspected of being contaminated and observing the resultant fluorescence with the unaided eye. This method suffers from being nonspecific and insensitive, requiring the survey to be made at night or, if indoors, with the lights out. A suggested improvement is to use a fabric-skirted box with a photovoltaic fluorescence detector, which would allow measurements in lighted areas.[2]

A prototype portable survey instrument for detecting PNA-contaminated surfaces is being developed at ORNL.[24] With an intensity-modulated (1 kHz) UV light source to illuminate the area being examined, this instrument, with appropriate filters, can be operated in lighted areas with good sensitivity. A synchronous demodulator circuit detects signals from the photomultiplier tube due to PNA fluorescence while rejecting signals due to background lighting. The prototype fluoroscope is expected to be able to detect nanogram quantities of many PNAs at close range and microgram quantities at a distance of 1 m. When completed, the instrument will be handheld with an auxiliary battery pack, allowing considerable freedom to an industrial hygienist performing a survey of contaminated surfaces.

A more advanced, second generation luminoscope being developed at ORNL[25] uses small monochromators instead of filters to obtain improved spectral resolution. Improved selectivity will come from the ability to select both the excitation and emission wavelength and to provide synchronous scanning. A prototype of the device is currently being assembled, and a working version of the luminoscope is expected by the end of 1979.

SUMMARY

There is a firmly established need for worker protection from potentially toxic exposures at coal conversion facilities. One method of protection used by the industrial hygienist is a real-time, portable, multicomponent area monitor. A variety of instruments is being developed for area monitoring to aid in the reduction of occupational exposures. No single instrument can, however, be expected to provide a complete analysis of all of the toxic chemicals that an industrial hygienist wishes to measure.

Although still undergoing evaluation and further development, the portable area monitors discussed in this paper should provide an important addition to the arsenal of field monitoring instruments available to industrial hygienists, environmental scientists, and field analytical chemists. These units have shown the capability of giving fast, selective, and sensitive measurements for a wide range of potentially hazardous compounds.

Further improvements of instrumentation through enhanced sensitivity, increased selectivity, improved data analysis capabilities, and increased portability can be expected.

REFERENCES

1. H. M. Braunstein, E. D. Copenhaver, and H. A. Pfuderer (ed.), *Environmental, Health, and Control Aspects of Coal Conversion: An Information Overview*, ORNL/EIS-94, Oak Ridge National Laboratory, Oak Ridge, Tenn., 1976.

2. *Recommended Health and Safety Guidelines for Coal Gasification Pilot Plants*, DHEW(NIOSH)-78-120, U.S. Department of Health, Education, and Welfare, Washington, D.C., 1978.

3. E. Bingham and H. L. Falk, "Environmental Carcinogens: The Modifying Effect of Co-carcinogens on the Threshold Response," *Arch. Environ. Health* 19, 779 (1969).

4. I. Schmeltz, J. Task, J. Hilfrish, N. Hirota, D. Hoffman, and E. L. Wynder, "Bioassays of Naphthalene and Alkylnaphthalenes for Co-carcinogenic Activity: Relation to Tobacco Carcinogenesis," in *Polynuclear Aromatic Hydrocarbons: Analysis, Chemistry, and Biology,* ed. P. W. Jones and R. I. Freudenthal, Vol. 3, p. 47, Raven Press, New York, 1978.

5. M. Dong, I. Schmeltz, E. LaVoie, and D. Hoffman, "Aza-Arenes in the Respiratory Environment: Analysis and Assays for Mutagenicity," in *Polynuclear Aromatic Hydrocarbons: Analysis, Chemistry, and Biology,* ed. P. W. Jones and R. I. Freudenthal, Vol. 3, p. 97, Raven Press, New York, 1978.

6. "Occupational Safety and Health Standards: Exposure to Coke Oven Emissions," *Federal Register* 41(206), 46741 (1976).

7. C. M. Judson, C. S. Josias, and J. L. Lawrence, "A Small Computerized Mass Spectrometer for the Automaic Determination of Trace Constituents in Gases," in *Pittsburgh Conference on Analytical Chemistry and Applied Spectroscopy* 28, 104 (1977).

8. J. E. Evans and J. T. Arnold, "Monitoring Organic Vapors," *Environ. Sci. Technol.* 9, 1134 (1975).

9. R. W. Meier, "A Field Portable Mass Spectrometer for Monitoring Organic Vapors," *Am. Ind. Hyg. Assoc. J.* 39, 233 (1978).

10. W. B. Telfair, A. C. Gilby, R. J. Syrjala, and P. A. Wilks, Jr., "A Microcomputer-Controlled Infrared Analyzer for Multicomponent Analysis," *Am. Lab.* 8, 91 (1976).

11. P. A. Wilks, Jr., "Infrared Monitors of Industrial Applications," in *Proceedings of the Second ORNL Workshop on Exposure to Polynuclear Aromatic Hydrocarbons in Coal Conversion Processes,* CONF-770361, p. 123, Oak Ridge National Laboratory, Oak Ridge, Tenn., 1977.

12. American Conference of Governmental Industrial Hygienists, *Threshold Limit Values for Chemi-*

cal Substances and Physical Agents in the Workroom Environment, ACGIH, Cincinnati, Ohio, 1976.

13. R. N. Hager, Jr., "Derivative Spectroscopy with Emphasis on Trace Gas Analysis," Anal. Chem. **45**, 1131A (1973).

14. T. C. O'Haver and G. L. Green, "Numerical Error Analysis of Derivative Spectrometry for the Quantitative Analysis of Mixtures," Anal. Chem. **48**, 312 (1975).

15. A. R. Hawthorne and J. H. Thorngate, "Improving Analysis from Second-Derivative UV-Absorption Spectrometry," Appl. Opt. **17**, 724 (1978).

16. A. R. Hawthorne and J. H. Thorngate, "Application of Second-Derivative UV-Absorption Spectrometry to PNA Analysis," Appl. Spectrosc. (in press).

17. A. R. Hawthorne, J. H. Thorngate, R. B. Gammage, and T. Vo-Dinh, "Trace Organic Analysis Using Second-Derivative UV-Absorption Spectroscopy," to be published in the Proceedings of the Third International Symposium on Polynuclear Aromatic Hydrocarbons, Columbus, Ohio, October 25–27, 1978.

18. S. C. Terry and J. B. Angell, "A Pocket-Sized Personal Air Contaminant Monitor," Paper 25, in Abstracts of 175th ACS National Meeting, March 13–17, 1978.

19. G. S. Hurst, M. H. Nayfeh, and J. P. Young, "One-Atom Detection Using Resonance Ionization Spectroscopy," Phys. Rev. A **15**, 2283 (1977).

20. T. Wakefield and R. E. Blank, "A Toxic Gas Monitor with PPB Sensitivity Using an Automated "Optoacoustic Spectrometer," Pittsburgh Conference on Analytical Chemistry and Applied Spectroscopy **29**, 634 (1978).

21. T. Vo-Dinh and J. D. Winefordner, "Room Temperature Phosphorimetry as a New Spectrochemical Method of Analysis," Appl. Spectrosc. Rev. **13**, 261 (1977).

22. T. Vo-Dinh and J. R. Hooyman, "Selective Triplet Emission Enhancement for Multicomponent Analysis by Room Temperature Phosphorimetry," submitted for publication in Anal. Chem.

23. T. Vo-Dinh, "Multicomponent Analysis by Synchronous Luminescence Spectrometry," Anal. Chem. **50**, 396 (1978).

24. D. D. Schuresko, "In-Plant Environmental Monitors," Fossil Energy Program Quarterly Progress Report for the Period Ending June 30, 1978, ORNL-5444, Oak Ridge National Laboratory, Oak Ridge, Tenn., 1978.

25. T. Vo-Dinh, Health and Safety Research Division, Oak Ridge National Laboratory, Oak Ridge, Tenn., personal communication (1978).

DISCUSSION

Dick Gammage, Oak Ridge National Laboratory: Since you are stressing the compound-selective capability of most of these monitoring techniques, and since the samples you deal with can contain tens of thousands of compounds, would you like to comment on the select compounds you might be thinking of monitoring?

Alan Hawthorne: This is indeed a problem. There are so many, many compounds to measure and so many of them are hazardous. Do you try to measure everything? The answer is obviously no. Do you select only a single compound to try to monitor it? I think once again the answer is no. Some small number of compounds should probably be measured and attempts made to correlate these with more detailed characterization data for the product streams. One of the things that we at Oak Ridge have been concentrating on is selection of a few of the more volatile PNAs to give an indication of areas where PNA exposure can build up and to detect leaks of tarry vapors. Instrumentation that we are trying to develop will be used to track down these leaks. We think a good candidate for indicator of PNA vapors is 2-methylnaphthalene. It is usually a very abundant compound in most of the oil and tar products from coal conversion facilities. I do discuss in the paper the concept of using proxy compounds. Jim Evans suggests that carbon monoxide is the best indicator substance for characterizing fugitive emissions of coal gasification plants and there is instrumentation currently available for that. There are a lot of good reasons for doing this. To get a handle on the heavier PNAs, the particulate PNA matter, we suggest that pyrene as well as benzo(a)pyrene be used as indicators.